SCHAUM'S OUTLINE OF

THEORY AND PROBLEMS

of

OPTICS

by

EUGENE HECHT, Ph.D.

Associate Professor of Physics
Adelphi University

SCHAUM'S OUTLINE SERIES

McGRAW-HILL BOOK COMPANY

New York   St. Louis   San Francisco   Auckland   Düsseldorf   Johannesburg
Kuala Lumpur   London   Mexico   Montreal   New Delhi   Panama
Paris   São Paulo   Singapore   Sydney   Tokyo   Toronto

07-027730-3

2 3 4 5 6 7 8 9 10 11 12 13 14 15 16 17 18 19 20   SH SH   7 9 8 7 6

Library of Congress Cataloging in Publication Data

Hecht, Eugene.

    Schaum's outline of theory and problems of optics.

    (Schaum's outline series)

    1. Optics.   I. Title.   II. Title: Theory and problems of optics.

QC355.2.H43   535   74-32228

ISBN 0-07-027730-3

Preface

This book is intended as a supplement to a first course in undergraduate optics. As is customary in this series, the book takes the form of a succinct outline of principles with numerous sets of solved and unsolved problems which elaborate and illustrate those principles.

Most of the treatment is in terms of the wave model of light, although the photon picture is used when essential to an understanding of the phenomenon. Accordingly, Chapter 1 gives a mathematical description of wave motion in general, while Chapter 2 relates that description to Maxwell's equations. Chapter 3 examines the laws of propagation and Chapter 4 applies them to the practical problems of geometrical optics. Diffraction theory is developed on the basis of the simple Huygens-Fresnel principle. The last chapter, Chapter 8, is concerned with an elementary discussion of Fourier optics.

The modern jargon of picoseconds, megahertz and nanometers — of coherence length, frequency stability and bandwidth — is extensively used, and the problems run the gamut from candles to lasers. Optics is a broad subject in the midst of a marvelous renaissance. Even though much of the material to be dealt with in an introductory treatment is quite traditional, I have tried to imbue it with the vitality and excitement of the contemporary scene.

A proper preface happily serves to extend thanks to those who have contributed to the effort; and so I do, indeed, thank all my students for their help and inspiration, particularly Patricia Fazio, John Ryan and Richard Deem. The entire manuscript was flawlessly typed by Miriam LaRosa, whose spelling is legend, and the work was meticulously edited by David Beckwith. Lastly, I extend sincere appreciation to my wife, Carolyn Eisen Hecht, for her rendition of the little green frog in Fig. 4-42, and I accordingly dedicate this book to her.

EUGENE HECHT

Adelphi University
January 1975

CONTENTS

CONTENTS

Chapter 1

Wave Motion

1.1 INTRODUCTION

Optics is the study of light or, more broadly, the study of the electromagnetic spectrum. For our purposes the wave aspects of light will be of paramount concern. Although light is an electromagnetic phenomenon much of optics can be understood without specifying the nature of the waves we are dealing with. For example, the extensive work of Fresnel (1788-1827), which is so highly useful even today, was derived within the framework of the elastic medium model, now long defunct.

1.2 THE DIFFERENTIAL WAVE EQUATION

A simple wave moving along a string has a great many properties in common with a light wave. The displacement of the string is perpendicular to the direction of motion of the disturbance, i.e. the wave propagates along the string while each element of the string itself merely moves back and forth. Waves of this sort are said to be *transverse*. Light is just such a transverse wave, with the electric and magnetic fields varying in directions perpendicular to the propagation direction.

The *differential wave equation*

$$\frac{\partial^2 \psi}{\partial x^2} \;=\; \frac{1}{v^2}\frac{\partial^2 \psi}{\partial t^2}$$

describes such phenomena (when there is only one space variable). The quantity $\psi(x,t)$, known as the *wave function,* represents the disturbance in space (x) and time (t), be it a string's displacement or the magnitude of a field. Here v is the speed of propagation of the wave.

One solution of the wave equation has the form

$$\psi(x,t) \;=\; f(x - vt)$$

wherein f is an arbitrary twice differentiable function of the variable $(x - vt)$. In other words, $(x - vt)$ may be squared or cubed or what have you, but it must appear as a unit. The shape of the disturbance, its *profile*, can be gotten by "photographing" the wave function at a given time. Mathematically this is equivalent to setting t equal to a constant; for example, at $t = 0$

$$\psi(x, 0) \;=\; f(x)$$

is the profile. Thus if $f(x)$ is the shape of a bump on a string, $f(x - vt)$ describes the bump moving with a speed v in the positive x-direction. In the same way, $g(x + vt)$ is a solution of the wave equation corresponding to an arbitrary profile $g(x)$ propagating in the negative x-direction.

1

SOLVED PROBLEMS

1.1.   Show that $f(x - vt)$ is a progressive wave moving in the positive x-direction with an unchanging profile.

Set up a coordinate system S′ moving to the right with the disturbance at a speed v, as in Fig. 1-1. At $t = 0$, the two systems S and S′ overlapped and so $x' = x - vt$. In S′ the wave function is independent of time. Therefore, in S′ the profile is unchanging and the wave function is given by

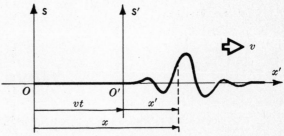

$$\psi(x,t) = f(x')$$
$$= f(x - vt)$$

Fig. 1-1

1.2.   Show that $\psi(x, t) = f(x \mp vt)$ is a solution of the one-dimensional differential wave equation.

Here f is a function of x', where $x' \equiv x \mp vt$ is in turn a function of x and t. Thus, using the chain rule,

$$\frac{\partial \psi}{\partial x} = \frac{\partial f}{\partial x'} \frac{\partial x'}{\partial x} = \frac{\partial f}{\partial x'} \quad \text{and} \quad \frac{\partial \psi}{\partial t} = \frac{\partial f}{\partial x'} \frac{\partial x'}{\partial t} = \mp v \frac{\partial f}{\partial x'}$$

and so

$$\frac{\partial^2 \psi}{\partial x^2} = \frac{\partial^2 f}{\partial x'^2} \quad \text{while} \quad \frac{\partial^2 \psi}{\partial t^2} = \frac{\partial}{\partial t}\left(\mp v \frac{\partial f}{\partial x'} \right) = \mp v \frac{\partial}{\partial x'}\left(\frac{\partial f}{\partial t} \right)$$

Hence

$$\frac{\partial^2 \psi}{\partial t^2} = \mp v \frac{\partial}{\partial x'}\left(\mp v \frac{\partial f}{\partial x'} \right) = v^2 \frac{\partial^2 f}{\partial x'^2} = v^2 \frac{\partial^2 \psi}{\partial x^2}$$

or

$$\frac{\partial^2 \psi}{\partial x^2} = \frac{1}{v^2} \frac{\partial^2 \psi}{\partial t^2}$$

1.3.   If $\psi_1(x, t)$ and $\psi_2(x, t)$ are both solutions of the differential wave equation, show that $\psi_1(x, t) + \psi_2(x, t)$ is also a solution.

Since both ψ_1 and ψ_2 are solutions,

$$\frac{\partial^2 \psi_1}{\partial x^2} = \frac{1}{v^2} \frac{\partial^2 \psi_1}{\partial t^2} \quad \text{and} \quad \frac{\partial^2 \psi_2}{\partial x^2} = \frac{1}{v^2} \frac{\partial^2 \psi_2}{\partial t^2}$$

Adding these together yields

$$\frac{\partial^2 \psi_1}{\partial x^2} + \frac{\partial^2 \psi_2}{\partial x^2} = \frac{1}{v^2}\left(\frac{\partial^2 \psi_1}{\partial t^2} + \frac{\partial^2 \psi_2}{\partial t^2} \right)$$

or

$$\frac{\partial^2}{\partial x^2}(\psi_1 + \psi_2) = \frac{1}{v^2} \frac{\partial^2}{\partial t^2}(\psi_1 + \psi_2)$$

The above result is the *superposition principle* for the one-dimensional wave equation. It follows that

$$\psi(x, t) = f(x - vt) + g(x + vt)$$

is the general solution to the equation.

1.4.   Given the profile

$$\psi(y, 0) = \frac{3}{2y^2 + 1}$$

(a) Write an expression for the corresponding progressive wave moving with a speed of 2 m/s in the increasing y-direction. (b) Sketch the profile at $t = 0$ and $t = 1$ s.

(a) Simply replace y by $y \mp vt$, or, in this particular case, by $y - 2t$.   Hence

$$\psi(y, t) \;=\; \frac{3}{2(y - 2t)^2 + 1}$$

(b) See Fig. 1-2.

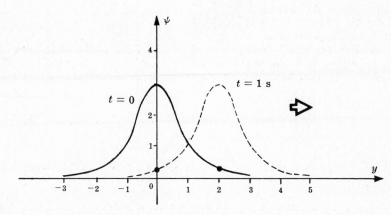

Fig. 1-2

1.5.   (a) Show that the expression $\psi(z, t) = A e^{-(2z + 3t)^2}$ is a progressive wave and (b) verify that it is a solution of the wave equation.

(a) The expression can be rewritten as

$$\psi \;=\; A e^{-4(z + 3t/2)^2}$$

which is a function of $z + vt$, with $v = 3/2$.   Accordingly, ψ represents a wave moving with speed 3/2 in the negative z-direction.

(b) By differentiation:

$$\frac{\partial \psi}{\partial z} \;=\; -8(z + 3t/2) A e^{-4(z + 3t/2)^2}, \qquad \frac{\partial^2 \psi}{\partial z^2} \;=\; [-8(z + 3t/2)]^2 \psi - 8\psi$$

$$\frac{\partial \psi}{\partial t} \;=\; [-8(z + 3t/2)(3/2)] \psi, \qquad \frac{\partial^2 \psi}{\partial t^2} \;=\; [-8(z + 3t/2)(3/2)]^2 \psi - 8(3/2)^2 \psi$$

and the wave equation becomes

$$\{[-8(z + 3t/2)]^2 - 8\} \psi \;=\; \frac{1}{v^2} (3/2)^2 \{[-8(z + 3t/2)]^2 - 8\} \psi$$

which is indeed satisfied for $v = 3/2$.

1.3 SINUSOIDAL WAVES

A wave which has as its profile a sinusoid (Fig. 1-3) is said to be *harmonic*. These waves are of particular interest because we can mathematically synthesize more complicated profiles out of sums of sine functions by Fourier methods.

If $\psi(x, 0) = A \sin kx$, then

$$\psi(x, t) \;=\; A \sin k(x \mp vt)$$

is a *progressive* harmonic wave. The argument of a sine function has to be unitless and to that end we have introduced the positive constant k, called the *propagation number*. The maximum value of the magnitude of $\psi(x, t)$ is A, the *amplitude*. Now, the wave function repeats itself after a *spatial period* or *wavelength* λ, i.e. $\psi(x, t) = \psi(x \pm \lambda, t)$. For this

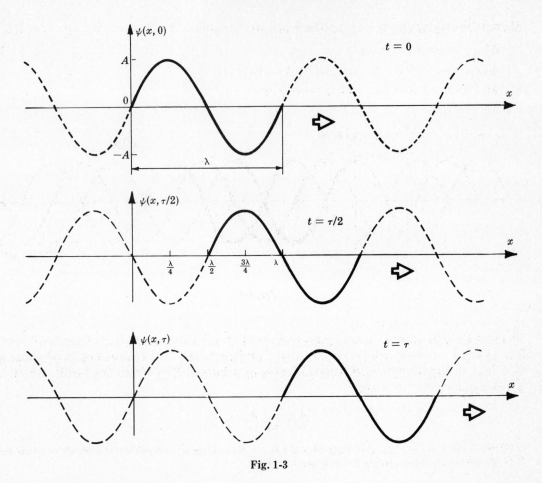

Fig. 1-3

to be the case the propagation number must be given by $k = 2\pi/\lambda$. Similarly, if the wave is to repeat itself after a *temporal period* τ, i.e. $\psi(x, t) = \psi(x, t \pm \tau)$, it follows that $\tau = \lambda/v$. The *period* is the number of units of time per wave, the reciprocal of which is the *frequency* ν, or the number of waves per unit of time. Thus

$$v = \nu\lambda$$

In analogy with mechanics we can introduce the *angular frequency* $\omega \equiv 2\pi/\tau$. Although nothing is actually revolving here, it is convenient to use a quantity such as ω having the units of *radians per second*. Accordingly, the wave function can be recast as

$$\psi(x, t) = A \sin(kx \mp \omega t)$$

The above harmonic waves range from $-\infty$ to $+\infty$ in both space and time, and they are therefore mathematical abstractions. Since they contain only a single frequency, the waves are referred to as *monochromatic*. No actual physical disturbance has this form, although ones approaching it to varying degrees exist and are said to be *quasimonochromatic*.

SOLVED PROBLEMS

1.6. Show that for a harmonic wave the repetitive nature in space, $\psi(x, t) = \psi(x \pm \lambda, t)$, requires that $k = 2\pi/\lambda$.

We know that the sine function repeats itself when the argument increases or decreases by 2π. Thus

$$A \sin k(x - vt) = A \sin k[(x \pm \lambda) - vt] = A \sin[k(x - vt) \pm 2\pi]$$

The second equality gives $|k\lambda| = 2\pi$, or since both k and λ are positive, $k = 2\pi/\lambda$.

1.7.   Sketch the wave $\psi(x, t) = A \cos(kx - \omega t)$ at the times $t = 0$, $t = \tau/4$, and $t = \tau/2$.

At $t = 0$, $\psi(x, 0) = A \cos kx$.

At $t = \tau/4 = 1/4\nu = \pi/2\omega$, $\psi(x, \tau/4) = A \cos(kx - \pi/2)$.

At $t = \tau/2 = 1/2\nu = \pi/\omega$, $\psi(x, \tau/2) = A \cos(kx - \pi)$.

See Fig. 1-4.

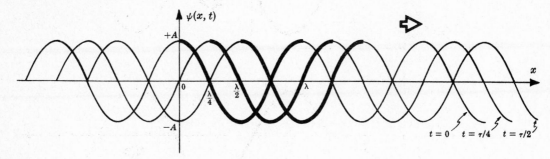

Fig. 1-4

1.8.   The wavelength of light is generally measured in units of *nanometers* (1 nm = 10^{-9} m). For example, yellow, which is just about midspectrum, has a wavelength of roughly 580 nm. Compare this with the thickness of a human hair (from the head), which is approximately 4×10^{-2} mm.

$$\frac{4 \times 10^{-5} \text{ m}}{580 \times 10^{-9} \text{ m}} = 69$$

Sixty-nine wavelengths per thickness of a hair. Light, although minute in wavelength, is still not that remote from things we're used to dealing with.

1.9.   Light ranges in wavelength roughly from violet at 390 nm to red at 780 nm. Its speed in vacuum is about 3×10^8 m/s, as is the case for all electromagnetic waves. Determine the corresponding frequency range.

Since $v = \nu\lambda$,

$$\nu_{\text{vio}} = \frac{3 \times 10^8 \text{ m/s}}{390 \times 10^{-9} \text{ m}} = 7.7 \times 10^{14} \text{ s}^{-1} \quad \text{and} \quad \nu_{\text{red}} = \frac{3 \times 10^8 \text{ m/s}}{780 \times 10^{-9} \text{ m}} = 3.8 \times 10^{14} \text{ s}^{-1}$$

The units are inverse seconds or cycles per second. Nowadays one uses the units of *hertz*, abbreviated Hz, instead of cps. The frequency range is then from 380 THz to 770 THz (1 terahertz = 10^{12} Hz = 1 THz).

1.10.   Verify that the harmonic wave function $\psi(x, t) = A \sin(kx - \omega t)$ is a solution of the one-dimensional differential wave equation.

The wave equation is given by

$$\frac{\partial^2 \psi}{\partial x^2} = \frac{1}{v^2} \frac{\partial^2 \psi}{\partial t^2}$$

Now $\quad\quad \dfrac{\partial \psi}{\partial x} = kA \cos(kx - \omega t), \quad\quad \dfrac{\partial^2 \psi}{\partial x^2} = -k^2 A \sin(kx - \omega t) = -k^2 \psi$

while $\quad\quad \dfrac{\partial \psi}{\partial t} = -\omega A \cos(kx - \omega t), \quad\quad \dfrac{\partial^2 \psi}{\partial t^2} = -\omega^2 A \sin(kx - \omega t) = -\omega^2 \psi$

The wave equation becomes

$$-k^2\psi \;=\; \frac{1}{v^2}(-\omega^2\psi)$$

Therefore, provided that $v = \omega/k = \nu\lambda$, which we know to be the case, $\psi(x,t)$ is a solution.

1.11. Prove that a progressive harmonic wave can be described alternatively by:

(a) $\psi \;=\; A\sin 2\pi\!\left(\dfrac{x}{\lambda}\mp\dfrac{t}{\tau}\right)$

(b) $\psi \;=\; A\sin 2\pi\nu\!\left(\dfrac{x}{v}\mp t\right)$

(c) $\psi \;=\; A\sin 2\pi(\kappa x\mp\nu t),$ where $\kappa\equiv 1/\lambda$

(a) Starting with $\psi = A\sin k(x\mp vt)$, use $k = 2\pi/\lambda$ to obtain

$$\psi \;=\; A\sin 2\pi\!\left(\frac{x}{\lambda}\mp\frac{vt}{\lambda}\right)$$

But $v/\lambda = \nu = 1/\tau$.

(b) From the result of (a):

$$\psi \;=\; A\sin 2\pi\nu\!\left(\frac{x}{\lambda\nu}\mp\frac{t}{\tau\nu}\right) \;=\; A\sin 2\pi\nu\!\left(\frac{x}{v}\mp t\right)$$

since $\lambda\nu = v$ and $\tau\nu = 1$.

(c) Substitute $\lambda = 1/\kappa$ and $\tau = 1/\nu$ in the result of (a).

1.12. Given the wave function (in SI units) for a light wave to be

$$\psi(x,t) \;=\; 10^3\sin\pi(3\times 10^6 x - 9\times 10^{14} t)$$

Determine (a) the speed, (b) wavelength, (c) frequency, (d) period, and (e) amplitude.

(a) By comparison with $\psi(x,t) = A\sin k(x-vt)$ the given wave function can be written as

$$\psi(x,t) \;=\; 10^3\sin 3\times 10^6\pi(x - 3\times 10^8 t)$$

whereupon we immediately have $v = 3\times 10^8$ m/s.

(b) By (a), $k = 3\times 10^6\pi$ m^{-1}. Therefore, $\lambda = 2\pi/k = 666$ nm.

(c) $$\nu \;=\; \frac{v}{\lambda} \;=\; \frac{3\times 10^8\text{ m/s}}{(2/3)\times 10^{-6}\text{ m}} \;=\; 4.5\times 10^{14}\text{ Hz}$$

(d) $$\tau \;=\; 1/\nu \;=\; 2.2\times 10^{-15}\text{ s}$$

(e) By (a), $A = 10^3$ V/m (i.e. volts per meter).

1.4 PHASE AND PHASE VELOCITY

One of the most important concepts we shall deal with is the *phase*, φ, of a harmonic wave, which is simply defined as the argument of the sine function:

$$\varphi \;=\; kx\mp\omega t$$

The wave function as written thus far is actually a special case, since at $t = 0$ and $x = 0$, $\psi(0,0) = 0$. There is no reason why the magnitude of a wave couldn't be anything you like at $t = 0$, $x = 0$. This can be accomplished by shifting the sine function, through the introduction of an *initial phase* ε such that

$$\varphi \;=\; kx\mp\omega t+\varepsilon$$

When we envision a harmonic wave sweeping by, we determine its speed by observing the motion of a point at which the magnitude of the disturbance remains constant. For such a point the phase must be constant as well. Thus the speed of the wave is the speed at which the condition of constant phase travels, i.e.

$$\left(\frac{\partial x}{\partial t}\right)_{\varphi} \;=\; -\frac{(\partial \varphi/\partial t)_x}{(\partial \varphi/\partial x)_t} \;=\; \pm\frac{\omega}{k} \;=\; \pm v$$

The positive quantity v, which is the speed of propagation of a harmonic wave, is also referred to as the *phase velocity*.

SOLVED PROBLEMS

1.13. Sketch the profile of the wave $\psi(x,t) = A \sin(kx - \omega t + \varepsilon)$, where the initial phase is given by each of the following: $\varepsilon = 0$, $\varepsilon = \pi/2$ and $\varepsilon = \pi$.

Setting $t = 0$, $\psi(x,0) = A \sin(kx + \varepsilon)$. Marking off the x-axis at intervals of $\lambda/4$, we obtain the profiles shown in Fig. 1-5.

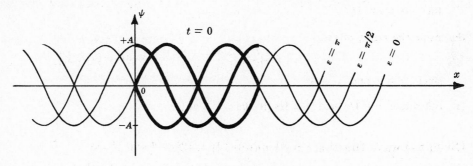

Fig. 1-5

If we were describing a wave generated along a rope by a hand at $x = 0$, the displacement would initially $(t = 0)$ be downward from zero for $\varepsilon = 0$ and upward from zero for $\varepsilon = \pi$.

1.14. What is the magnitude of the wave function $\psi(x,t) = A \cos(kx - \omega t + \pi)$ at $x = 0$ when $t = 0$, $t = \tau/4$, $t = \tau/2$, $t = 3\tau/4$ and $t = \tau$?

By substituting directly into $\psi(x,t)$ we get

$$\psi(0,0) \;=\; A \cos(0 - 0 + \pi) \;=\; -A$$

Moreover, since $\tau = 2\pi/\omega$, $\tau/4 = \pi/2\omega$ and therefore

$$\psi(0,\tau/4) \;=\; A \cos(0 - \pi/2 + \pi) \;=\; 0$$

Similarly,
$$\psi(0,\tau/2) \;=\; A \cos(0 - \pi + \pi) \;=\; +A$$

$$\psi(0,3\tau/4) \;=\; A \cos(0 - 3\pi/2 + \pi) \;=\; 0$$

$$\psi(0,\tau) \;=\; A \cos(0 - 2\pi + \pi) \;=\; -A$$

1.15. By examining the phase, determine the direction of motion of the progressive waves represented by

$$\psi_1(y,t) \;=\; A \cos 2\pi\left(\frac{t}{\tau} + \frac{y}{\lambda} - \varepsilon\right), \qquad \psi_2(z,t) \;=\; A \cos \pi 10^{15}\left(t - \frac{z}{v} + \varepsilon\right)$$

Maintenance of the condition of constant phase, i.e.

$$\varphi \;=\; 2\pi\left(\frac{t}{\tau} + \frac{y}{\lambda} - \varepsilon\right) \;=\; \text{constant}$$

requires that y be decreasing, since t is positive and increasing. In other words, for φ to be constant ψ_1 must be a wave moving in the negative y-direction. Similarly, ψ_2 is a wave moving in the increasing or positive z-direction. The sign of ε is irrelevant to the direction of motion.

1.16. Using the fact that $v = \left(\dfrac{\partial x}{\partial t}\right)_\varphi$, compute the speed of the wave

$$\psi(x, t) = 10^3 \sin \pi(3 \times 10^6 x - 9 \times 10^{14} t)$$

and compare your answer with that of Problem 1.12, again assuming SI units.

The condition $\varphi = \text{constant}$ is equivalent to

$$0 = \frac{d\varphi}{dt} = \left(\frac{\partial \varphi}{\partial t}\right)_x + \left(\frac{\partial \varphi}{\partial x}\right)_t \left(\frac{\partial x}{\partial t}\right)_\varphi = \left(\frac{\partial \varphi}{\partial t}\right)_x + \left(\frac{\partial \varphi}{\partial x}\right)_t v$$

Hence
$$v = -\frac{(\partial \varphi / \partial t)_x}{(\partial \varphi / \partial x)_t} = -\frac{-\pi 9 \times 10^{14}}{\pi 3 \times 10^6} = +3 \times 10^8 \text{ m/s}$$

Recall that v is a positive quantity.

1.17. Write an expression for the profile $(t = 0)$ of a harmonic wave moving in the $+x$-direction such that at $x = 0$, $\psi = 10$; at $x = \lambda/6$, $\psi = 20$; and at $x = 5\lambda/12$, $\psi = 0$.

Since $t = 0$, $\psi(x, 0) = A \sin(kx + \varepsilon)$. Substituting the data, we have

$$\psi(0, 0) = A \sin \varepsilon = 10$$

$$\psi(\lambda/6, 0) = A \sin\left(\frac{\pi}{3} + \varepsilon\right) = 20$$

$$\psi(5\lambda/12, 0) = A \sin\left(\frac{5\pi}{6} + \varepsilon\right) = 0$$

Combining the first and second of these yields

$$10 \sin\left(\frac{\pi}{3} + \varepsilon\right) = 20 \sin \varepsilon$$

or
$$\sin\frac{\pi}{3} \cos \varepsilon + \cos\frac{\pi}{3} \sin \varepsilon = 2 \sin \varepsilon$$

from which
$$\tan \varepsilon = \frac{\sin \pi/3}{2 - \cos \pi/3} = \frac{1}{\sqrt{3}}$$

Thus $\varepsilon = \pi/6$ radians and $A = 20$. Hence

$$\psi(x, 0) = 20 \sin(kx + \pi/6)$$

1.18. For a wave with an unchanging profile propagating in the positive x-direction with a speed v, we can expect that $\psi(x, t) = \psi(x + v\,\Delta t, t + \Delta t)$. (This just says that a point on the wave having a given phase will move a distance $v\,\Delta t$ in a time Δt.) Show that the wave function $f(x - vt)$ satisfies this condition.

Substituting $x + v\,\Delta t$ for x and $t + \Delta t$ for t, the given wave function becomes

$$f[(x + v\,\Delta t) - v(t + \Delta t)] = f(x + v\,\Delta t - vt - v\,\Delta t) = f(x - vt)$$

1.5 COMPLEX NUMBER REPRESENTATION

The trigonometric expressions we'll have to deal with can be simplified by use of complex exponentials. Recall that a *complex number z* has the form

$$z = x + iy$$

where $i = \sqrt{-1}$, and x and y are the *real part* and the *imaginary part* of z, respectively. Note that both x and y are themselves real numbers. This can be rewritten as

$$z = A(\cos \varphi + i \sin \varphi)$$

using the fact that $x = A \cos \varphi$ and $y = A \sin \varphi$, as is evident in the Argand diagram of Fig. 1-6. Euler's formula

$$e^{i\varphi} = \cos \varphi + i \sin \varphi$$

then allows us to write

$$z = Ae^{i\varphi}$$

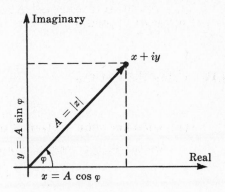

Fig. 1-6

wherein A is the magnitude and φ the phase of the complex quantity z. The *complex conjugate* z^* is obtained by changing the sign of i wherever it appears in z. Hence

$$zz^* = (Ae^{i\varphi})(Ae^{-i\varphi}) = A^2$$

and so the magnitude of z is just

$$A = (zz^*)^{1/2}$$

Multiplication of complex numbers is easy when they are expressed as exponentials. The product of

$$z_1 = A_1 e^{i\varphi_1} \quad \text{and} \quad z_2 = A_2 e^{i\varphi_2}$$

is just

$$z_1 z_2 = A_1 A_2 e^{i(\varphi_1 + \varphi_2)}$$

Notice that if we write

$$\psi(x, t) = Ae^{i(kx - \omega t + \varepsilon)}$$

the real part is $A \cos \varphi$ and the imaginary part is $A \sin \varphi$, where, of course, $\varphi = kx - \omega t + \varepsilon$. Consequently, we can manipulate exponentials in any calculation and then get back to either the cosine or sine form of the wave by taking the real or imaginary part of the answer.

SOLVED PROBLEMS

1.19. Show that the real part of the complex number z is given by

$$\text{Re}\,(z) = \tfrac{1}{2}(z + z^*)$$

Since $z = x + iy$, we can write

$$\tfrac{1}{2}(z + z^*) = \tfrac{1}{2}[(x + iy) + (x - iy)] = x$$

which is indeed the real part of z.

1.20. Derive the expressions

$$\cos \varphi = \frac{e^{i\varphi} + e^{-i\varphi}}{2} \qquad \sin \varphi = \frac{e^{i\varphi} - e^{-i\varphi}}{2i}$$

from Euler's formula

$$e^{i\varphi} = \cos \varphi + i \sin \varphi$$

In Euler's formula,

$$e^{i\varphi} = \cos\varphi + i\sin\varphi$$

replace φ by $-\varphi$, obtaining

$$e^{-i\varphi} = \cos\varphi - i\sin\varphi$$

since $\cos(-\varphi) = \cos\varphi$, $\sin(-\varphi) = -\sin\varphi$.   Adding these two forms yields

$$e^{i\varphi} + e^{-i\varphi} = 2\cos\varphi$$

while subtracting them results in

$$e^{i\varphi} - e^{-i\varphi} = 2i\sin\varphi$$

The first result is equivalent to that of Problem 1.19; the second result is equivalent to

$$\text{Im}(z) = (z - z^*)/2i$$

1.21. Writing the wave function as $\psi = Ae^{i\varphi}$, show that ψ is unchanged when its phase is increased or decreased by 2π.

When the phase is changed by $\pm 2\pi$, the wave function takes on the value

$$\psi' = Ae^{i(\varphi \pm 2\pi)} = Ae^{i\varphi}e^{\pm 2\pi i} = \psi e^{\pm 2\pi i}$$

But by Euler's formula

$$e^{\pm 2\pi i} = \cos(\pm 2\pi) + i\sin(\pm 2\pi) = 1 + i0 = 1$$

Therefore, $\psi' = \psi$.

1.22. Show that multiplying a complex wave function by $\pm i$ is equivalent to shifting its phase by $\pm\pi/2$.

If $\psi = Ae^{i\varphi}$, then $\pm i\psi = \pm iAe^{i\varphi}$.   But from Euler's formula

$$e^{\pm i\pi/2} = \cos(\pm\pi/2) + i\sin(\pm\pi/2) = \pm i$$

and so

$$\pm iAe^{i\varphi} = Ae^{\pm i\pi/2}e^{i\varphi} = Ae^{i(\varphi \pm \pi/2)}$$

1.23. Imagine that we have two waves of the same amplitude, speed and frequency overlapping in some region of space such that the resultant disturbance is

$$\psi(y, t) = A\cos(ky + \omega t) + A\cos(ky - \omega t + \pi)$$

Using complex exponentials show that

$$\psi(y, t) = -2A\sin ky \sin \omega t$$

This is known as a *standing wave*.

Keeping in mind that we are interested only in the *real part*, we can recast the wave function as

$$\psi(y, t) = A\left[e^{i(ky + \omega t)} + e^{i(ky - \omega t + \pi)}\right]$$

$$= Ae^{iky}\left[e^{i\omega t} + e^{i\pi}e^{-i\omega t}\right]$$

Now, from Euler's formula, $e^{i\pi} = \cos\pi + i\sin\pi = -1$; hence

$$\psi(y, t) = Ae^{iky}\left[e^{i\omega t} - e^{-i\omega t}\right]$$

It follows from the second result of Problem 1.20 that

$$\psi(y, t) = Ae^{iky}\, 2i\sin\omega t$$

and so
$$\psi(y, t) = A(2i \cos ky \sin \omega t - 2 \sin ky \sin \omega t)$$

The real part of this wave function is $\psi(y, t) = -2A \sin ky \sin \omega t$.

Had we begun with sine waves, i.e.
$$\psi(y, t) = A \sin (ky + \omega t) + A \sin (ky - \omega t + \pi)$$

the treatment would have been identical up until the last step, where, this time, the imaginary part would be taken to give
$$\psi(y, t) = 2A \cos ky \sin \omega t$$

One can indicate explicitly the use of the real (Re) or imaginary (Im) part by writing, for example,
$$x = \text{Re}(z) \quad \text{or} \quad y = \text{Im}(z)$$

In calculations such as that above we shall omit these designations; if you like, you can include them — it's a matter of taste as long as there is no ambiguity.

1.6 THREE-DIMENSIONAL WAVES

We can generalize the differential wave equation to three dimensions by noting that the space variables should appear symmetrically. That is, the equation should not change if we interchange the space variables, as long as the coordinate system remains right-handed. In any event
$$\frac{\partial^2 \psi}{\partial x^2} + \frac{\partial^2 \psi}{\partial y^2} + \frac{\partial^2 \psi}{\partial z^2} = \frac{1}{v^2} \frac{\partial^2 \psi}{\partial t^2}$$

is the appropriate three-dimensional form in Cartesian coordinates. The particular solutions of most concern to us in our study of optics are those associated with plane and spherical waves.

We now write the equation for a plane passing through an arbitrary point (x_0, y_0, z_0) and perpendicular to a given direction delineated by the *propagation vector* **k**, as in Fig. 1-7. The vector **r** − **r**$_0$ will sweep out the desired plane provided that

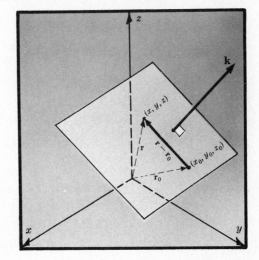

$$(\mathbf{r} - \mathbf{r}_0) \cdot \mathbf{k} = 0$$

or
$$\mathbf{k} \cdot \mathbf{r} = \text{constant}$$

This is the equation of a plane and so
$$\psi(\mathbf{r}) = A \sin (\mathbf{k} \cdot \mathbf{r})$$

is a function defined on a family of planes all perpendicular to **k**. Over each of these $\mathbf{k} \cdot \mathbf{r} =$ constant, and so $\psi(\mathbf{r})$ is a constant. As we move from plane to plane, $\psi(\mathbf{r})$ varies sinusoidally. As before, to convert this into a progressive *harmonic plane wave*, we simply rewrite it as

Fig. 1-7

$$\psi(\mathbf{r}, t) = A \sin (\mathbf{k} \cdot \mathbf{r} \mp \omega t)$$

or
$$\psi(\mathbf{r}, t) = A e^{i(\mathbf{k} \cdot \mathbf{r} \mp \omega t)}$$

The minus sign corresponds to motion in the positive **k**-direction, the plus sign to motion in the negative **k**-direction.

The form of the *harmonic spherical wave* is most easily arrived at by solving the differential wave equation in spherical coordinates. That procedure leads to

$$\psi(r, t) \,=\, \frac{\mathcal{A}}{r} \sin k(r \mp vt)$$

or

$$\psi(r, t) \,=\, \frac{\mathcal{A}}{r} e^{ik(r \mp vt)}$$

where the constant $\mathcal{A}$ is known as the *source strength*. Observe that the amplitude $\mathcal{A}/r$ varies inversely with distance from the origin. This is a requirement of energy conservation. Again, the minus and plus signs in the phase respectively correspond to waves diverging from and converging toward the origin. The expression at any instant represents a cluster of concentric spheres, over each of which r is constant and therefore $\psi(r, t)$ is constant.

Instead of a harmonic wave we could equally well have considered a spherical or planar *pulse*. For example, imagine a point source which, rather than oscillating harmonically, just turns on, builds up, and then shuts off. The disturbance, although short-lived, would move out in all directions as a spherical pulse of some sort.

SOLVED PROBLEMS

1.24. We found that $\mathbf{k} \cdot \mathbf{r} = \text{constant}$ is the equation of a plane normal to $\mathbf{k}$ and passing through some point (x_0, y_0, z_0). Determine the form of the constant and write out the harmonic wave function in Cartesian coordinates.

The equation of the plane is

$$(\mathbf{r} - \mathbf{r}_0) \cdot \mathbf{k} \,=\, 0$$

In Cartesian coordinates $\mathbf{r} = [x, y, z]$, $\mathbf{r}_0 = [x_0, y_0, z_0]$ and $\mathbf{k} = [k_x, k_y, k_z]$. Hence

$$(x - x_0)k_x + (y - y_0)k_y + (z - z_0)k_z \,=\, 0$$

or

$$xk_x + yk_y + zk_z \,=\, x_0k_x + y_0k_y + z_0k_z$$

The left side of this last equation is $\mathbf{k} \cdot \mathbf{r}$, while the right side is the constant in question. The harmonic wave function $\psi(\mathbf{r}, t) = A \sin(\mathbf{k} \cdot \mathbf{r} \mp \omega t)$ becomes

$$\psi(x, y, z, t) \,=\, A \sin(xk_x + yk_y + zk_z \mp \omega t)$$

1.25. Because of the spatial repetition in a harmonic plane wave, we can expect that

$$\psi(\mathbf{r}, 0) \,=\, \psi(\mathbf{r} + \lambda\mathbf{k}/k, 0)$$

In other words, the profile at one location in space is identical to the profile a distance λ farther along in the direction of the unit propagation vector, $\mathbf{k}/k$. Use this and the exponential form of $\psi(\mathbf{r}, 0)$ to show that $|\mathbf{k}| = k = 2\pi/\lambda$.

In terms of exponentials, $\psi(\mathbf{r}, 0) = \psi(\mathbf{r} + \lambda\mathbf{k}/k, 0)$ becomes

$$A e^{i\mathbf{k} \cdot \mathbf{r}} \,=\, A e^{i\mathbf{k} \cdot (\mathbf{r} + \lambda\mathbf{k}/k)} \,=\, A e^{i\mathbf{k} \cdot \mathbf{r}} e^{i\lambda \mathbf{k} \cdot \mathbf{k}/k}$$

But $\mathbf{k} \cdot \mathbf{k}/k = |\mathbf{k}|^2/k = k$ and so

$$A e^{i\mathbf{k} \cdot \mathbf{r}} \,=\, A e^{i\mathbf{k} \cdot \mathbf{r}} e^{i\lambda k}$$

This implies that $e^{i\lambda k} = 1$. Since λ is the minimum repetition-distance and since (Problem 1.21) $e^{i2\pi} = 1$, we must have $\lambda k = 2\pi$ or $k = 2\pi/\lambda$.

1.26. (a) Draw a sketch of a planar wavefront propagating in the positive x-direction.
(b) Write an expression for a harmonic plane wave of this sort moving along the x-axis.

(a) See Fig. 1-8.

(b) The wave function in general is $\psi(\mathbf{r}, t) = A \sin(\mathbf{k} \cdot \mathbf{r} - \omega t)$ or in Cartesian coordinates (see Problem 1.24)

$$\psi(\mathbf{r}, t) = A \sin(k_x x + k_y y + k_z z - \omega t)$$

But here $\mathbf{k}$ is along x, so that $k_x = k$, $k_y = k_z = 0$ and

$$\psi(\mathbf{r}, t) = A \sin(kx - \omega t)$$

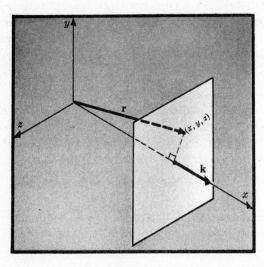

Fig. 1-8   Fig. 1-9

1.27. Write the planar harmonic wave function in Cartesian coordinates in terms of direction cosines α, β, γ, where

$$k_x = \alpha k \qquad k_y = \beta k \qquad k_z = \gamma k$$

and $\alpha^2 + \beta^2 + \gamma^2 = 1$. Then show that the function is a solution of the three-dimensional differential wave equation.

Beginning with

$$\psi(\mathbf{r}, t) = A \sin(k_x x + k_y y + k_z z - \omega t)$$

we replace k_x, k_y and k_z by the corresponding direction cosine terms. As shown in Fig. 1-9,

$$k_x = k \cos \theta_1 = k\alpha$$
$$k_y = k \cos \theta_2 = k\beta$$
$$k_z = k \cos \theta_3 = k\gamma$$

where

$$k_x^2 + k_y^2 + k_z^2 = k^2(\alpha^2 + \beta^2 + \gamma^2) = k^2$$

Hence

$$\psi(\mathbf{r}, t) = A \sin[k(\alpha x + \beta y + \gamma z) - \omega t]$$

Now to check that this is a solution of the wave equation take the appropriate derivatives:

$$\frac{\partial^2 \psi}{\partial x^2} = -\alpha^2 k^2 \psi \qquad \frac{\partial^2 \psi}{\partial y^2} = -\beta^2 k^2 \psi \qquad \frac{\partial^2 \psi}{\partial z^2} = -\gamma^2 k^2 \psi$$

and

$$\frac{\partial^2 \psi}{\partial t^2} = -\omega^2 \psi$$

Adding the first three equations and making use of $\alpha^2 + \beta^2 + \gamma^2 = 1$ and $k = 2\pi/\lambda = \omega/v$ (see Problem 1.25), we obtain

$$\frac{\partial^2 \psi}{\partial x^2} + \frac{\partial^2 \psi}{\partial y^2} + \frac{\partial^2 \psi}{\partial z^2} \;=\; -k^2 \psi \;=\; -\frac{\omega^2}{v^2}\psi$$

Expressing ψ on the right side in terms of the second time-derivative yields the wave equation.

1.28.   Show that

$$\psi(r, t) \;=\; \frac{f(r - vt)}{r}$$

is the solution of the three-dimensional wave equation which corresponds to a spherical disturbance centered at the origin and moving out from it with a speed v.   Here $f(r - vt)$ is an arbitrary twice differentiable function.

The wave equation in rectangular or Cartesian coordinates is

$$\frac{\partial^2 \psi}{\partial x^2} + \frac{\partial^2 \psi}{\partial y^2} + \frac{\partial^2 \psi}{\partial z^2} \;=\; \frac{1}{v^2}\frac{\partial^2 \psi}{\partial t^2}$$

By using the coordinates r, θ and ϕ as indicated in Fig. 1-10:

$$x \;=\; r \sin\theta \cos\phi \qquad y \;=\; r \sin\theta \sin\phi \qquad z \;=\; r \cos\theta$$

the wave equation can be recast as

$$\frac{\partial^2 \psi}{\partial r^2} + \frac{2}{r}\frac{\partial \psi}{\partial r} + \frac{1}{r^2 \sin\theta}\frac{\partial}{\partial \theta}\left(\sin\theta \frac{\partial \psi}{\partial \theta}\right) + \frac{1}{r^2 \sin^2\theta}\frac{\partial^2 \psi}{\partial \phi^2} \;=\; \frac{1}{v^2}\frac{\partial^2 \psi}{\partial t^2}$$

We are looking for the simple solution possessing spherical symmetry.   That is, we require the wave function to be independent of θ and ϕ:

$$\psi(r, \theta, \phi, t) \;=\; \psi(r, t)$$

Thus, the partial derivatives with respect to θ and ϕ will drop out of the wave equation, leaving only

$$\frac{\partial^2 \psi}{\partial r^2} + \frac{2}{r}\frac{\partial \psi}{\partial r} \;=\; \frac{1}{v^2}\frac{\partial^2 \psi}{\partial t^2}$$

This can be restated as

$$\frac{1}{r}\frac{\partial^2 (r\psi)}{\partial r^2} \;=\; \frac{1}{v^2}\frac{\partial^2 \psi}{\partial t^2}$$

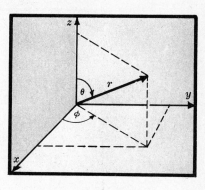

Fig. 1-10

The independent variable r is not a function of t so that

$$r\frac{\partial^2 \psi}{\partial t^2} \;=\; \frac{\partial^2 (r\psi)}{\partial t^2}$$

and the wave equation becomes

$$\frac{\partial^2 (r\psi)}{\partial r^2} \;=\; \frac{1}{v^2}\frac{\partial^2 (r\psi)}{\partial t^2}$$

Now this has the same form as the one-dimensional wave equation, whose general solution was found to be (Problem 1.3)

$$\psi(x, t) \;=\; f(x - vt) + g(x + vt)$$

Here r rather than x is the space variable, while $r\psi(r, t)$ is the unknown function rather than $\psi(x, t)$. Hence, the general solution is

$$r\psi(r, t) \;=\; f(r - vt) + g(r + vt)$$

and so for an outgoing wave

$$\psi(r, t) \;=\; \frac{f(r - vt)}{r}$$

1.29. Thus far we have considered *scalar wave functions* of the form $\psi(x, t) = A \sin(kx - \omega t)$. Suppose that you actually tried to set up a transverse wave of this sort; perhaps a disturbance on a string. It would become evident that since you do not know the direction of the displacement the scalar wave function does not adequately specify the wave. (*a*) How can this deficiency be corrected? (*b*) If the disturbance resides in a plane, known as the *plane of vibration*, the wave is said to be *plane polarized* or *linearly polarized*. Write an expression for a linearly polarized harmonic plane wave.

(*a*) The direction of the displacement in a transverse wave can be fixed by making the amplitude a vector:

$$\boldsymbol{\psi}(x, t) \;=\; \mathbf{A} \sin(kx - \omega t)$$

where now $\boldsymbol{\psi}(x, t)$ is spoken of as the *wave vector*. The vectors $\mathbf{A}$ and $\mathbf{k}$ determine the plane of vibration at any instant in time.

(*b*) For a linearly polarized harmonic plane wave $\mathbf{A}$ is constant in time and

$$\boldsymbol{\psi}(\mathbf{r}, t) \;=\; \mathbf{A} e^{i(\mathbf{k \cdot r} \mp \omega t + \varepsilon)}$$

Figure 1-11 shows several planar wavefronts normal to $\mathbf{k}$. It depicts a harmonic plane wave, so that $\boldsymbol{\psi}(\mathbf{r}, t)$ varies sinusoidally from one plane to the next. Furthermore it's linearly polarized, so that at all points on any planar wavefront the amplitude vector is identical and the corresponding planes of vibration are all parallel. If the amplitude vector is a function of time which varies sufficiently rapidly and randomly, the wave is said to be *unpolarized*. In that case the scalar wave function will generally suffice; hence its interest to us here.

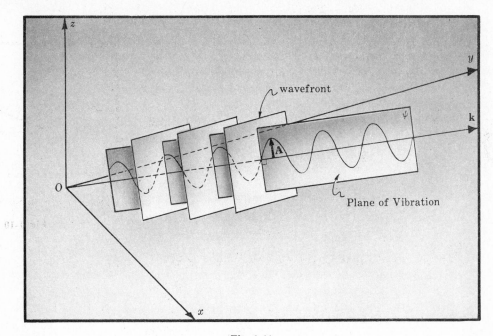

Fig. 1-11

1.7 WAVEFRONTS

Quite generally, a surface over which the phase of a wave is constant is called a *wavefront*. Clearly, for a plane wave the wavefronts are planar surfaces for which $\mathbf{k \cdot r} = \text{constant}$. Similarly, a spherical wave has spherical wavefronts where $r = \text{constant}$. If the wave function is constant over the wavefront, i.e. if the amplitude is constant, the wave is said to be *homogeneous*. This is often the case, but there are many instances of interest (such as in frustrated total internal reflection or a laser beam in the TEM_{00} mode) where the amplitude varies over a wavefront and the wave is *inhomogeneous*.

SOLVED PROBLEMS

1.30.   The wavefronts of starlight reaching your eye or a telescope are essentially planar. Similarly, to photograph a distant terrestrial object you would set the camera lens to focus at ∞, i.e. to receive incoming plane waves.   Discuss the operative phenomenon.

   The wavefronts arising from a distant point source are spheres having a very large radius and therefore very little curvature.   Over the relatively tiny area of a remote detector these wavefronts appear planar, as in Fig. 1-12.

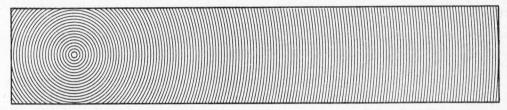

Fig. 1-12

Supplementary Problems

THE WAVE EQUATION

1.31.   Show that $\psi(z, t) = A \sin^2 4\pi(t + z)$ is a solution of the one-dimensional differential wave equation.

1.32.   Which of the following functions describe progressive waves?   Here A, B and C are constants.

$$\psi_1(x, t) = A(x - t)^2 \qquad\qquad \psi_3(z, t) = A \sin B(z^2 - Ct^2)$$
$$\psi_2(y, t) = A(y + t + B) \qquad\qquad \psi_4(x, t) = A/(Bx^2 - t)$$

   Ans. ψ_1

1.33.   What is the speed and direction of propagation of each of the following waves?

$$\psi_1(y, t) = A(y - t)^2$$
$$\psi_2(x, t) = A(Bx + Ct + D)^2$$
$$\psi_3(z, t) = A \exp(Bz^2 + BC^2t^2 - 2BCzt)$$

   Here A, B, C and D are constants.   [We shall sometimes use the notation $\exp u$ for e^u.]

   Ans. $v_1 = 1$ in the positive y-direction, $v_2 = C/B$ in the negative x-direction, $v_3 = C$ in the positive z-direction.

1.34.   Show that $g(x + vt)$ is a progressive wave moving in the negative x-direction with an unchanging profile.

1.35.   Is $\psi(x, t) = A(x + Bt + D)^2 + A \exp(Cx^2 + B^2Ct^2 - 2BCxt)$,   where A, B, C and D are constants, a solution of the one-dimensional differential wave equation?   What is the speed of the wave if $\psi(x, t)$ is in fact a wave function?

   Ans. $\psi(x, t)$ has the form $g(x + Bt) + f(x - Bt)$ and so it is a solution for which $v = B$

1.36.   Show that $h\left(t - \dfrac{x}{v}\right)$, which is an arbitrary twice differentiable function, is a solution of the wave equation.

Ans. $h\left(t - \dfrac{x}{v}\right) = h\left(-\dfrac{x - vt}{v}\right) = f(x - vt)$

1.37.   Prove that the rate of change of the wave function $\psi(x, t)$ with respect to t equals its rate of change with respect to x, to within a multiplicative constant.

Ans. $\dfrac{\partial \psi}{\partial t} = \mp v \dfrac{\partial \psi}{\partial x}$

SINUSOIDAL WAVES

1.38.   Show that the repetitive nature in time of a harmonic wave, i.e. $\psi(x, t) = \psi(x, t \pm \tau)$, requires that $\tau = \lambda/v$.

1.39.   What is the wavelength of a harmonic electromagnetic wave having a frequency of 100 Hz? Since the length of an antenna should be roughly comparable to the wavelength, this calculation immediately suggests using a high frequency carrier for radio signals. What frequency would provide 1 m waves?

Ans. $\lambda = 3 \times 10^6$ m, 3×10^8 Hz or 300 MHz

1.40.   It is evident from Fig. 1-5 that $\sin(kx - \omega t + \pi/2) = \cos(kx - \omega t)$. Show that this is actually true analytically.

1.41.   Given the wave

$$\psi(x, t) = 10 \cos 2\pi\left(\frac{x}{2 \times 10^{-7}} - 1.5 \times 10^{15}t\right)$$

determine the speed, wavelength and frequency. Use SI units.

Ans. $\lambda = 200$ nm, $\nu = 1.5 \times 10^{15}$ Hz, $v = 3 \times 10^8$ m/s

1.42.   Given a harmonic disturbance of amplitude 10 units which is described by a wave function $\psi(x, t)$ such that $\psi(0, 0) = 0$. If the wave has an angular frequency of $\pi/2$ and moves with a speed of 10 m/s, determine its magnitude at $t = 3$ s at a point 20 m from the origin.

Ans. $\psi(20, 3) = 10$ units

1.43.   Plot the function $y(0, t)$, where

$$y(x, t) = y_0 \sin(kx + \omega t + \varepsilon)$$

and $y_0 = 3$, $k = 2\pi$, $\omega = \pi/4$ and $\varepsilon = -\pi$.

1.44.   Imagine that you have a photograph of a wave at $t = 0$ showing its shape to have the mathematical form $\psi(x, 0) = 5 \sin(\pi x/25)$. If the wave is moving in the negative x-direction at a rate of 2 m/s, write an expression for the disturbance at $t = 4$ s.

Ans. $\psi(x, 4) = 5 \sin\left[\dfrac{\pi}{25}(x + 8)\right]$

1.45.   Envision a wave of the form $y(x, t) = 10^2 \sin(2\pi x - 4\pi t)$ and locate two detectors to measure the disturbances at points $x_1 = 2$ and $x_2 = 10$. What will be the magnitude of the disturbance at x_2 at the instant t' when $y(x_1, t') = 10^2$?

Ans. $y(x_2, 7/8) = 10^2$

1.46.   In Problem 1.45, suppose that $x_1 = B$, $y(B, T) = C$. Determine the value of $y(D, T)$ at $x_2 = D$, provided that D and B are whole numbers. What is the value of $y(D, T + 1)$?

Ans. $y(D, T) = C$, since $\lambda = 1$; $y(D, T + 1) = C$, since $\tau = 1/2$

1.47.   Is

$$\psi(x, t) = \frac{A \sin^2 2\pi 10^{15}\left(\dfrac{x}{3 \times 10^8} + t\right)}{[(2 \times 10^7)(x + 3 \times 10^8 t)]^2}$$

a solution of the wave equation? Make a sketch of its profile and determine its speed of propagation if it is a wave.

Ans.   It has the form $g(x + vt)$ and is therefore a solution of the wave equation; $v = 3 \times 10^8$ in the negative x-direction.

PHASE AND PHASE VELOCITY

1.48.   Given a sinusoidal wave $E(x, t)$ of amplitude 20 V/m. If $E(0, 0) = -20$ V/m, what is the initial phase of the wave? (E can be thought of as the electric field component of an electromagnetic wave.)

Ans. $\varepsilon = +3\pi/2$

1.49.   A sinusoidal wave at $t = 0$ has its maximum magnitude at $x = 0$. What is its initial phase?

Ans. $\varepsilon = \pi/2$

1.50.   A harmonic wave is moving in the positive x-direction and you set yourself up at some point to observe the variations in the phase with time. Discuss what results you might anticipate.

Ans. $\left(\dfrac{\partial \varphi}{\partial t}\right)_x = \mp \omega$

1.51.   Imagine that you have a photograph of a harmonic wave on a string stretched along the x-axis. Discuss how the phase of the wave varies with changing x.

Ans. $\left(\dfrac{\partial \varphi}{\partial x}\right)_t = k$

1.52.   An electromagnetic wave of frequency 3×10^{14} Hz sweeps across the room with a speed essentially equal to that in vacuum. If the phase angle difference between two points at a given instant is 60°, what can you say about their separation in space?

Ans. $\lambda/6$, $\lambda + \lambda/6$, $2\lambda + \lambda/6$, etc.; or 166 nm, 1166 nm, 2166 nm, etc.

1.53.   An orange light wave of frequency 500 THz exists in a region of space. (a) By how much would the phase vary in a billionth of a second? (b) How long would be the wave train corresponding to that interval in time? (Because of the exceedingly high frequency of light there is no existent means of measuring the instantaneous values of either the magnitude or phase of a light wave.)

Ans.   (a) 5×10^5 cycles or $\pi \times 10^6$ rad,   (b) 0.3 m

1.54.   The rate at which the phase of a sinusoidal wave changes in time, at any given point in space, is $12\pi \times 10^{14}$ rad/s and the rate at which the phase changes with distance x, at any given time, is $4\pi \times 10^6$ rad/m. Write an expression for the wave function provided that the initial phase is $\pi/3$, the amplitude is 10 and the wave advances in the positive x-direction. What is its speed?

Ans. $\psi(x, t) = 10 \sin (4\pi \times 10^6 x - 12\pi \times 10^{14} t + \pi/3)$, $v = 3 \times 10^8$ m/s

COMPLEX NUMBER REPRESENTATION

1.55.   What is the complex conjugate of each of the following?

$$z = \frac{1 - 4i}{2i} \qquad z = 2e^{i\omega t}e^{-ikx} \qquad z = \frac{1}{5i} - (4i)^2 + \frac{i}{4}$$

$$Ans. \quad z^* = \frac{1+4i}{-2i}, \quad 2e^{-i\omega t}e^{ikx}, \quad -\frac{1}{5i} - (-4i)^2 - \frac{i}{4}$$

1.56. Determine the real part of $z = (1-4i)/2i$ and $z = 2e^{i\omega t}e^{-ikx}$.

Ans. -2, $2\cos(\omega t - kx)$

1.57. Determine the imaginary part of

$$z = 5e^{ikx}e^{i\omega t}e^{i\varepsilon}, \quad z = \left(\frac{Ae^{i\omega t}}{Be^{ikx}}\right)e^{i\varepsilon}, \quad z = \frac{e^{i\omega t} + e^{-i\omega t}}{2}$$

Ans. $5\sin(kx+\omega t+\varepsilon)$, $\dfrac{A}{B}\sin(\omega t - kx + \varepsilon)$, 0

1.58. Find the magnitudes of the complex quantities

$$\psi(x,t) = e^{ikx}e^{-i\omega t}e^{i\varepsilon} \qquad \psi(y,t) = 2e^{iky}e^{i\omega t} + 4e^{iky}e^{-i\omega t}$$

Ans. 1, $2[5 + 4\cos 2\omega t]^{1/2}$

1.59. It's quite usual in many branches of physics, and certainly in optics, to compute the square of some sort of harmonic function, e.g. the kinetic energy if v is sinusoidal. When this is done in the complex representation, *the greatest of care must be used* to avoid an extremely common error. To examine this point in detail, determine $\psi^2(x,t)$, where $\psi(x,t) = A\cos(kx-\omega t)$, using the complex representation. Where lies the rub?

Ans. $\psi^2 = A^2\cos^2(kx-\omega t)$. If ψ is written as $Ae^{i(kx-\omega t)}$ we must evaluate $[\mathrm{Re}\,(\psi)]^2$, which is not equal to $\mathrm{Re}\,(\psi\,\psi^*)$ but rather to $[(\psi+\psi^*)/2]^2$.

THREE-DIMENSIONAL WAVES

1.60. In light of the results of Problem 1.27 show that $\psi(x,y,z,t) = f[k(\alpha x + \beta y + \gamma z) - \omega t]$ is a plane wave solution of the three-dimensional differential equation, where f is an arbitrary twice differentiable function.

1.61. Given a harmonic plane wave of wavelength λ, propagating with a speed v in a direction given by the unit vector $(\hat{\mathbf{i}}+\hat{\mathbf{j}})/\sqrt{2}$ in Cartesian coordinates. Write an expression for the wave function. Here $\hat{\mathbf{i}}$, $\hat{\mathbf{j}}$ and $\hat{\mathbf{k}}$ are the usual unit basis vectors.

Ans. $\psi(x,y,t) = A\sin\left(\dfrac{k}{\sqrt{2}}x + \dfrac{k}{\sqrt{2}}y - \omega t\right)$, where $k = \dfrac{2\pi}{\lambda}$

1.62. In Problem 1.51 we saw that for one-dimensional harmonic waves $\left(\dfrac{\partial\varphi}{\partial x}\right)_t = k$. For three-dimensional harmonic plane waves, determine the value of

$$(\nabla\varphi)_t \equiv \hat{\mathbf{i}}\left(\frac{\partial\varphi}{\partial x}\right)_t + \hat{\mathbf{j}}\left(\frac{\partial\varphi}{\partial y}\right)_t + \hat{\mathbf{k}}\left(\frac{\partial\varphi}{\partial z}\right)_t$$

i.e. the gradient of the phase holding the time constant.

Ans. $(\nabla\varphi)_t = \mathbf{k}$

1.63. Determine the direction of propagation of the plane wave

$$\psi(x,y,z,t) = A\sin\left(\frac{k}{\sqrt{14}}x + \frac{2k}{\sqrt{14}}y + \frac{3k}{\sqrt{14}}z - \omega t\right)$$

Ans. Along the unit vector $\dfrac{1}{\sqrt{14}}\hat{\mathbf{i}} + \dfrac{2}{\sqrt{14}}\hat{\mathbf{j}} + \dfrac{3}{\sqrt{14}}\hat{\mathbf{k}}$

1.64. Write the expression in Cartesian coordinates for a harmonic plane wave for which $k = 2\pi/\lambda$ and $\mathbf{k}$ is directed along a line from the origin through the point $(2,2,3)$.

Ans. $\psi(x,y,z,t) = A\sin\left(\dfrac{2k}{\sqrt{17}}x + \dfrac{2k}{\sqrt{17}}y + \dfrac{3k}{\sqrt{17}}z - \omega t\right)$

Chapter 2

Electromagnetic Waves and Photons

2.1 MAXWELL'S EQUATIONS AND ELECTROMAGNETIC WAVES

In 1865 Maxwell unified and extended the laws of Faraday, Gauss and Ampère, forming a set of encompassing expressions which have since become known as *Maxwell's equations*. They interrelate the spatial and temporal variations of the *electric field intensity* **E** and the *magnetic induction* **B**. In free space, and using Cartesian coordinates, Maxwell's equations can be written in differential form as follows:

$$\frac{\partial E_z}{\partial y} - \frac{\partial E_y}{\partial z} = -\frac{\partial B_x}{\partial t}$$

$$\frac{\partial E_x}{\partial z} - \frac{\partial E_z}{\partial x} = -\frac{\partial B_y}{\partial t}$$

$$\frac{\partial E_y}{\partial x} - \frac{\partial E_x}{\partial y} = -\frac{\partial B_z}{\partial t}$$

$$\frac{\partial B_z}{\partial y} - \frac{\partial B_y}{\partial z} = \epsilon_0 \mu_0 \frac{\partial E_x}{\partial t}$$

$$\frac{\partial B_x}{\partial z} - \frac{\partial B_z}{\partial x} = \epsilon_0 \mu_0 \frac{\partial E_y}{\partial t}$$

$$\frac{\partial B_y}{\partial x} - \frac{\partial B_x}{\partial y} = \epsilon_0 \mu_0 \frac{\partial E_z}{\partial t}$$

$$\frac{\partial B_x}{\partial x} + \frac{\partial B_y}{\partial y} + \frac{\partial B_z}{\partial z} = 0$$

$$\frac{\partial E_x}{\partial x} + \frac{\partial E_y}{\partial y} + \frac{\partial E_z}{\partial z} = 0$$

Here the electric and magnetic properties of the medium, in this case vacuum, are represented by the constants ϵ_0 and μ_0, the *permittivity* and *permeability*, respectively. By manipulating these expressions, Maxwell was able to show that each component of the electric and magnetic fields obeys the differential wave equation (Section 1.6). Explicitly,

$$\frac{\partial^2 E_x}{\partial x^2} + \frac{\partial^2 E_x}{\partial y^2} + \frac{\partial^2 E_x}{\partial z^2} = \epsilon_0 \mu_0 \frac{\partial^2 E_x}{\partial t^2}$$

with identical relations existing for E_y, E_z, B_x, B_y and B_z. Thus the electric and magnetic fields can couple together as an electromagnetic wave traveling through space at a speed $v = 1/\sqrt{\epsilon_0 \mu_0}$. Maxwell, using numerical values of ϵ_0 and μ_0, determined that $v \approx 3 \times 10^8$ m/s, in fine agreement with Fizeau's measurement of the speed of light. The conclusion was inescapable – light was an electromagnetic wave.

The customary symbol for the speed of light in vacuum is c, and its presently accepted value is 2.997924562×10^8 m/s $\pm$ 1.1 m/s or 186,282.3960 miles/s $\pm$ 3.6 ft/s.

SOLVED PROBLEMS

2.1.   A plane electromagnetic wave is one where the electric and magnetic fields are constant on a plane perpendicular to the direction of propagation. Show that such a wave must have its electric field *transverse* to the propagation direction.

If the wave propagates in the z-direction, the electric field must be independent of x and y; that is, $\mathbf{E} = \mathbf{E}(z, t)$. The last of Maxwell's equations

$$\frac{\partial E_x}{\partial x} + \frac{\partial E_y}{\partial y} + \frac{\partial E_z}{\partial z} = 0$$

then leads to

$$\frac{\partial E_z}{\partial z} = 0$$

since $\mathbf{E}$ is not a function of either x or y. This means that $E_z = $ constant and is therefore of no interest. We are concerned only with the electromagnetic wave, which must vary along z. Thus the wave can possess only x- and y-components, and so $\mathbf{E}$ is transverse.

2.2.   Suppose that we have a linearly polarized electromagnetic plane wave (see Fig. 1-11, page 15) whose electric field is of the form $\mathbf{E} = E_x(z, t)\,\hat{\mathbf{i}}$. Show that $\mathbf{B} = B_y(z, t)\,\hat{\mathbf{j}}$.

The electric field is polarized along the x-axis, as indicated by the presence of only an $\hat{\mathbf{i}}$-term. Moreover, the wave propagates in the z-direction. Since $E_y = E_z = 0$ and $E_x = E_x(z, t)$, the first three of Maxwell's equations reduce to

$$0 = -\frac{\partial B_x}{\partial t} \qquad \frac{\partial E_x}{\partial z} = -\frac{\partial B_y}{\partial t} \qquad 0 = -\frac{\partial B_z}{\partial t}$$

This means that B_x and B_z are constant in time and of no concern to us. Hence B_y is the only time-varying term, and so $\mathbf{B} = B_y(z, t)\,\hat{\mathbf{j}}$ is the wave's magnetic component. Note that $\mathbf{E}$ and $\mathbf{B}$ are perpendicular to each other and to the propagation direction as well.

2.3.   Given a harmonic plane electromagnetic wave whose $\mathbf{E}$-field has the form

$$E_z(y, t) = E_{0z} \sin\left[\omega\left(t - \frac{y}{c}\right) + \epsilon\right]$$

determine the corresponding $\mathbf{B}$-field and make a sketch of the wave.

Since $E_x = E_y = 0$, the first of Maxwell's equations yields

$$\frac{\partial E_z}{\partial y} = -\frac{\partial B_x}{\partial t}$$

or

$$\frac{\partial B_x}{\partial t} = \frac{\omega}{c} E_{0z} \cos\left[\omega\left(t - \frac{y}{c}\right) + \epsilon\right]$$

Integrating both sides with respect to t leads to

$$B_x(y, t) = \frac{1}{c} E_{0z} \sin\left[\omega\left(t - \frac{y}{c}\right) + \epsilon\right] = \frac{1}{c} E_z(y, t)$$

The electric and magnetic fields are orthogonal, their magnitudes are related by $E = cB$, and both are normal to the propagation direction (Fig. 2-1, page 22).

2.4.   Quite generally, an electromagnetic wave propagates in a direction given by the cross product $\mathbf{E} \times \mathbf{B}$. Prove that this is true for a plane harmonic wave moving in the positive x-direction, whose $\mathbf{E}$-field is $E(x, t) = E_z(x, t)$.

Maxwell's first three component equations lead to

$$-\frac{\partial E_z}{\partial x} = -\frac{\partial B_y}{\partial t}$$

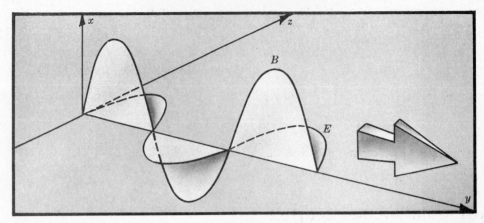

Fig. 2-1

where $E_x = E_y = 0$. This, upon integration, yields $E_z(x, t) = -cB_y(x, t)$; the **E**- and **B**-fields are out of phase. In vector form we have $\mathbf{E} = E_z \hat{\mathbf{k}}$, $\mathbf{B} = -(E_z/c)\hat{\mathbf{j}}$ and $\mathbf{E} \times \mathbf{B} = -(E_z^2/c)\hat{\mathbf{k}} \times \hat{\mathbf{j}}$. Since $-\hat{\mathbf{k}} \times \hat{\mathbf{j}} = \hat{\mathbf{i}}$, all is in agreement.

2.5. Imagine an electromagnetic plane wave in vacuum whose **E**-field (in SI units) is given by

$$E_x = 10^2 \sin \pi(3 \times 10^6 z - 9 \times 10^{14} t) \qquad E_y = 0 \qquad E_z = 0$$

Determine the speed, frequency, wavelength, period, initial phase, **E**-field amplitude and polarization.

The wave function has the basic form

$$E_x(z, t) = E_{0x} \sin k(z - vt)$$

(see Section 1.3). Consequently, it can be reformulated as

$$E_x = 10^2 \sin [3 \times 10^6 \pi(z - 3 \times 10^8 t)]$$

whereupon we see that $k = 3 \times 10^6 \pi$ m^{-1} and $v = 3 \times 10^8$ m/s. Since $k = 2\pi/\lambda = 3 \times 10^6 \pi$, $\lambda = 666$ nm. Furthermore,

$$\nu = \frac{v}{\lambda} = \frac{3 \times 10^8}{(2/3) \times 10^{-6}} = 4.5 \times 10^{14} \text{ Hz}$$

The period τ is $\tau = 1/\nu = 2.2 \times 10^{-15}$ s, while the initial phase is evidently zero. The field amplitude is just $E_{0x} = 10^2$ V/m. The wave is linearly polarized in the x-direction and propagates along the z-axis. This wave corresponds to red light.

2.6. Write an expression for the magnetic field associated with the wave of Problem 2.5.

The wave propagates in the z-direction while the **E**-field oscillates along x. In other words, the **E**-field resides in the xz-plane. Accordingly, since **B** is normal to both **E** and the propagation direction, it must reside in the yz-plane. Thus, $B_x = 0$, $B_z = 0$ and $\mathbf{B} = B_y(z, t)\hat{\mathbf{j}}$. As we saw in Problem 2.3, $E = cB$, the application of which leads to

$$B_y(z, t) = 0.33 \times 10^{-6} \sin \pi(3 \times 10^6 z - 9 \times 10^{14} t)$$

Here $B_{0y} = E_{0x}/c = 10^2/(3 \times 10^8)$, the unit of which is the *tesla* (T), where 1 T $= 1$ kg s^{-1} C^{-1}.

2.7. A plane electromagnetic harmonic wave of frequency 600×10^{12} Hz (green light), propagating in the positive x-direction in vacuum, has an electric field amplitude of 42.42 V/m. The wave is linearly polarized such that the plane of vibration of the electric field is at $45°$ to the xz-plane. Write expressions for **E** and **B**.

The amplitude of the **E**-field, E_0, is equal to $(E_{0y}^2 + E_{0z}^2)^{1/2}$, where, because of the polarization at 45°, $E_{0y} = E_{0z}$. Thus $E_0 = 42.42 = \sqrt{2}\, E_{0y}$ and $E_{0y} = E_{0z} = 30$ V/m. Writing the phase in the form $\omega(t - x/v)$, the electric field becomes

$$E_x = 0 \qquad E_y = E_z = 30 \sin\left[2\pi 600 \times 10^{12}\left(t - \frac{x}{3 \times 10^8}\right)\right]$$

where, of course, $\omega = 2\pi\nu$. Inasmuch as $E = cB$,

$$B_x = 0 \qquad B_z = -B_y = 10^{-7} \sin\left[2\pi 600 \times 10^{12}\left(t - \frac{x}{3 \times 10^8}\right)\right]$$

Notice that B_z is perpendicular to E_y, as is $-B_y$ to E_z. Check this against Problem 2.4.

2.2 THE INDEX OF REFRACTION

Maxwell's theoretical treatment resulted in a predicted propagation velocity of $c = 1/\sqrt{\epsilon_0 \mu_0}$ for electromagnetic waves in vacuum. In contrast, a wave moving through a material medium travels at a speed $v = 1/\sqrt{\epsilon \mu}$. Here ϵ and μ are the permittivity and permeability of the medium. The *absolute index of refraction n* is then defined by

$$n \equiv \frac{c}{v} = \sqrt{\frac{\epsilon \mu}{\epsilon_0 \mu_0}}$$

Generally the magnetic properties of the media have little effect on v, since in materials of concern to us $\mu \approx \mu_0$.

An incoming electromagnetic wave applies an electric field to the medium, which as a result becomes electrically polarized. That, in turn, contributes to ϵ, which then determines n. All of this is dependent on the driving frequency of the incident wave. Figure 2-2 illustrates the frequency dependence of n for various substances of interest.

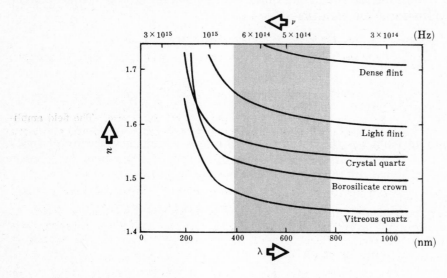

Fig. 2-2

SOLVED PROBLEMS

2.8. Light having a free space wavelength of $\lambda_0 = 500$ nm passes from vacuum into diamond ($n_d = 2.4$). Under ordinary circumstances the frequency is unaltered as light traverses different substances. Assuming this to be the case, compute the wave's speed and wavelength in the diamond.

Since $n = c/v$, $v = c/n = 3 \times 10^8/2.4 = 1.25 \times 10^8$ m/s. As for the wavelength,

$$n \;=\; \frac{c}{v} \;=\; \frac{\lambda_0 \nu_0}{\lambda \nu} \;=\; \frac{\lambda_0}{\lambda}$$

inasmuch as $\nu_0 = \nu$. Hence, $\lambda = 500/2.4 = 208$ nm.

2.9. Suppose a light wave propagates from a point A to another point B, and we introduce into its path a glass plate ($n_g = 1.5$) of thickness $\ell = 1$ mm. By how much will that alter the phase of the wave at B if $\lambda_0 = 500$ nm?

The refractive index of air (1.000293 at 0 °C and 1 atmosphere) is assumed equal to one. The number of waves in air over the distance $\overline{AB}$ is just $\overline{AB}/\lambda_0$. The associated phase shift is $2\pi(\overline{AB}/\lambda_0)$. With the glass inserted there are $(\overline{AB} - \ell)/\lambda_0$ waves in air and ℓ/λ waves in glass. The phase difference is then

$$\Delta\varphi \;=\; \frac{2\pi(\overline{AB} - \ell)}{\lambda_0} + \frac{2\pi\ell}{\lambda} - \frac{2\pi\overline{AB}}{\lambda_0} \;=\; 2\pi\ell\!\left(\frac{1}{\lambda} - \frac{1}{\lambda_0}\right)$$

But $1/\lambda = n/\lambda_0$ and so

$$\Delta\varphi \;=\; \frac{2\pi\ell}{\lambda_0}(n - 1)$$

In this particular case

$$\Delta\varphi \;=\; \frac{2\pi\, 10^{-3}}{500 \times 10^{-9}}(1.5 - 1) \;=\; 2\pi\, 10^3 \text{ rad}$$

2.10. A plane harmonic infrared wave traveling through a transparent medium is given by

$$E_x(y, t) \;=\; E_{0x} \sin 2\pi\!\left(\frac{y}{5 \times 10^{-7}} - 3 \times 10^{14}t\right)$$

in SI as usual. Determine the refractive index of the medium at that frequency, and the vacuum wavelength of the disturbance.

The phase is familiar in the form $k(y - vt)$; accordingly, we rewrite the above as

$$\varphi \;=\; \frac{2\pi}{5 \times 10^{-7}}(y - 15 \times 10^7 t)$$

Clearly, $\lambda = 5 \times 10^{-7}$ m and $v = 1.5 \times 10^8$ m/s. Thus $n = c/v = 2$ and $\lambda_0 = n\lambda = 1000$ nm.

2.11. Light from a sodium lamp ($\lambda_0 = 589$ nm) passes through a tank of glycerin (of index 1.47) 20 m long in a time t_1. If it takes a time t_2 to traverse the same tank when filled with carbon disulfide (of index 1.63), determine the difference $t_2 - t_1$.

Since $v = c/n$,

$$t_1 = \frac{20}{c/n} = \frac{20(1.47)}{c} \qquad \text{and} \qquad t_2 = \frac{20(1.63)}{c}$$

Accordingly,

$$t_2 - t_1 \;=\; \frac{20}{c}(1.63 - 1.47) \;=\; 1.07 \times 10^{-8} \text{ s}$$

2.3 IRRADIANCE

A light wave flashing through space at 186,000 miles per second carries electromagnetic energy and thereby can interact with a detector, be it a film, retina or photocell. Energy flows in the direction in which the wave advances, i.e. (see Problem 2.4) in the direction of **E** x **B**. Accordingly, the energy per unit area per unit time flowing perpendicularly into a surface in free space is given by the *Poynting vector* **S**, where

$$\mathbf{S} = c^2 \epsilon_0 \, \mathbf{E} \times \mathbf{B}$$

Energy over time is power, so that the SI units of **S** are W/m². At optical frequencies **E**, **B** and **S** all oscillate at exceedingly rapid rates and it remains impractical to measure an instantaneous value of **S** directly. Instead, one determines its average value $\langle S \rangle$ over a convenient time interval. This, in turn, is known as the *radiant flux density*. When energy emerges from a surface, the flux density is spoken of as *exitance*; when energy is incident, the flux density is called *irradiance*, symbolized by $I \equiv \langle S \rangle$.

SOLVED PROBLEMS

2.12. A laser emits a 2-mm diameter beam of highly collimated light at a power level, or *radiant flux*, of 100 mW. Neglecting any divergence of the beam, compute its irradiance.

The cross-sectional area of the beam is $\pi (10^{-3})^2$ and so

$$I = \frac{100 \times 10^{-3}}{\pi (10^{-3})^2} = 31.8 \times 10^3 \text{ W/m}^2$$

2.13. A harmonic electromagnetic wave in free space is described by $\mathbf{E} = \mathbf{E}_0 \cos (kx - \omega t)$. Show that $I = (c \epsilon_0 / 2) E_0^2$.

The **B**-field has the form $\mathbf{B} = \mathbf{B}_0 \cos (kx - \omega t)$ and therefore

$$\mathbf{S} = c^2 \epsilon_0 \, \mathbf{E} \times \mathbf{B} = c^2 \epsilon_0 \, \mathbf{E}_0 \times \mathbf{B}_0 \cos^2 (kx - \omega t)$$

Hence

$$\langle S \rangle = c^2 \epsilon_0 \, |\mathbf{E}_0 \times \mathbf{B}_0| \, \langle \cos^2 (kx - \omega t) \rangle$$

Calculating the average over a time interval of length T, we find:

$$\langle \cos^2 (kx - \omega t) \rangle = \frac{1}{T} \int_t^{t+T} \cos^2 (kx - \omega t') \, dt'$$

$$= \frac{1}{2} - \frac{1}{4 \omega T} \{ \sin [2kx - 2\omega(t + T)] - \sin 2(kx - \omega t) \}$$

When $T \gg \tau$, $\omega T \gg 1$ and $\langle \cos^2 (kx - \omega t) \rangle = 1/2$. Consequently, since $E_0 = cB_0$,

$$I = \frac{c \epsilon_0}{2} E_0^2$$

or, if you like,

$$I = c \epsilon_0 \langle \mathbf{E}^2 \rangle$$

2.14. A plane electromagnetic wave moving through free space has an **E**-field (also referred to as the *optical field*) given by $E_x = 0$, $E_y = 0$ and

$$E_z = 100 \sin \left[8\pi \times 10^{14} \left(t - \frac{x}{3 \times 10^8} \right) \right]$$

Calculate the corresponding flux density.

From Problem 2.13, $I = (c \epsilon_0 / 2) E_0^2$. Then, since $\epsilon_0 = 8.8542 \times 10^{-12}$ C² N⁻¹ m⁻² and $E_0 = 100$ V/m,

$$I = \frac{(3 \times 10^8)(8.85 \times 10^{-12})(100)^2}{2} = 13.3 \text{ W/m}^2$$

2.15. Envision a plane harmonic electromagnetic wave propagating in space along the y-axis. If the **E**-field is linearly polarized in the yz-plane and if $\lambda_0 = 500$ nm, write an expression for the corresponding **B**-field when the irradiance is 53.2 W/m².

We can determine E_0 from the irradiance:

$$I = c\epsilon_0 E_0^2/2$$

$$53.2 = (3 \times 10^8)(8.85 \times 10^{-12})E_0^2/2$$

$$E_0 = 200 \text{ V/m}$$

Then, from $B_0 = E_0/c = 66.7 \times 10^{-8}$ T, it follows that

$$B_x = 66.7 \times 10^{-8} \sin \frac{2\pi}{500 \times 10^{-9}}(y - 3 \times 10^8 t) \qquad B_y = B_z = 0$$

2.16. A 60-W monochromatic point source radiating equally in all directions in vacuum is being monitored at a distance of 2.0 m. Using the fact that $\mu_0 = 4\pi \times 10^{-7}$ N s^2 C^{-2}, determine the amplitude of the **E**-field at the detector.

If A is the area of a sphere of radius r surrounding the source and I is the irradiance at that distance, then the power radiated by the source is given by $IA = I(4\pi r^2)$, or equivalently, $\langle S \rangle (4\pi r^2)$. Thus

$$60 = \left(\frac{c\epsilon_0}{2}E_0^2\right)4\pi r^2 = \frac{4\pi r^2}{2\mu_0 c}E_0^2$$

Hence

$$E_0 = \left(\frac{30\mu_0 c}{\pi r^2}\right)^{1/2} = \left[\frac{30(4\pi \times 10^{-7})(3 \times 10^8)}{\pi (2)^2}\right]^{1/2} = 30 \text{ V/m}$$

2.4 PHOTONS — ENERGY AND MOMENTUM

The photon picture portrays the emission and absorption of radiant energy ($\mathcal{E}$) in the form of quanta having the value

$$\mathcal{E} = h\nu$$

In other words, a photon has an energy $\mathcal{E}$ which is proportional to its frequency ν. The constant of proportionality $h = 6.6256 \times 10^{-34}$ J s is known as *Planck's constant*. It can be shown — even classically — that electromagnetic energy and momentum (p) are related by

$$p = \frac{\mathcal{E}}{c}$$

Inasmuch as $\mathcal{E} = h\nu$, we then have

$$p = \frac{h}{\lambda}$$

In vector form:

$$\mathbf{p} = \hbar\mathbf{k} \quad \text{where} \quad \hbar \equiv \frac{h}{2\pi}$$

SOLVED PROBLEMS

2.17. Compute the frequency, vacuum wavelength and energy in joules of a photon having an energy of 2 electron volts (2 eV).

Since 1 eV $= 1.6021 \times 10^{-19}$ J, $\mathcal{E} = 2$ eV $= 3.2 \times 10^{-19}$ J. Knowing that $\mathcal{E} = h\nu$, we have

$$3.2 \times 10^{-19} = 6.6 \times 10^{-34}\nu \quad \text{or} \quad \nu = 4.8 \times 10^{14} \text{ Hz}$$

Finally, because $c = \lambda\nu$, $\lambda = 3 \times 10^8/(4.8 \times 10^{14}) = 625$ nm.

2.18. What is the momentum of a single photon of red light ($\nu = 400 \times 10^{12}$ Hz) moving through free space?

The momentum is given by $p = h/\lambda = h\nu/c$. Hence

$$p = \frac{(6.6 \times 10^{-34})(400 \times 10^{12})}{3 \times 10^8} = 8.8 \times 10^{-28} \text{ kg m s}^{-1}$$

2.19. A stream of photons slamming normally into a completely absorbing screen exerts a pressure $\mathcal{P}$ in newtons per meter squared. Show that $\mathcal{P} = I/c$ when the screen is in vacuum.

The time rate of change of momentum equals the force, i.e. $\Delta p/\Delta t = F$. The force per unit area, F/A, is the pressure, and so

$$\mathcal{P} = \frac{1}{A}\frac{\Delta p}{\Delta t}$$

But $\mathcal{E} = cp$ in vacuum, which means that $\Delta \mathcal{E} = c\,\Delta p$ and

$$\mathcal{P} = \frac{1}{Ac}\frac{\Delta \mathcal{E}}{\Delta t}$$

Since irradiance by definition is energy per unit area per unit time, we obtain $\mathcal{P} = I/c$.

2.20. A collimated beam of light of flux density 3×10^4 W/m² is incident normally on a 1.0-cm² completely absorbing screen. Using the results of Problem 2.19, determine both the pressure exerted on and the momentum transferred to the screen during a 1000-s interval.

The pressure is simply

$$\mathcal{P} = \frac{I}{c} = \frac{3 \times 10^4}{3 \times 10^8} = 10^{-4} \text{ N m}^{-2}$$

and

$$\Delta p = \mathcal{P}A\,\Delta t = (10^{-4})(10^{-4})(10^3) = 10^{-5} \text{ kg m s}^{-1}$$

(One could actually build an interplanetary sailboat using solar pressure.)

2.21. How many red photons ($\lambda = 663$ nm) must strike a totally *reflecting* screen per second, at normal incidence, if the exerted force is to be 0.225 lb?

We know that $F = \Delta p/\Delta t$ and, in this case, Δp is *twice* the incident momentum. Thus, if N is the number of incoming photons per second,

$$F = N\frac{2h}{\lambda}$$

Happily, 0.225 lb = 1 newton, and so

$$N = \frac{\lambda}{2h} = \frac{663 \times 10^{-9}}{2(6.63 \times 10^{-34})} = 5 \times 10^{26} \text{ photons per second}$$

This quantity is referred to as the *photon flux*.

2.22. Imagine a source emitting 100 W of green light at a wavelength of 500 nm. How many photons per second are emerging from the source?

The power multiplied by the given time interval is the emitted energy, i.e. (100 W)(1 s) = 100 J. Denoting the photon flux by N, we have

$$N = \frac{100}{h\nu} = \frac{100\lambda}{hc} = \frac{100(500 \times 10^{-9})}{(6.6 \times 10^{-34})(3 \times 10^8)} = 25 \times 10^{19}$$

2.5 THE ELECTROMAGNETIC-PHOTON SPECTRUM

The radiant energy spectrum ranges from gigantic radio waves millions of kilometers long to the minutest γ-rays millions of times smaller than a nucleus. The spectrum is generally divided into somewhat overlapping classifications designated by such familiar terms as microwaves, ultraviolet, infrared, etc. This is done more for historical than pedagogical reasons, and one should always bear in mind the single nature of all radiant energy. At the low-frequency end of the spectrum, wave characteristics predominate; at the high-frequency end, the corpuscular properties prevail – but it's all electromagnetic energy.

Radio-frequency waves, which were first generated by Hertz in 1887, range from a few Hz up to about 10^9 Hz. This includes radiation from power lines, AM and FM radio, and TV.

Microwaves extend from 10^9 Hz to approximately 3×10^{11} Hz, i.e. from roughly 1/3 m to 1 mm. The region is of interest in communications, radar work and radio astronomy.

Infrared radiant energy corresponds to the frequency band from about 3×10^{11} Hz to 5×10^{14} Hz. Alternatively, the wavelengths of IR go from 1.0 mm down to 780 nm.

Light is the very small electromagnetic spectral region extending from 780 nm to 390 nm, as indicated in Table 2-1. This is the band we shall define as light, despite the fact that the human eye is capable of responding to a somewhat broader range.

Table 2-1

Light		
Color	Vacuum Wavelength (nm)	Frequency (THz)
Red	780–622	384–482
Orange	622–597	482–503
Yellow	597–577	503–520
Green	577–492	520–610
Blue	492–455	610–659
Violet	455–390	659–769

1 nm = 10^{-9} m,   1 THz = 10^{12} Hz

Ultraviolet picks up, as its name would imply, at the violet end of the visible region ($\approx 7.7 \times 10^{14}$ Hz) and extends to about 3×10^{17} Hz. A UV photon has an energy in the range from 3.2 eV to 1.2 keV and the corpuscular properties of radiant energy begin to obtrude.

X-rays are the still higher-frequency manifestation of electromagnetic energy. Discovered by Roentgen in 1895, they occupy the band from 3×10^{17} Hz to 5×10^{19} Hz.

Gamma rays correspond to photon energies of from 10^4 eV to 10^{19} eV and beyond.

SOLVED PROBLEMS

2.23. Determine the vacuum wavelength corresponding to a γ-ray energy of 10^{19} eV.

Converting to joules   (1 eV = 1.6×10^{-19} J),   we have
$$\mathcal{E} = (1.6 \times 10^{-19})(10^{19}) = 1.6 \text{ J}$$
and since $\mathcal{E} = h\nu = hc/\lambda$, $\lambda = hc/\mathcal{E}$. Hence
$$\lambda = \frac{(6.6 \times 10^{-34})(3 \times 10^8)}{1.6} = 1.2 \times 10^{-25} \text{ m}$$

2.24. What is the photon energy in joules corresponding to a 60-Hz wave emitted from a power line? How does this compare with the energy range for light?

Using the expression $\mathcal{E} = h\nu$, we have
$$\mathcal{E} = (6.6 \times 10^{-34})(60) = 39.6 \times 10^{-33} \text{ J}$$

as the energy of a 60-Hz photon. Light extends from 3.8×10^{14} Hz to 7.7×10^{14} Hz, or, from 25.1×10^{-20} J to 50.8×10^{-20} J. Thus, 60 Hz is less energetic than light by a factor of about 10^{13}.

2.25. What is the largest momentum we can expect for a microwave photon?

Microwave frequencies go up to 3×10^{11} Hz. Therefore, since $p = h/\lambda = h\nu/c$,

$$p = \frac{(6.6 \times 10^{-34})(3 \times 10^{11})}{3 \times 10^8} = 6.6 \times 10^{-31} \text{ kg m s}^{-1}$$

Supplementary Problems

MAXWELL'S EQUATIONS AND ELECTROMAGNETIC WAVES

2.26. Given a planar electromagnetic wave in vacuum whose **B**-field is denoted by

$$B_x = 0 \qquad B_y = 66.7 \times 10^{-8} \sin 4\pi 10^6 (z - 3 \times 10^8 t) \qquad B_z = 0$$

Write an expression for the **E**-field. What are the wavelength, speed and direction of motion of the disturbance?

Ans. $E_x = 200 \sin 4\pi 10^6 (z - 3 \times 10^8 t)$, $E_y = E_z = 0$; $\lambda_0 = 500$ nm; $v = 3 \times 10^8$ m/s in the positive z-direction

2.27. Figure 2-3 depicts the electric field (in the y-direction) of a planar electromagnetic wave traveling in the positive x-direction in vacuum. Determine the corresponding **B**-field.

Ans. $B_x = B_y = 0$, $B_z = 6.6 \times 10^{-8} \cos 2\pi 10^3 (x - 3 \times 10^8 t)$

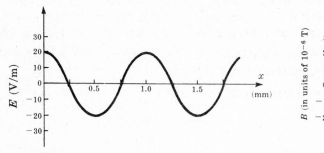

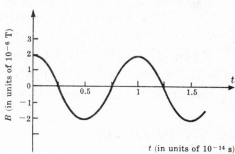

Fig. 2-3   Fig. 2-4

2.28. The time-variation of the magnetic field of a planar electromagnetic wave is represented in Fig. 2-4. The wave propagates in the x-direction in vacuum and the **B**-field is along the z-axis. Determine the corresponding **E**-field.

Ans. $\mathbf{E} = \hat{\mathbf{j}} 600 \cos \frac{2}{3}\pi 10^6 (x - 3 \times 10^8 t)$. (The little cap on the $\hat{\mathbf{j}}$ indicates a *unit vector*; here, of course, it's a Cartesian base vector along the y-axis.)

2.29. Consider a harmonic planar light wave of wavelength 500 nm propagating in vacuum along the positive y-direction. If its **B**-field is confined to the xy-plane and the radiant flux density is 1.197 W/m², determine the **E**-field.

Ans. $\mathbf{E} = \hat{\mathbf{k}} 30 \sin 4\pi 10^6 (y - 3 \times 10^8 t)$. The wave travels along $\hat{\mathbf{j}}$, **E** is along $\hat{\mathbf{k}}$ and **B** is along $\hat{\mathbf{i}}$.

THE INDEX OF REFRACTION

2.30. The vacuum wavelength of a light beam is 600 nm. What is its propagation number in a medium of refractive index 1.5?

Ans. 1.57×10^7 rad/m

2.31. We wish to compare the times of flight of two light beams; one in a tank of carbon tetrachloride ($n = 1.46$), the other in air ($n = 1.0$). If the path lengths are made identical, how long must the tank be when a millionth of a second difference in the transit times is required?

Ans.   650 m

2.32. Light of vacuum wavelength $\lambda_0 = 589$ nm is associated with a refractive index of 2.417 for diamond (C) and 1.923 for zircon ($ZrO_2 \cdot SiO_2$). Compute the ratio of its wavelength in diamond to that in zircon.

Ans.   0.79

2.33. What is the index of refraction of a glass if its dielectric constant K_e, defined to be ϵ/ϵ_0, has the value 2.31?

Ans.   1.52

IRRADIANCE

2.34. A collimated light beam of flux density 10 W/cm² impinges normally on a perfectly absorbing planar surface of area 1 cm². If this occurs for 1000 s, how much energy is imparted to the surface?

Ans. 10^4 J

2.35. A focused CO_2 laser beam putting out a continuous wave ($\lambda_0 = 10,600$ nm) of 3 kW is capable of burning a hole through a quarter-inch thick stainless steel slab in about 10 seconds. Determine the irradiance when such a beam has a focused spot area of 10^{-5} cm². What is the electric field amplitude?

Ans. $I = 3 \times 10^{12}$ W/m², $E_0 = 4.8 \times 10^7$ V/m

2.36. Show that the flux density for harmonic electromagnetic waves in vacuum is given by

$$I = (1.33 \times 10^{-3} \text{ W/V}^2)E_0^2$$

2.37. An isotropic point source radiates equally in all directions. If the electric field amplitude at 10 meters from the source is measured to be 10 V/m, determine the radiant flux (power).

Ans.   167.6 W

2.38. With Problem 2.13 in mind, show that the irradiance corresponding to an **E**-field of the form

$$\mathbf{E} = \mathbf{E}_0 \sin{(kx - \omega t)}$$

is given by $I = (c\epsilon_0/2)E_0^2$.

PHOTONS — ENERGY AND MOMENTUM

2.39. Show that

$$\mathcal{E} = \frac{1239}{\lambda}$$

where energy is in electron volts (1 eV = 1.6021×10^{-19} J) and λ is in nanometers. (There will not be much error if you remember the above number as 1234.)

2.40. The electromagnetic flux density impinging normally on a surface just outside the earth's atmosphere is about 2 cal cm^{-2} min^{-1}. Assuming perfect reflection, determine the corresponding radiation pressure from the sun. (1 J = 0.239 cal)

Ans. $\mathcal{P} = 9.32 \times 10^{-6}$ N/m^2, or about 10^{-10} atm

2.41. The fact that photons striking a metal will liberate electrons is the quintessence of the photoelectric effect. If the minimum energy required to free a photoelectron from sodium is 1.8 eV, what is the longest-wavelength light that should be used?

Ans. $\lambda_0 = 688.3$ nm

2.42. A flashlight emits 1 mW of collimated light. What average thrust is exerted by the flashlight?

Ans. $\frac{1}{3} \times 10^{-11}$ N, which is about a billionth of the weight of a 1-gram mass

2.43. Tiny glass spheres have been suspended in midair on a beam of laser light. What is the force on a perfectly reflecting 9×10^{-2}-cm^2 surface arising from a 600-W collimated laser beam having a cross-sectional area of 4 mm^2?

Ans. $F = 4 \times 10^{-6}$ N

THE ELECTROMAGNETIC-PHOTON SPECTRUM

2.44. Radiation from interstellar clouds of hydrogen is detected at a 21-cm wavelength. What class of electromagnetic waves are these? Determine their frequency and photon energy.

Ans. Microwaves; $\nu = 1.4 \times 10^9$ Hz, $\mathcal{E} = 9.24 \times 10^{-25}$ J

2.45. In an article "The Longest Electromagnetic Waves" (*Sci. Am.* March 1962, 128) J. R. Heirtzler describes the detection of waves 18,600,000 miles long. What kind of electromagnetic waves are these? Determine their period, and their photon energy in eV.

Ans. Radio-frequency waves; $\tau = 100$ s, 4.1×10^{-17} eV

2.46. Consider radiant energy of wavelengths $\lambda_0 = 10^{-12}$ m (γ-ray), $\lambda_0 = 500$ nm (green light) and $\lambda_0 = 1$ cm (microwave). How many photons of each are needed to carry 1 erg of energy? (1 J = 10^7 ergs)

Ans. 5×10^5, 3×10^{11}, 5×10^{15}

2.47. Compare the photon energy of 10-cm microwaves to that of a He-Ne laser beam ($\lambda_0 = 632.9$ nm).

Ans. 1.98×10^{-24} J and 3.1×10^{-19} J

Chapter 3

Reflection and Transmission

3.1 INTRODUCTION

This chapter deals, for the most part, with reflection at, or transmission through, an interface separating two different materials. In theory, at least, one could use Maxwell's formalism to trace an electromagnetic wave through a system, but there are other, generally simpler methods. Accordingly, Snell's law, the law of reflection and Fermat's principle — all of them at least three hundred years old — describe various aspects of the behavior of light with no concern as to its actual nature. Going beyond a simple determination of the direction of travel, the Fresnel equations make it possible to calculate just how much light will be reflected and how much transmitted at each interface.

3.2 THE LAWS OF REFLECTION AND REFRACTION

Figure 3-1(a) depicts an incident plane wave arriving at the interface between two media of refractive indices n_i and n_t. Most generally, a portion of the incoming light is reflected back into the incident medium, while the remainder propagates into the transmitting medium. The latter portion is often referred to as the *refracted wave*. Here the angles θ_i, θ_r and θ_t relate to the incident, reflected and transmitted waves, respectively. Figure 3-1(b) is the associated ray diagram. *A ray is a line in the direction of flow of radiant energy* and in isotropic media it simply corresponds to a normal to the wavefronts. Clearly, in such media, rays are parallel to the wave's propagation vector, **k**.

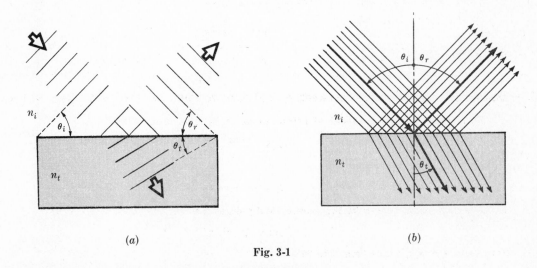

(a)   (b)

Fig. 3-1

The three basic laws of reflection and refraction are as follows:

(1) The incident, reflected and transmitted rays all reside in a plane, known as the *plane of incidence*, which is normal to the interface.

(2) The angle of incidence equals the angle of reflection: $\theta_i = \theta_r$.

32

(3) The incident and transmitted ray directions are related by *Snell's law:*

$$n_i \sin \theta_i = n_t \sin \theta_t$$

SOLVED PROBLEMS

3.1. A beam of collimated light (i.e. having parallel rays) traveling in air makes an angle of 30° to the normal to a glass plate. If the index of the glass is $n_g = 3/2$, determine the direction of the transmitted beam within the plate.

Snell's law,

$$n_i \sin \theta_i = n_t \sin \theta_t$$

gives us the relationship between the incident and transmitted angles. Here $n_i = 1$ for air, $\theta_i = 30°$ and $n_t = 3/2$; hence

$$(1) \quad \sin 30° = \frac{3}{2} \sin \theta_t$$

Inasmuch as $\sin 30° = 1/2$, $\sin \theta_t = 0.333$ and so $\theta_t = 19.5°$. This is the angle made with the normal, as in Fig. 3-1(*b*).

3.2. Envision the interface between two regions, one of glass ($n_g = 1.5$) and the other of water ($n_w = 1.33$). A ray traveling in the glass impinges on the interface at 45° and refracts into the water. What is the transmission angle?

We apply Snell's law

$$n_i \sin \theta_i = n_t \sin \theta_t$$

The incident medium is glass, $n_i = 1.5$, and $n_t = 1.33$. This leads to

$$1.5 \sin 45° = 1.33 \sin \theta_t$$

$$\frac{(1.5)(0.707)}{1.33} = \sin \theta_t$$

$$0.794 = \sin \theta_t$$

Accordingly, $\theta_t = 52.6°$.

3.3. Describe the relationship between θ_t and θ_i both when $n_t > n_i$ and when $n_i > n_t$.

By defining the *relative index of refraction* as $n_{ti} \equiv n_t/n_i$, Snell's law,

$$\sin \theta_i = \frac{n_t}{n_i} \sin \theta_t$$

becomes

$$\sin \theta_i = n_{ti} \sin \theta_t$$

If the transmitting medium is the more optically dense, i.e. $n_t > n_i$, then $n_{ti} > 1$ and

$$\sin \theta_i > \sin \theta_t$$

Since both θ_i and θ_t range from 0 to 90°,

$$\theta_i > \theta_t$$

Similarly, when $n_i > n_t$, $\theta_i < \theta_t$.

Compare these results with the results of Problems 3.1 and 3.2.

3.4. (*a*) Prove that a ray incident at θ_i to a planar glass plate immersed in air will emerge from the plate at the same angle. (*b*) Derive an expression for the displacement a of the ray if the thickness of glass is d.

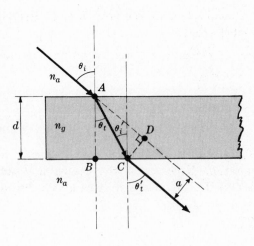

Fig. 3-2

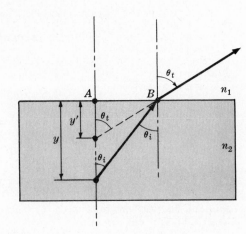

Fig. 3-3

(a) We must show that $\theta_i = \theta_t'$ in Fig. 3-2. Accordingly, Snell's law yields

$$n_a \sin \theta_i = n_g \sin \theta_t$$

where $\theta_t = \theta_i'$. Furthermore, at the second surface

$$n_g \sin \theta_i' = n_a \sin \theta_t'$$

Comparison of these two expressions leads to

$$n_a \sin \theta_i = n_g \sin \theta_i' = n_a \sin \theta_t'$$

and so $\theta_i = \theta_t'$; the incident and emerging rays are parallel.

(b) In Fig. 3-2, $\angle CAD = \theta_i - \theta_t$ and therefore in triangle CAD

$$a = \overline{AC} \sin (\theta_i - \theta_t)$$

But $\overline{AC} = d/(\cos \theta_t)$, from which it follows that

$$a = \frac{d \sin (\theta_i - \theta_t)}{\cos \theta_t}$$

3.5. Imagine that we have two media (of indices n_1 and n_2) separated by a planar interface. An object in the more dense medium (n_2) is a distance y below the interface. An observer above the boundary will see the object as if it were at a distance y' below it. Write an expression for y' in terms of y and the refractive indices, assuming the line of sight to be nearly normal to the interface.

The geometry is illustrated in Fig. 3-3. We know that

$$n_2 \sin \theta_i = n_1 \sin \theta_t$$

and from the figure

$$\overline{AB} = y \tan \theta_i = y' \tan \theta_t$$

Dividing these equations leads to

$$\frac{n_2 \cos \theta_i}{y} = \frac{n_1 \cos \theta_t}{y'}$$

In this case where θ_i, and therefore also θ_t, is small, $\cos \theta_i \approx \cos \theta_t \approx 1$ and

$$y' = y \frac{n_1}{n_2}$$

The expression is a lot more complicated when θ_i is not small.

3.6.   A fish appears to be 2 m below the surface of a pond when viewed from almost directly
above by an angler.   What is the actual depth of the fish?

From Problem 3.5, $y' = y n_1/n_2$.   In this instance $y' = 2$ m, $n_1 = 1$ and $n_2 = 1.33$.   Substitution then gives

$$y = \frac{1.33}{1}(2) = 2.66 \text{ m}$$

as the actual depth.

3.7.   Imagine a stratified system consisting of planar layers of transparent materials of
different thicknesses.   Show that the propagation direction of the emerging beam
is determined by only the incident direction and the refractive indices of the initial
and final layers (n_1 and n_f).

Referring to Fig. 3-4, we obtain from Snell's law:

$$n_1 \sin \theta_{i1} = n_2 \sin \theta_{t2}$$
$$n_2 \sin \theta_{i2} = n_3 \sin \theta_{t3}$$
$$\cdots\cdots\cdots\cdots\cdots\cdots\cdots$$
$$n_\ell \sin \theta_{i\ell} = n_f \sin \theta_{tf}$$

Because $\theta_{t2} = \theta_{i2}$, $\theta_{t3} = \theta_{i3}$, etc., these equations lead to

$$n_1 \sin \theta_{i1} = n_2 \sin \theta_{i2} = \cdots = n_\ell \sin \theta_{i\ell} = n_f \sin \theta_{tf}$$

whence
$$n_1 \sin \theta_{i1} = n_f \sin \theta_{tf}$$

Notice that if $n_1 = n_f$, as for a stack of plates immersed in air, $\theta_{i1} = \theta_{tf}$ and the incoming and
outgoing rays are parallel.

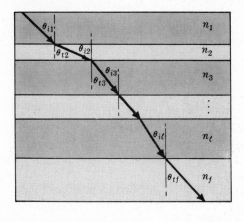

Fig. 3-4

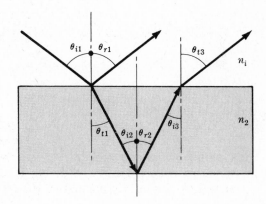

Fig. 3-5

3.8.   A collimated laser beam shines on a tank of water.   Part of the beam reflects off the
top surface and part off the bottom surface, as shown in Fig. 3-5.   Show that the
two beams heading back into the incident medium are parallel.

We have, by the fact that alternate interior angles of parallel lines are equal,

$$\theta_{t1} = \theta_{i2} \qquad \theta_{r2} = \theta_{i3}$$

The law of reflection states that

$$\theta_{i1} = \theta_{r1} \qquad \theta_{i2} = \theta_{r2}$$

Hence $\theta_{t1} = \theta_{i3}$.   Snell's law now gives

$$n_1 \sin \theta_{i1} = n_2 \sin \theta_{t1} \qquad n_2 \sin \theta_{i3} = n_1 \sin \theta_{t3}$$

Substituting $\theta_{t1} = \theta_{i3}$, we obtain $n_1 \sin \theta_{i1} = n_1 \sin \theta_{t3}$, and hence

$$\theta_{i1} = \theta_{r1} = \theta_{t3}$$

3.3 FERMAT'S PRINCIPLE

About two thousand years ago, Hero of Alexandria asserted that *a ray of light leaving a point S, reflecting off a mirror and then arriving at some point P traverses the shortest possible path in space* (Fig. 3-6). Although true for reflection in a homogeneous material, this would obviously not obtain for a ray leaving S, *refracting* at an interface, and arriving at P in the transmitting medium (Fig. 3-7). There the shortest distance is a straight line from S to P, and that is certainly not the path taken by light. In 1657 Fermat generalized Hero's observation as follows: *A ray of light in traversing a route from any one point to another follows the path which takes the least time to negotiate.* Although often correct as it stands, this statement is not the whole truth and will need some modification.

Suppose a ray in going from S to P traverses distances $s_1, s_2, s_3, \ldots, s_m$ in media of indices $n_1, n_2, n_3, \cdots, n_m$, respectively. The total time of flight is then

$$t = \sum_{i=1}^{m} \frac{s_i}{v_i} = \frac{1}{c} \sum_{i=1}^{m} n_i s_i$$

This last summation is referred to as the *optical path length,* or O.P.L. Fermat's principle can then be reworded as: *A ray traverses a route which corresponds to the shortest optical path length.*

To give the modern and most general statement of Fermat's principle, we recall the notion of a *stationary value* of a function. The function $f(x)$ is said to have a stationary value at $x = x_0$ if its derivative, df/dx, vanishes at $x = x_0$. A stationary value could correspond to a maximum, a minimum or a point of inflection with a horizontal tangent. In any case, $f(x)$ varies slowly in the vicinity of a stationary value $f(x_0)$, so that $f(x) \approx f(x_0)$ for $x \approx x_0$.

We may now express Fermat's principle as: *A ray of light in going from any one point to another follows, regardless of the media involved, a route which corresponds to a stationary value of the optical path length.* This applies equally well for inhomogeneous media, for which we have

$$\text{O.P.L.} = \int_{S}^{P} n(s)\, ds$$

The actual path is again the one for which the derivative of the O.P.L. is zero.

Physically, Fermat's principle can be interpreted as a statement of the effects of constructive interference. More about this in Chapter 6.

SOLVED PROBLEMS

3.9.   Justify that an alternative statement of Fermat's principle is: *The actual route taken by a ray of light is the one whose O.P.L. is very nearly equal (i.e. within a first approximation) to the optical path lengths of possible nearby routes.*

The validity of the statement follows from the remark that if x_0 is a point at which a function $f(x)$ is stationary, then

$$f(x_0) \approx f(x)$$

for values of x such that $x \approx x_0$. Here the function is the O.P.L. Thus the actual route taken by light has a stationary O.P.L. whose value is nearly equal to the optical path lengths of adjacent routes.

Another way to think about this is in terms of the Taylor series expansion of $f(x)$. Taken about x_0 where

$$\frac{df}{dx} = 0$$

the Taylor series yields

$$f(x) \approx f(x_0) \quad \text{for} \quad x \approx x_0$$

just as expected. The phrase "within a first approximation" arises from the fact that the second-derivative term in the series need not be zero. Were we to examine a possible but unreal path (i.e. one not taken by light), adjacent routes would be found to differ greatly in their O.P.L. values.

3.10. Use Fermat's principle (calculus version) to arrive at the law of reflection.

As shown in Fig. 3-6, a ray leaves S, strikes the interface at an unspecified point B, and reflects off to P. Assuming the medium to be homogeneous and of index n, we have

$$\text{O.P.L.} = n\overline{SB} + n\overline{BP}$$
$$= n(h^2 + x^2)^{1/2} + n[b^2 + (a-x)^2]^{1/2}$$

Here the O.P.L. is a function of the variable x, and light will take only the route for which

$$\frac{d(\text{O.P.L.})}{dx} = 0$$

i.e.

$$nx(h^2 + x^2)^{-1/2} - n(a-x)[b^2 + (a-x)^2]^{-1/2} = 0$$

But this is equivalent to

$$n \sin \theta_i - n \sin \theta_r = 0$$

and so $\theta_i = \theta_r$. Thus, if a ray goes from S to P via reflection at B, Fermat's principle demands that B be located such that the angle of incidence equals the angle of reflection.

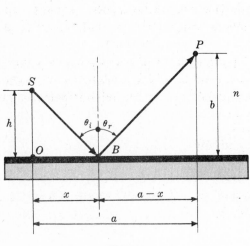

Fig. 3-6

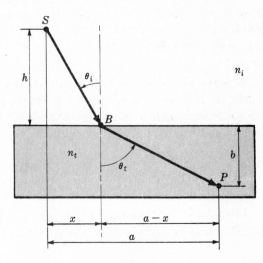

Fig. 3-7

3.11. Apply Fermat's principle (calculus version) to the case of refraction in order to derive Snell's law.

In Fig. 3-7 a ray goes from S to P via refraction at point B on the interface. The scheme is to locate B such that the derivative of the O.P.L. is zero. Thus we write

$$\text{O.P.L.} = n_i \overline{SB} + n_t \overline{BP}$$
$$= n_i(h^2 + x^2)^{1/2} + n_t[b^2 + (a-x)^2]^{1/2}$$

The variable is x and so

$$\frac{d(\text{O.P.L.})}{dx} = 0 = n_i x (h^2 + x^2)^{-1/2} - n_t(a - x)[b^2 + (a - x)^2]^{-1/2}$$

This has the form $0 = n_i \sin \theta_i - n_t \sin \theta_t$, which is obviously equivalent to Snell's law. The value of x corresponding to a stationary O.P.L. is the one for which Snell's law applies. Other locations of B mean different values of x, none of them corresponding to a stationary value of the O.P.L.

3.12. Use the alternative formulation of Fermat's principle given in Problem 3.9 to arrive at Snell's law without calculus.

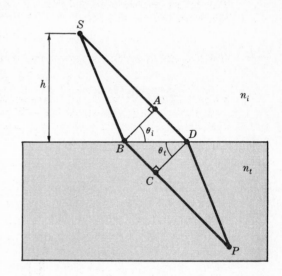

Figure 3-8 shows two rays going from S to P. The optical path lengths along these two routes for an actual light ray will be nearly equal if they are adjacent to each other. Accordingly, assuming $\angle BSD$ and $\angle CPD$ to be small, it follows that $\overline{SB} \approx \overline{SA}$ and $\overline{PC} \approx \overline{PD}$. If the optical path lengths

$$n_i \overline{SB} + n_t \overline{BP}$$

and

$$n_i \overline{SD} + n_t \overline{DP}$$

are to be approximately equal it is necessary that

$$n_i \overline{SB} + n_t \overline{BC} + n_t \overline{CP}$$
$$\approx n_i \overline{SA} + n_i \overline{AD} + n_t \overline{DP}$$

This means that $n_t \overline{BC}$ must nearly equal $n_i \overline{AD}$. If we imagine $\overline{BA}$ and $\overline{CD}$ to correspond to segments of planar wavefronts as in Fig. 3-1(a), then $\overline{BC} \approx \overline{BD} \sin \theta_t$ and $\overline{AD} \approx \overline{BD} \sin \theta_i$. This approximation is good provided $\overline{BD}$ is quite small. Finally, then,

Fig. 3-8

$$n_i \overline{BD} \sin \theta_i = n_t \overline{BD} \sin \theta_t$$

or

$$n_i \sin \theta_i = n_t \sin \theta_t$$

The above treatment is perhaps the easiest mathematically, but it verges on being simplistic.

3.13. A spherical wave diverging from a point S is to enter some arbitrary optical system from which it is to emerge as a wave converging to a point P (as in Fig. 3-9). What does Fermat's principle tell us about the optical path lengths for the various rays going from S to P?

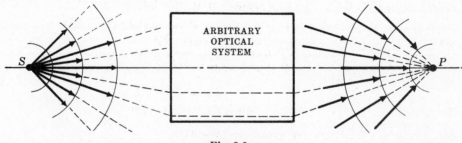

Fig. 3-9

Rays from S to P will presumably traverse a great many different paths through the system. Suppose that one such path corresponds to the minimum O.P.L. between S and P. Fermat's principle implies that light would traverse that minimum O.P.L. route and no other. But other routes must obviously be taken because rays leave S in many directions. It follows that a minimum (or maximum) O.P.L. cannot be *uniquely* attained. In other words, *all rays from S through the system to P must traverse identical optical path lengths*. This is true for all sorts of focusing systems (such as lenses and mirrors).

3.14. A collimated beam incident parallel to the symmetry axis of a certain concave **mirror** is reflected into a converging beam. Use Fermat's principle to show that the **mirror** is paraboloidal.

Figure 3-10 depicts, in cross section, parallel rays corresponding to a plane wave Σ incident on mirror M. The reflected rays converge on point F. The optical path lengths of all routes to F must be the same (compare Problem 3.13, with $S = \infty$ and $P = F$); hence

$$n_i(\overline{AB} + \overline{BF}) \;=\; n_i(\overline{EG} + \overline{GF}) \;=\; \cdots \;=\; n_i(\overline{XY} + \overline{YF})$$

Now let the line segments $\overline{AB}$, $\overline{EG}$, $\ldots$, $\overline{XY}$ be prolonged through the mirror to points $C, H, \ldots, Z$, which are chosen such that

$$\overline{BC} = \overline{BF}, \quad \overline{GH} = \overline{GF}, \quad \ldots, \quad \overline{YZ} = \overline{YF}$$

The two sets of equalities above imply that

$$\overline{AB} + \overline{BC} \;=\; \overline{EG} + \overline{GH} \;=\; \cdots \;=\; \overline{XY} + \overline{YZ}$$

which tells us that the distance between Σ and the line Σ' through $C, H, \ldots, Z$ is constant. We have thus constructed a *straight* line Σ' such that the points of M are equidistant from it and from the point F. By definition, then, M is a parabola (with *focus* F and *directrix* Σ').

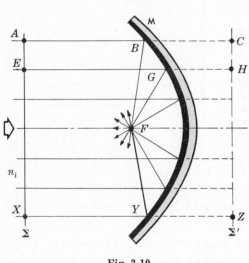

Fig. 3-10

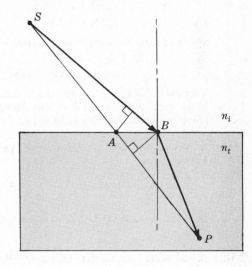

Fig. 3-11

3.15. Imagine a ray of light leaving a point S in air and refracting to a point P in water as shown in Fig. 3-11. Using Fermat's principle but not Snell's law, show that $v_t < v_i$.

It is observed that rays leaving S arrive at P via only one route, $\overline{SBP}$. Accordingly the corresponding O.P.L. must be a maximum or a minimum. But it isn't a maximum, since a ray traveling way out to the side and back to P would have a much greater O.P.L. Thus, $n_i\overline{SB} + n_t\overline{BP}$ is the minimum O.P.L. The straight-line path $\overline{SAP}$ therefore has a larger O.P.L. From the diagram $\overline{SB} > \overline{SA}$, $\overline{AP} > \overline{BP}$ and

$$\overline{SB} + \overline{BP} \;>\; \overline{SA} + \overline{AP}$$

Hence

$$\overline{SB} - \overline{SA} \;>\; \overline{AP} - \overline{BP}$$

both sides of which are positive quantities. Moreover,

$$n_i\overline{SA} + n_t\overline{AP} \;>\; n_i\overline{SB} + n_t\overline{BP}$$

and so

$$n_t(\overline{AB} - \overline{BP}) \;>\; n_i(\overline{SB} - \overline{SA})$$

which means that $n_t > n_i$ and $v_t < v_i$.

3.16. Referring back to Fig. 3-6, derive the fact that $\theta_i = \theta_r$ using Fermat's principle, but this time let θ_i itself be the space variable and not x.

As in Problem 3.10, O.P.L. $= n\overline{SB} + n\overline{BP}$. Here, however, we note that

$$\overline{SB} = \frac{h}{\cos \theta_i} \qquad \overline{BP} = \frac{b}{\cos \theta_r}$$

Consequently,

$$\text{O.P.L.} = \frac{nh}{\cos \theta_i} + \frac{nb}{\cos \theta_r}$$

and differentiating,

$$\frac{d(\text{O.P.L.})}{d\theta_i} = \frac{nh \sin \theta_i}{\cos^2 \theta_i} + \frac{nb \sin \theta_r}{\cos^2 \theta_r} \frac{d\theta_r}{d\theta_i}$$

Hence

$$\frac{h \sin \theta_i}{\cos^2 \theta_i} = -\frac{b \sin \theta_r}{\cos^2 \theta_r} \frac{d\theta_r}{d\theta_i} \qquad (1)$$

It follows from the geometry that θ_i and θ_r are related via

$$h \tan \theta_i + b \tan \theta_r = a$$

and differentiation with respect to θ_i yields

$$h \sec^2 \theta_i + b \sec^2 \theta_r \frac{d\theta_r}{d\theta_i} = 0$$

or

$$\frac{h}{\cos^2 \theta_i} = -\frac{b}{\cos^2 \theta_r} \frac{d\theta_r}{d\theta_i} \qquad (2)$$

From (1) and (2), $\sin \theta_i = \sin \theta_r$; that is, $\theta_i = \theta_r$.

3.4 THE FRESNEL EQUATIONS

Augustin Jean Fresnel, about one hundred and fifty years ago, derived a set of expressions which allow us to calculate the amount of light reflected and transmitted at an interface. Envision a planar harmonic light wave incident on the interface between two dielectric media, with the E-field normal to the plane of incidence (Fig. 3-12). The boundary conditions require that we match the phases of the incident, reflected and transmitted waves [i.e. $\mathbf{E}_i(\mathbf{r}, t)$, $\mathbf{E}_r(\mathbf{r}, t)$ and $\mathbf{E}_t(\mathbf{r}, t)$]. This leads to the laws of reflection and refraction, with which we are already familiar. There are still other boundary conditions on the E- and B-fields and these, in turn, yield the Fresnel relationships. Thus, with E_{0i}, E_{0r} and E_{0t} denoting the amplitudes of the incident, reflected and transmitted waves, we find

$$r_\perp \equiv \left(\frac{E_{0r}}{E_{0i}}\right)_\perp = \frac{n_i \cos \theta_i - n_t \cos \theta_t}{n_i \cos \theta_i + n_t \cos \theta_t}$$

$$t_\perp \equiv \left(\frac{E_{0t}}{E_{0i}}\right)_\perp = \frac{2n_i \cos \theta_i}{n_i \cos \theta_i + n_t \cos \theta_t}$$

as the desired expressions for the *amplitude coefficients of reflection and transmission*, respectively. A similar matching of boundary conditions when the E-field is parallel to the plane of incidence (Fig. 3-13) results in another set of amplitude coefficients:

$$r_{||} \equiv \left(\frac{E_{0r}}{E_{0i}}\right)_{||} = \frac{n_t \cos \theta_i - n_i \cos \theta_t}{n_i \cos \theta_t + n_t \cos \theta_i}$$

$$t_{||} \equiv \left(\frac{E_{0t}}{E_{0i}}\right)_{||} = \frac{2n_i \cos \theta_i}{n_i \cos \theta_t + n_t \cos \theta_i}$$

In addition to these field amplitude ratios, one can define the *reflectance, R*, as the ratio of the reflected to the incident flux (or power) and the *transmittance, T*, as the ratio of the transmitted to the incident flux. In other words,

$$R = \left(\frac{E_{0r}}{E_{0i}}\right)^2 = r^2$$

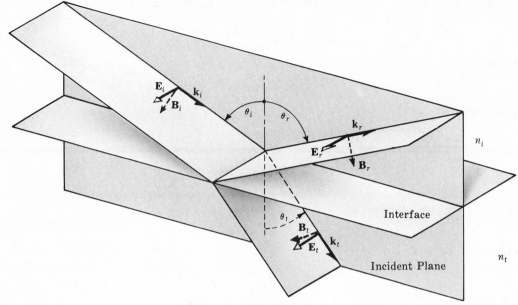

Fig. 3-12

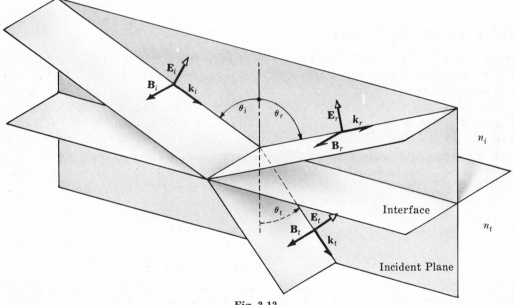

Fig. 3-13

and
$$T = \frac{n_t \cos \theta_t}{n_i \cos \theta_i}\left(\frac{E_{0t}}{E_{0i}}\right)^2 = \left(\frac{n_t \cos \theta_t}{n_i \cos \theta_i}\right)t^2$$

are the reflected and transmitted *power coefficients*. While R is simply the ratio of reflected to incident irradiance, T is somewhat more involved in its form. This results from the fact that the cross-sectional area of the transmitted beam differs from that of the other two, and while power is independent of area, irradiance is not.

SOLVED PROBLEMS

3.17. Rewrite the expressions for the amplitude reflection coefficients as functions of θ_i and θ_t only, i.e. get rid of the explicit dependence of $r_\perp$ and $r_{||}$ on n_i and n_t.

Beginning with the amplitude coefficient expression

$$r_\perp = \frac{n_i \cos \theta_i - n_t \cos \theta_t}{n_i \cos \theta_i + n_t \cos \theta_t}$$

divide by n_i and use Snell's law $(\sin \theta_i / \sin \theta_t = n_t/n_i)$ to get

$$r_\perp = \frac{\cos \theta_i - (\sin \theta_i \cos \theta_t / \sin \theta_t)}{\cos \theta_i + (\sin \theta_i \cos \theta_t / \sin \theta_t)}$$

This can be consolidated into

$$r_\perp = -\frac{\sin (\theta_i - \theta_t)}{\sin (\theta_i + \theta_t)}$$

using the identity $\sin (\alpha \pm \beta) = \sin \alpha \cos \beta \pm \cos \alpha \cos \beta$.

Similarly,
$$r_{\parallel} = \frac{n_t \cos \theta_i - n_i \cos \theta_t}{n_i \cos \theta_t + n_t \cos \theta_i}$$

can be reformulated as

$$r_{\parallel} = \frac{\sin \theta_i \cos \theta_i - \sin \theta_t \cos \theta_t}{\sin \theta_t \cos \theta_t + \sin \theta_i \cos \theta_i}$$

via Snell's law. But this last equation is equivalent to

$$r_{\parallel} = \frac{(\sin \theta_i \cos \theta_t - \sin \theta_t \cos \theta_i)(\cos \theta_i \cos \theta_t - \sin \theta_i \sin \theta_t)}{(\sin \theta_i \cos \theta_t + \sin \theta_t \cos \theta_i)(\cos \theta_i \cos \theta_t + \sin \theta_i \sin \theta_t)}$$

which, in turn, leads to

$$r_{\parallel} = \frac{\sin (\theta_i - \theta_t) \cos (\theta_i + \theta_t)}{\sin (\theta_i + \theta_t) \cos (\theta_i - \theta_t)}$$

and finally

$$r_{\parallel} = \frac{\tan (\theta_i - \theta_t)}{\tan (\theta_i + \theta_t)}$$

3.18. Derive expressions for the amplitude coefficients at normal incidence and compute their numerical values at an air-glass interface where $n_t = 1.5$.

The key point here is that $\theta_i \approx \theta_t \approx 0$. Hence, $\cos \theta_i \approx \cos \theta_t \approx 1$ and

$$r_{\parallel} = \frac{n_t \cos \theta_i - n_i \cos \theta_t}{n_i \cos \theta_t + n_t \cos \theta_i} = \frac{n_t - n_i}{n_i + n_t}$$

$$r_\perp = \frac{n_i \cos \theta_i - n_t \cos \theta_t}{n_i \cos \theta_i + n_t \cos \theta_t} = \frac{n_i - n_t}{n_i + n_t}$$

In summary,

$$[r_{\parallel}]_{\theta_i=0} = [-r_\perp]_{\theta_i=0} = \frac{n_t - n_i}{n_t + n_i}$$

For the case of external reflection $(n_i < n_t)$ at an air-glass interface $(n_i = 1, n_t = 1.5)$,

$$[r_{\parallel}]_{\theta_i=0} = [-r_\perp]_{\theta_i=0} = \frac{1.5 - 1}{1.5 + 1} = \frac{1/2}{5/2} = 0.2$$

In much the same way,

$$[t_{\parallel}]_{\theta_i=0} = [t_\perp]_{\theta_i=0} = \frac{2n_i}{n_i + n_t}$$

which, for air and glass, has the value 0.8.

3.19. Express the amplitude reflection coefficients in terms of θ_i and n_{ti}.

Starting with

$$r_\perp = \frac{n_i \cos \theta_i - n_t \cos \theta_t}{n_i \cos \theta_i + n_t \cos \theta_t}$$

divide by n_i and substitute $n_{ti} \equiv n_t/n_i$ to get

$$r_\perp = \frac{\cos \theta_i - n_{ti} \cos \theta_t}{\cos \theta_i + n_{ti} \cos \theta_t}$$

Snell's law can be written as $(\sin \theta_i)/n_{ti} = \sin \theta_t$. Since

$$\cos^2 \theta_t = 1 - \sin^2 \theta_t = 1 - \frac{\sin^2 \theta_i}{n_{ti}^2}$$

the amplitude coefficient becomes

$$r_\perp = \frac{\cos \theta_i - n_{ti}(1 - \sin^2 \theta_i/n_{ti}^2)^{1/2}}{\cos \theta_i + n_{ti}(1 - \sin^2 \theta_i/n_{ti}^2)^{1/2}}$$

More simply,

$$r_\perp = \frac{\cos \theta_i - (n_{ti}^2 - \sin^2 \theta_i)^{1/2}}{\cos \theta_i + (n_{ti}^2 - \sin^2 \theta_i)^{1/2}}$$

Similarly,

$$r_{||} = \frac{n_{ti}^2 \cos \theta_i - (n_{ti}^2 - \sin^2 \theta_i)^{1/2}}{n_{ti}^2 \cos \theta_i + (n_{ti}^2 - \sin^2 \theta_i)^{1/2}}$$

3.20. Determine the values of the amplitude reflection coefficients for light incident at 30° on an air-glass interface, $n_{ti} = 1.50$.

From Problem 3.19

$$r_\perp = \frac{\cos \theta_i - (n_{ti}^2 - \sin^2 \theta_i)^{1/2}}{\cos \theta_i + (n_{ti}^2 - \sin^2 \theta_i)^{1/2}}$$

or, since $\cos 30° = 0.866$ and $\sin 30° = 0.5$,

$$r_\perp = \frac{0.866 - (9/4 - 1/4)^{1/2}}{0.866 + (9/4 - 1/4)^{1/2}} = \frac{0.866 - 1.414}{0.866 + 1.414}$$

$$= -\frac{0.548}{2.280} = -0.240$$

Similarly,

$$r_{||} = \frac{(9/4)(0.866) - 1.414}{(9/4)(0.866) + 1.414} = \frac{1.949 - 1.414}{1.949 + 1.414} = 0.159$$

The minus sign on the perpendicular coefficient means that the reflected field points in the opposite direction to that shown in Fig. 3-12. In other words, the perpendicular component of the **E**-field is shifted in phase by 180° upon reflection.

3.21. Determine the value of $t_\perp$ for light incident at 30° on an air-glass interface, $n_{ti} = 1.50$, and show that $t_\perp + (-r_\perp) = 1$ for this case.

Using Snell's law, $\sin \theta_i = n_{ti} \sin \theta_t$, we find

$$\frac{1}{2} = \frac{3}{2} \sin \theta_t$$

whence $\sin \theta_t = 0.333$ and therefore $\cos \theta_t = 0.9428$. Thus

$$t_\perp = \frac{2n_i \cos \theta_i}{n_i \cos \theta_i + n_t \cos \theta_t} = \frac{2(0.866)}{0.866 + 1.5(0.943)} = 0.759$$

and, using the results of Problem 3.20,

$$t_\perp + (-r_\perp) = 0.999$$

which is close enough to 1 for our purposes.

The equation $t_\perp + (-r_\perp) = 1$ actually holds for all θ_i, whereas $t_{||} + r_{||} = 1$ obtains only when $\theta_i \approx 0$.

3.22. Suppose that a linearly polarized wave impinges on an interface such that the plane of vibration makes an angle of γ_i with the plane of incidence. If the reflectance of the components parallel and perpendicular to the plane of incidence are $R_{||}$ and $R_\perp$, respectively, write an expression for the total reflectance, R.

Since the cross-sectional areas of the incident and reflected beams are equal, we can simply work with their irradiances. Accordingly $R = I_r / I_i$, where $I_r = I_{r||} + I_{r\perp}$, and so

$$R = \frac{I_{r||} + I_{r\perp}}{I_i} = \frac{I_{r||}}{I_i} + \frac{I_{r\perp}}{I_i}$$

Inasmuch as the field components are expressible as

$$[E_{0i}]_{||} = E_{0i} \cos \gamma_i \qquad [E_{0i}]_\perp = E_{0i} \sin \gamma_i$$

it follows from $I = c\epsilon_0 \langle E^2 \rangle$ that

$$I_{i||} = I_i \cos^2 \gamma_i \qquad I_{i\perp} = I_i \sin^2 \gamma_i$$

Substituting back into the reflectance yields

$$R = \frac{I_{r||}}{I_i} \cos^2 \gamma_i + \frac{I_{r\perp}}{I_{i\perp}} \sin^2 \gamma_i$$

or

$$R = R_{||} \cos^2 \gamma_i + R_\perp \sin^2 \gamma_i$$

The transmittance has the same form:

$$T = T_{||} \cos^2 \gamma_i + T_\perp \sin^2 \gamma_i$$

3.23. Write expressions for (a) the reflectance and (b) the transmittance at near-normal incidence. (c) Determine the percentage of light lost in reflection at an air-glass ($n_g = 1.5$) interface.

(a) From Problem 3.22

$$R = R_{||} \cos^2 \gamma_i + R_\perp \sin^2 \gamma_i$$

where $R_{||} = r_{||}^2$ and $R_\perp = r_\perp^2$. At normal incidence, we know from Problem 3.18 that

$$[R_{||}]_{\theta_i=0} = \left(\frac{n_t - n_i}{n_t + n_i} \right)^2 = [R_\perp]_{\theta_i=0}$$

in which case

$$[R]_{\theta_i=0} = [R_{||}]_{\theta_i=0} = [R_\perp]_{\theta_i=0} = \left(\frac{n_t - n_i}{n_t + n_i} \right)^2$$

(b) The transmittance is given by

$$T = \frac{n_t \cos \theta_t}{n_i \cos \theta_i} t^2$$

and Problem 3.18 led to

$$[t_{||}]_{\theta_i=0} = [t_\perp]_{\theta_i=0} = \frac{2n_i}{n_i + n_t}$$

Hence, keeping in mind that

$$T = T_{||} \cos^2 \gamma_i + T_\perp \sin^2 \gamma_i$$

and

$$[T_{||}]_{\theta_i=0} = \frac{n_t}{n_i} \left(\frac{2n_i}{n_i + n_t} \right)^2 = [T_\perp]_{\theta_i=0}$$

we finally arrive at

$$[T]_{\theta_i=0} \;=\; [T_{||}]_{\theta_i=0} \;=\; [T_\perp]_{\theta_i=0} \;=\; \frac{4n_t n_i}{(n_i+n_t)^2}$$

(c) Substitution of $n_i = 1$ and $n_t = 1.5$ into the result of (a) yields a loss of

$$[R]_{\theta_i=0} \;=\; 4\%$$

at a single air-glass interface.

3.24. Natural or unpolarized light is such that the azimuthal angle γ_i of Problem 3.22 changes rapidly and randomly, as does the field amplitude. Derive an expression for R_n, the reflectance of natural light, in terms of $R_{||}$ and $R_\perp$, bearing in mind that these, in turn, are both functions of θ_i and θ_t.

Our aim is to rewrite

$$R_n \;=\; \frac{I_{r||} + I_{r\perp}}{I_i}$$

while feeding in the information that we're dealing with unpolarized light (via I_i).

Looking back at Problem 3.22, notice that we found $I_{i||}$ and $I_{i\perp}$ in the usual way by squaring and time averaging the field components. Now, however, γ_i is a function of time and (compare Problem 2.13)

$$\langle \cos^2 \gamma_i(t)\rangle \;=\; \langle \sin^2 \gamma_i(t)\rangle \;=\; \frac{1}{2}$$

Thus for natural light $I_{i||} = I_i/2$, $I_{i\perp} = I_i/2$ and we can write

$$I_{r||} \;=\; \frac{I_{r||} I_i}{2 I_{i||}} \;=\; \frac{1}{2} R_{||} I_i$$

$$I_{r\perp} \;=\; \frac{I_{r\perp} I_i}{2 I_{i\perp}} \;=\; \frac{1}{2} R_\perp I_i$$

Consequently,

$$R_n \;=\; \frac{1}{2}(R_{||} + R_\perp) \;=\; \frac{1}{2}(r_{||}^2 + r_\perp^2)$$

3.25. What percentage of the incoming irradiance is reflected at an air-glass ($n_g = 1.5$) interface for a beam of natural light incident at $70°$?

From Problem 3.24

$$R_n \;=\; \frac{1}{2}(r_{||}^2 + r_\perp^2)$$

while Problem 3.19 led to

$$r_\perp \;=\; \frac{\cos\theta_i - (n_{ti}^2 - \sin^2\theta_i)^{1/2}}{\cos\theta_i + (n_{ti}^2 - \sin^2\theta_i)^{1/2}}$$

$$r_{||} \;=\; \frac{n_{ti}^2 \cos\theta_i - (n_{ti}^2 - \sin^2\theta_i)^{1/2}}{n_{ti}^2 \cos\theta_i + (n_{ti}^2 - \sin^2\theta_i)^{1/2}}$$

With $\theta_i = 70°$, $\cos\theta_i = 0.34$, $\sin\theta_i = 0.94$ and $n_{ti} = 1.5$, we obtain

$$r_\perp \;=\; \frac{0.34 - (2.25 - 0.88)^{1/2}}{0.34 + (2.25 - 0.88)^{1/2}} \;=\; -0.55$$

$$r_{||} \;=\; \frac{2.25(0.34) - (2.25 - 0.88)^{1/2}}{2.25(0.34) + (2.25 - 0.88)^{1/2}} \;=\; -0.21$$

Substitution into the expression for the reflectance yields $R_n = 17\%$.

3.5 THE CRITICAL ANGLE

For the most part we have concerned ourselves with the case of *external reflection* $(n_t > n_i)$. The opposite situation, in which $n_i > n_t$, is known as *internal reflection* and is also of considerable practical interest. Recall from Problem 3.17 that

$$r_\perp = -\frac{\sin(\theta_i - \theta_t)}{\sin(\theta_i + \theta_t)}$$

$$r_{||} = \frac{\tan(\theta_i - \theta_t)}{\tan(\theta_i + \theta_t)}$$

When $n_i > n_t$ Snell's law demands that $\theta_t > \theta_i$. Therefore $r_\perp$ is positive, rising with increasing θ_i from 0.2 to a value of 1.0. On the other hand, $r_{||}$ ranges from -0.2 to 1.0. Both coefficients reach a value of 1.0 at an incident angle referred to as θ_c, the *critical angle*. When $\theta_i = \theta_c$, $\theta_t = 90°$; beyond this, the amplitude coefficients become complex.

Envision a beam of light impinging on an interface between two transparent media where $n_i > n_t$. At normal incidence $(\theta_i = 0)$ most of the incoming light is transmitted into the less dense medium. As θ_i increases, more and more light is reflected back into the denser medium, while θ_t increases. When $\theta_t = 90°$, θ_i is defined to be θ_c and the transmittance becomes zero. For $\theta_i > \theta_c$ all of the light is *totally internally reflected*, remaining in the incident medium.

SOLVED PROBLEMS

3.26. Use Snell's law to derive an expression for θ_c. Compute the value of θ_c for a water-air interface $(n_w = 1.33)$.

Rewrite
$$n_i \sin \theta_i = n_t \sin \theta_t$$

as
$$\sin \theta_i = n_{ti} \sin \theta_t$$

where $n_{ti} < 1$. Requiring that $\theta_t = 90°$ for $\theta_i = \theta_c$ leads to
$$\sin \theta_c = n_{ti}$$

At a water-air interface
$$\theta_c = \sin^{-1}(1/1.33) = \sin^{-1} 0.752 = 48.8°$$

3.27. A tank of water is covered with a 1-cm thick layer of linseed oil $(n_o = 1.48)$ above which is air. What angle must a beam of light, originating in the tank, make at the water-oil interface if no light is to escape?

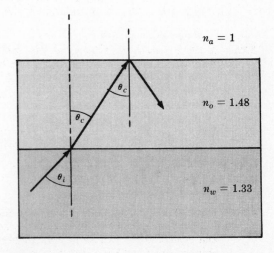

The appropriate geometry is indicated in Fig. 3-14. Total internal reflection can occur only at the oil-air interface since $n_w < n_o$. There
$$\theta_c = \sin^{-1}(1/1.48) = 42.5°$$

Snell's law applied to the water-oil interface is
$$1.33 \sin \theta_i = 1.48 \sin \theta_c = 1$$

and $\theta_i = 48.8°$ In other words, for incident angles equal to or greater than 48.8° the beam will be reflected back into the water. Notice that this same angle would obtain even without the oil layer.

Fig. 3-14

3.28. Imagine yourself lying on the floor of a pool filled with water, looking straight upwards. How large a plane angle does the field of view beyond the pool apparently subtend?

Rays striking the air-water interface from above at glancing incidence will enter the water at a transmission angle equal to θ_c. The plane angle subtended at the observer is therefore $2\theta_c$. Here,

$$\sin \theta_c = \frac{1}{1.33}$$

whence $\theta_c = 48.8°$ and $2\theta_c = 97.6°$.

3.29. The sparkling appearance of a gem-cut diamond arises from total internal reflection. Light entering from above is reflected back out toward the viewer, re-emerging through the top facets. Determine the critical angle ($n_d = 2.417$) and compare it to that of glass ($n_g = 1.5$).

By Problem 3.26

$$\sin \theta_c = n_{ti} = \frac{1}{2.417}$$

Hence

$$\theta_c = \sin^{-1} 0.4137 = 24.4°$$

This is a good deal smaller than the corresponding value of $41.8°$ for glass-air.

3.30. Determine the critical angle for a water ($n_w = 1.33$)-glass ($n_g = 1.50$) interface.

We have

$$\sin \theta_c = n_{ti}$$

or

$$\theta_c = \sin^{-1}\frac{1.33}{1.50} = \sin^{-1} 0.887 = 62.5°$$

Supplementary Problems

THE LAWS OF REFLECTION AND REFRACTION

3.31. Show that the displacement of a beam on passing through a parallel plate of index n_t (Problem 3.4) can be expressed as

$$a = d(\sin \theta_i)\left(1 - \frac{n_i \cos \theta_i}{n_t \cos \theta_t}\right)$$

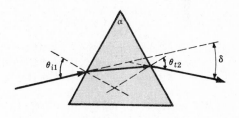

Fig. 3-15

3.32. A ray traverses a prism as in Fig. 3-15. Show that the ray undergoes a deviation δ given by

$$\delta = \theta_{i1} + \theta_{t2} - \alpha$$

3.33. Use the law of reflection to show that the deviation δ produced by the two mirrors in Fig. 3-16 is given by

$$\delta \equiv \alpha + \beta = 4\pi - 2\gamma$$

3.34. A ray enters the mirrors of Fig. 3-17, page 48, and is reflected several times, ultimately retracing its path and emerging. Write an expression for the relationship which must exist between θ_i and α.

Ans. $\theta_i = 3\alpha$

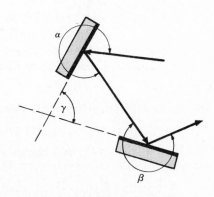

Fig. 3-16

3.35. Figure 3-18 depicts a scheme for graphically constructing the path of a refracted ray. First an arc is swept out having a radius equal to n_i and then another equal to n_t. The incident ray is extended to meet the first circular segment, a perpendicular to the interface is dropped and its intersection with the second arc locates the refracted ray. This is a very handy procedure for ray-tracing — prove that it's correct.

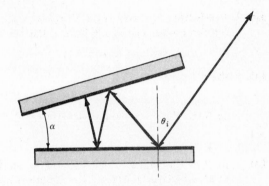

Fig. 3-17

FERMAT'S PRINCIPLE

3.36. Envision a concave ellipsoidal mirror whose axis of revolution passes through its two foci. Show, using Fermat's principle, that the rays diverging from a point source at one focus converge to the other focus after reflection from the inner surface of the ellipsoid. (Here the O.P.L. has neither a maximum nor a minimum.)

3.37. Referring back to Fig. 3-7, derive Snell's law where θ_i is the space variable rather than x. Remember that $h \tan \theta_i + b \tan \theta_t = a$, where where h, b and a are constants.

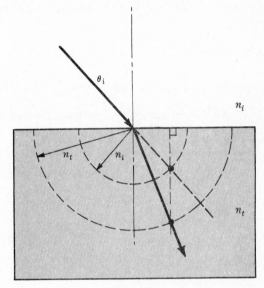

Fig. 3-18

3.38. As shown in Fig. 3-19, two possible adjacent paths from S to P are $\overline{SAP}$ and $\overline{SBP}$. Show that the difference in optical path lengths (which must be vanishingly small if $\overline{SAP}$ is to be an actual ray trajectory) is given by

$$\frac{n_i h \sin \theta_i}{\cos^2 \theta_i} d\theta_i + \frac{n_t b \sin \theta_t}{\cos^2 \theta_t} d\theta_t$$

Go on to derive Snell's law.

3.39. Use Fermat's principle to show that the incident and reflected rays at a planar interface must be in a common plane, viz, the plane of incidence.

THE FRESNEL EQUATIONS

3.40. A linearly polarized beam of light whose **E**-field is normal to the plane of incidence impinges in air at 45° on an air-glass interface. Assuming $n_g = 1.5$, determine the amplitude coefficients of reflection and transmission.

Ans. $r_\perp = -0.3034$, $t_\perp = 0.6966$

3.41. Show that the amplitude transmission coefficients can be reformulated without an explicit dependence on n_i or n_t, as follows:

$$t_\perp = \frac{2 \sin \theta_t \cos \theta_i}{\sin (\theta_i + \theta_t)}$$

$$t_{||} = \frac{2 \sin \theta_t \cos \theta_i}{\sin (\theta_i + \theta_t) \cos (\theta_i - \theta_t)}$$

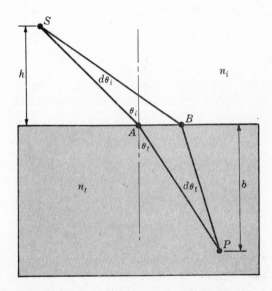

Fig. 3-19

3.42. Verify that $t_\perp + (-r_\perp) = 1$ for all values of θ_i. (It is probably simplest to use the expressions for the coefficients which depend only on θ_i and θ_t.) Show, as well, that $n_{ti}t_{||} + (-r_{||}) = 1$.

3.43. Show that

$$[R]_{\theta_i=0} + [T]_{\theta_i=0} = 1$$

In fact, $R + T = 1$ independently of θ_i.

3.44. In Problem 3.18 we calculated the amplitude coefficients at normal incidence for external reflection at an air-glass interface. Recompute these coefficients with the light incident in the glass.

Ans. $[-r_{||}]_{\theta_i=0} = [r_\perp]_{\theta_i=0} = 0.2$ and $[t_{||}]_{\theta_i=0} = [t_\perp]_{\theta_i=0} = 1.2$. (This is correct; E_{0t} is greater than E_{0i} here because the field on one side of the boundary must equal its value on the other side; i.e. $\mathbf{E}_i + \mathbf{E}_r = \mathbf{E}_t$, and $\mathbf{E}_i$ and $\mathbf{E}_r$ are in phase when $\theta_i = 0$. Energy is, of course, conserved.)

3.45. Rewrite the reflectance and transmittance at *normal incidence* as

$$[R]_{\theta_i=0} = \left(\frac{n_{it} - 1}{n_{it} + 1}\right)^2 \qquad [T]_{\theta_i=0} = \frac{4n_{it}}{(n_{it} + 1)^2}$$

(using the results of Problem 3.23) and verify several of the points on Fig. 3-20. Solve for the values of n_{it} such that $[R]_{\theta_i=0} = [T]_{\theta_i=0}$.

Ans. $n_{it} = 1$ corresponds to no interface at all, in which case $[T]_{\theta_i=0} = 1$, $[R]_{\theta_i=0} = 0$; for $n_{it} = 1/1.5 = 0.66$, $[R]_{\theta_i=0} = 0.04$, $[T]_{\theta_i=0} = 0.96$. The reflectance equals the transmittance when n_{it} equals 5.8 and 0.17.

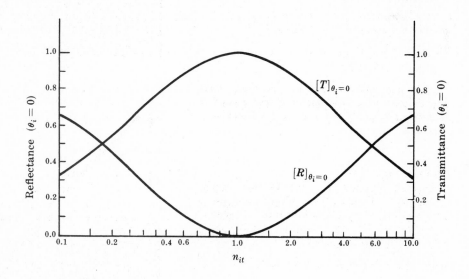

Fig. 3-20

THE CRITICAL ANGLE

3.46. Three transparent materials of indices $n_A < n_B < n_C$ form a layered structure with n_A on top and n_C on bottom. If the critical angles at both the A-B and B-C interfaces are 45°, determine n_{AC}.

Ans. $n_{AC} = \frac{1}{2}$

3.47. Light is to be totally internally reflected by the prism of Fig. 3-21. Determine the corresponding minimum index of refraction of the prism if it is to be imbedded in air.

Ans. 1.414

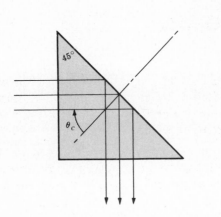

Fig. 3-21

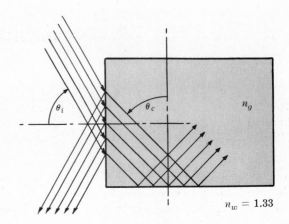

Fig. 3-22

3.48. What is the refractive index of the glass block in Fig. 3-22 if an incident angle of $\theta_i = 45°$ results in total internal reflection at the bottom surface? (This setup is the basis for several refractometers.)

Ans. 1.63

3.49. Imagine that we have a clear liquid in an open container and we determine that $\theta_c = 45°$. Now if we shine light from above at varying values of θ_i an orientation will be found ($\theta_i = \theta_p$) where the reflected light is linearly polarized, i.e. $r_{||} = 0$. Compute θ_p, the polarization angle.

Ans. 54.7°

3.50. Figure 3-23 depicts a thin glass fiber (n_f) surrounded by a lower-density cladding layer (n_c). There is a maximum incident angle $\theta_i = \theta_{max}$ such that any ray impinging on the face at $\theta_i > \theta_{max}$ will arrive at an internal wall at an angle less than θ_c and will not be totally internally reflected. Show that

$$\sin \theta_{max} = \frac{1}{n_0}(n_f^2 - n_c^2)^{1/2}$$

Light caught within the cylinder will be multiply reflected down its length. This is the basis for what has come to be known as *fiber optics*.

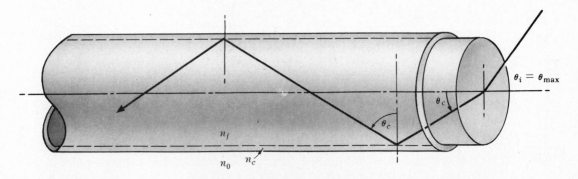

Fig. 3-23

Chapter 4

Geometrical Optics

4.1 INTRODUCTION

This chapter concerns itself with the basic techniques used to collect and reshape wavefronts, most often with the intent of forming some sort of image. Accordingly, we examine the configurations of various reflecting and refracting devices (lenses, mirrors, etc.) which will effect the wavefront alterations needed for specific purposes.

The domain of geometrical optics, as distinct from physical optics, is limited to situations where diffraction effects arising from the inherent wave nature of light are negligible. This simplification is tantamount to requiring rectilinear propagation in homogeneous media —that is, rays are assumed to traverse straight lines.

4.2 ASPHERICAL REFRACTING SURFACES

We first examine how to reconfigure a wavefront by causing it to propagate through a curved interface separating two transparent media. Once having done that it is simple to determine the effect of a series of such surfaces (a lens).

Suppose that we have a segment of a spherical wave diverging from a point source S and we wish to reshape it into a spherical wave converging to a point P, as in Fig. 4-1 (see

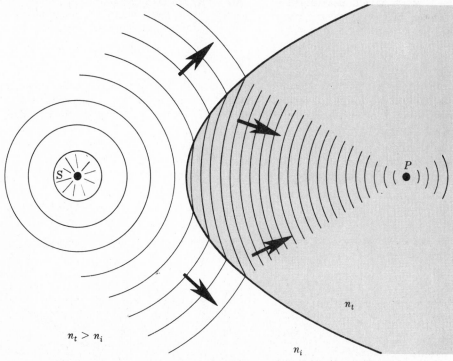

Fig. 4-1

51

Problem 3.13).   In effect, the central portion of the wavefront must be held back with respect to its edges.   Since the wave travels slower in a medium of higher index, this suggests elongating the interface in the vicinity of the SP-axis.   The edges of the wave would then travel longer in the optically less dense (faster) medium and in so doing overtake and pass the central region of the wave, thereby inverting the wavefront.   This particular interface configuration is known as a *Cartesian ovoid*.   Similarly, if we wanted the waves in the transmitting medium to be planar, the interface would now have to be a bit flatter so that the edges of the wavefront overtake, but this time do not pass, the front's central region.   It turns out that such an interface would have to be hyperboloidal with S at a focus.

Many aspherical surfaces are of practical interest but they all share the common drawback of being difficult to fabricate.   Even so, precision aspherical elements are being used where their high cost can be justified (e.g. in reconnaissance cameras).

SOLVED PROBLEMS

4.1.   Derive an expression for the Cartesian ovoid depicted in Fig. 4-2.   Use Fermat's principle, being careful to explain its application.

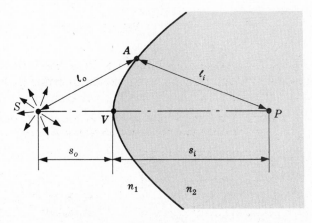

Fig. 4-2

Fermat's principle states that the optical path length (O.P.L.) along each ray from S to P must be stationary.   Here, a portion of a spherical wave leaves S and is to converge at P, so that there are many allowed ray paths.   No one O.P.L. can be a maximum or minimum, which means that they are all equal, i.e.

$$\ell_o n_1 + \ell_i n_2 = s_o n_1 + s_i n_2$$

regardless of the location of A.   Once having chosen the *object* and *image distances*, s_o and s_i, the expression for the ovoid becomes

$$\ell_o n_1 + \ell_i n_2 = \text{constant}$$

4.2.   Write an expression for the particular Cartesian ovoid whose object and image distances are 8 cm and 10 cm respectively.   Assume it to be made of glass ($n_g = 1.5$) and surrounded by air ($n_a = 1$).   Sketch the interface.

The ovoid is described by

$$\ell_o n_1 + \ell_i n_2 = \text{constant} = s_o n_1 + s_i n_2$$

(see Problem 4.1).   In this instance

$$\text{constant} = (8)(1) + (10)(1.5) = 23$$

and therefore

$$\ell_o + 1.5\,\ell_i = 23$$

Keeping $\ell_o > s_o$, let $\ell_o = 9$; then $\ell_i = 9.33$.   Similarly, when $\ell_o = 10$, $\ell_i = 8.66$.   Figure 4-3 is a sketch of this ovoid.

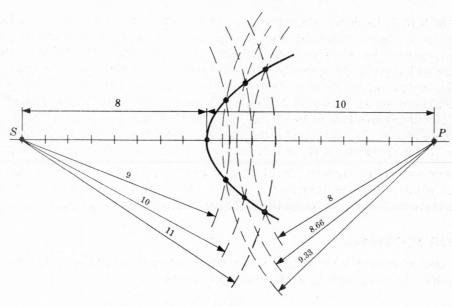

Fig. 4-3

4.3. Figure 4-4 shows a point source imbedded in a medium $(n_1 > n_2)$. Qualitatively discuss the shape which the interface must have in order that the wavefronts emerge planar.

To flatten out the wavefront you'll want a portion to advance more the farther it is off-axis. Since $n_1 > n_2$ this means that the interface must be convex to the right, as in Fig. 4-5.

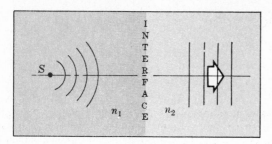

Fig. 4-4

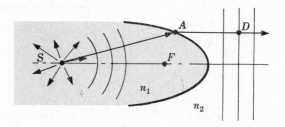

Fig. 4-5

4.4. Show that the desired interface in Problem 4.3 is actually an ellipsoid of revolution with an eccentricity of $e = n_{21}$.

Refer to Fig. 4-5. From Fermat's principle

$$(\overline{SA})n_1 + (\overline{AD})n_2 = \text{constant}$$

Dividing by n_1, this becomes

$$\overline{SA} + (\overline{AD})n_{21} = \text{constant}$$

If the curve is assumed to be an ellipse with S and F as the foci, then

$$\overline{SA} + \overline{FA} = \text{constant}$$

Furthermore if point D is on the directrix of the ellipse then $\overline{FA} = (\overline{AD})e$. If D is on a wavefront not corresponding to the directrix, $\overline{FA} = (\overline{AD})e + \text{constant}$. In either case

$$\overline{SA} + (\overline{AD})e = \text{constant}$$

Clearly, then, the interface is elliptical with $e = n_{21} < 1$. In contrast, when $n_1 < n_2$, $e > 1$ and the curve of the boundary is a hyperbola.

4.3 SPHERICAL REFRACTING SURFACES

Precise spherical surfaces are particularly easy to fabricate and, therefore, of considerable practical interest. Figure 4-6 depicts a spherical boundary surface of radius R centered at C. We know from Section 4.2 that a broad cone of rays corresponding to a spherical wave segment diverging from S would converge to P if we were dealing with a Cartesian ovoid. Even if the interface is spherical, it can be shown, using Fermat's principle, that a *narrow* cone of incident rays will arrive at P. In that case, where A is nearby V (i.e. $\ell_o \approx s_o$, $\ell_i \approx s_i$) we have

$$\frac{n_1}{s_o} + \frac{n_2}{s_i} = \frac{n_2 - n_1}{R}$$

Rays of this sort that make shallow angles with the optical axis are referred to as *paraxial*.

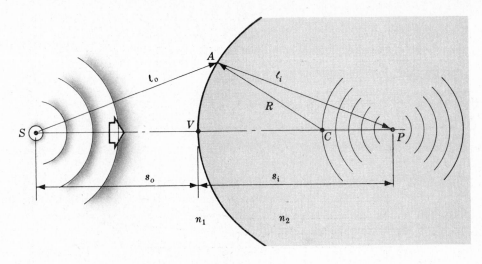

Fig. 4-6

There is a special object distance, $s_o \equiv f_o$, known as the *object focal length,* for which $s_i = \infty$ and the waves in the transmitting medium are planar (Fig. 4-7). Straight substitution yields

$$f_o = \frac{n_1}{n_2 - n_1} R$$

Similarly, when s_o is made infinite, $s_i \equiv f_i$ and the incident waves are planar (Fig. 4-8). The *image focal length* is then given by

$$f_i = \frac{n_2}{n_2 - n_1} R$$

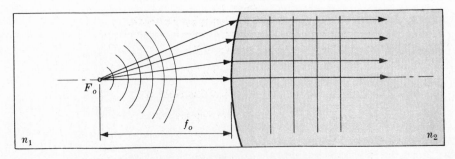

Fig. 4-7

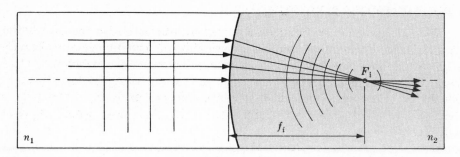

Fig. 4-8

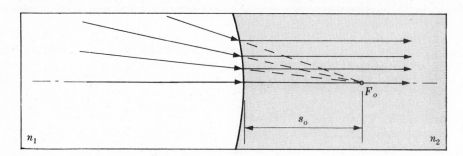

Fig. 4-9

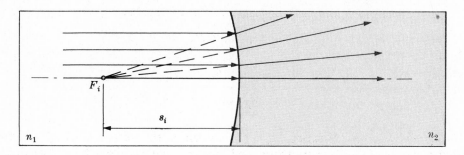

Fig. 4-10

An *object* is said to be *real* when light diverges from it (Fig. 4-7) and *virtual* when light converges toward it (Fig. 4-9).   Similarly, an *image* is said to be *real* when light converges toward it (Fig. 4-8) and *virtual* when light diverges from it (Fig. 4-10). Table 4-1 summarizes the sign convention we will adhere to — in all cases light is assumed to enter from the left. Notice that a negative value of s_o or s_i means a virtual object or a virtual image, respectively.

Table 4-1
Sign Convention for Spherical Surfaces

s_o, f_o	+ left of V
s_i, f_i	+ right of V
R	+ when C is right of V
y_o, y_i	+ above optical axis
x_o	+ left of F_o
x_i	+ right of F_i

SOLVED PROBLEMS

4.5. Suppose that we have a glass rod $(n_g = 1.50)$ surrounded by air with the left end ground to a convex hemisphere of 2-cm radius. If a point source is located 6 cm to the left of the hemisphere's vertex, where will its image appear?

Since $n_1 = 1$ and $n_2 = 1.5$,

$$\frac{1}{s_o} + \frac{1.5}{s_i} = \frac{1.5 - 1}{R}$$

In this case $s_o = 6$ cm and $R = 2$ cm, so that

$$\frac{1}{6} + \frac{3}{2s_i} = \frac{0.5}{2}$$

and $s_i = 18$ cm. The image is real and lies on the right side of V within the glass.

4.6. If the glass rod of Problem 4.5 is immersed in water $(n_w = 1.33)$ determine the new location of the image of the point source.

This time $n_1 = 1.33$ while $n_2 = 1.50$ and so

$$\frac{1.33}{s_o} + \frac{1.50}{s_i} = \frac{1.50 - 1.33}{R}$$

or

$$\frac{1.33}{6} + \frac{1.50}{s_i} = \frac{0.17}{2}$$

Hence $s_i = -10.98$ cm; the image is virtual, to the left of the vertex, and light diverges from it.

4.7. Consider the block of glass shown in Fig. 4-11. If the point source S is 30 cm from the vertex of the hemispherical end and if the latter has a radius of 10 cm, locate the image seen by the observer.

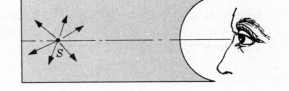

Here $n_1 = 1.5$, $n_2 = 1$, $s_o = 30$ cm and $R = +10$ cm. Therefore

$$\frac{1.5}{s_o} + \frac{1}{s_i} = \frac{1 - 1.5}{R}$$

or

$$\frac{1}{s_i} = \frac{-0.5}{10} - \frac{1.5}{30} = -\frac{1}{10}$$

Fig. 4-11

Thus $s_i = -10$ cm. Accordingly, the image is virtual, i.e. the rays diverge from it, and 10 cm to the left of the vertex in the glass.

4.8. If the interface in Fig. 4-6, page 54, has a radius of 5 cm and separates air on the left from glass $(n_g = 1.5)$ on the right, determine f_o and f_i. (These are also often referred to as the *first* and *second focal lengths*, respectively.)

By convention light enters from the left; consequently, $n_1 = 1$, $n_2 = 1.5$ and R, which is positive, equals 5 cm. Hence

$$f_o = \frac{1}{1.5 - 1}(5) = 10 \text{ cm} \qquad f_i = \frac{1.5}{1.5 - 1}(5) = 15 \text{ cm}$$

4.9. What must be the radius of curvature of the rod's right end in Fig. 4-12 if the parallel bundle of rays is to come to a focus 100 cm from the vertex? The glass rod $(n = 1.46)$ is immersed in ethyl alcohol $(n = 1.36)$.

The image focal length is given by

$$f_i = \frac{n_2}{n_2 - n_1} R$$

so that

$$R = \frac{f_i(n_2 - n_1)}{n_2}$$

$$= \frac{100(1.36 - 1.46)}{1.46} = -6.85 \text{ cm}$$

The minus sign occurs because the center of curvature is left of the vertex.

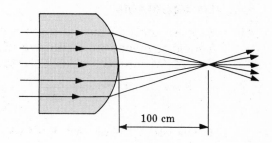

Fig. 4-12

4.4 THE THIN LENS EQUATION

A *lens* is a refracting system consisting of two or more interfaces, at least one of which is curved. Only lenses of uniform refractive index will be considered in this book. The rod in Fig. 4-12 qualifies as a *simple lens*, i.e. it consists of only one element, which in turn means that it has just two refracting interfaces. A *compound lens* is formed of two or more simple lenses. A *thin lens*, compound or simple, is one where the thickness of the elements plays no significant role and as such is negligible. Figure 4-13 illustrates the nomenclature associated with a thin simple spherical lens. Light can be traced through both its interfaces; provided that the thickness ($\overline{V_1 V_2}$) is indeed negligible and further that we are limited to paraxial rays, it can be shown that

$$\frac{1}{s_o} + \frac{1}{s_i} = (n_{\ell m} - 1)\left(\frac{1}{R_1} - \frac{1}{R_2}\right)$$

where, as usual, $n_{\ell m} = n_\ell/n_m$. This is the so-called *thin lens equation*, which is also referred to as the *lensmaker's formula*. Notice that if $s_o = \infty$, $1/f_i$ equals the quantity on the right and the same is true of $1/f_o$ when $s_i = \infty$. In other words, $f_o = f_i = f$, where

$$\frac{1}{f} = (n_{\ell m} - 1)\left(\frac{1}{R_1} - \frac{1}{R_2}\right)$$

The thin lens equation can then be rewritten as the *Gaussian lens formula*:

$$\frac{1}{s_o} + \frac{1}{s_i} = \frac{1}{f}$$

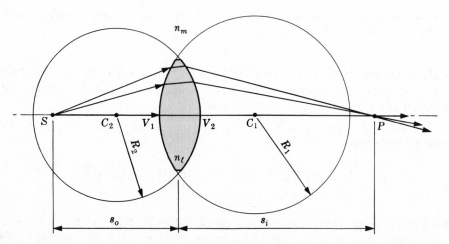

Fig. 4-13

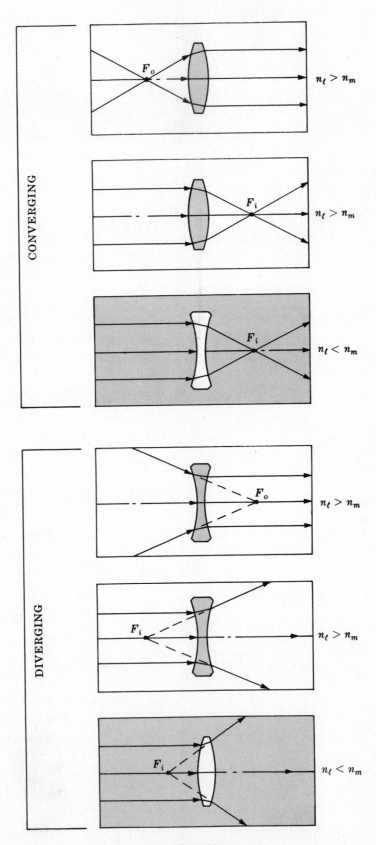

Fig. 4-14

A spherical wave leaving S in Fig. 4-13 impinges on a *positive lens*, i.e. one which is thicker at its center than at its edge. The central zone of the wavefront is slowed down more than its outer regions and the front inverts itself, thereafter converging to P. Quite reasonably, an element of this sort is said to be a *converging lens* and light rays are bent toward the central axis by the lens. As shown in Fig. 4-14 the above description presumes that the index of the medium, n_m, is less than n_ℓ. However, if $n_m > n_\ell$ a converging lens would certainly be thinner at its center. Generally speaking $(n_m < n_\ell)$, a lens which is in fact thinner at its center is variously known as a *negative, concave* or *diverging lens*. Light passing through such a lens tends to bend away from the central axis, at least more so than when it entered.

SOLVED PROBLEMS

4.10. Derive the thin lens equation using geometrical arguments and the fact that a spherical wave enters and a spherical wave leaves the lens.

Referring to Fig. 4-15(a), bear in mind that the wavefront ABL must be bent into wavefront EFG. That means that the optical path lengths between corresponding points on the fronts must be equal, i.e.

$$\overline{AD} + \overline{DE} = n_\ell \overline{BF}$$

Restricting ourselves to paraxial rays, $\overline{AD} \approx \overline{HO}$, $\overline{DE} \approx \overline{OI}$ and

$$\overline{HO} + \overline{OI} = n_\ell \overline{BF}$$

or

$$\overline{HB} + \overline{BO} + \overline{OF} + \overline{FI} = n_\ell \overline{BO} + n_\ell \overline{OF}$$

$$\overline{HB} + \overline{FI} = (n_\ell - 1)(\overline{BO} + \overline{OF}) \qquad (1)$$

Notice that each length $\overline{HB}$, $\overline{FI}$, $\overline{BO}$ and $\overline{OF}$ is the radial distance from a chord to the circumference of a circle (each is often called the *sagitta*).

Using Fig. 4-15(b), we obtain $(\overline{JH})(\overline{HB}) = (\overline{AH})(\overline{HC})$ or, in general,

$$(2R - x)x = h^2$$

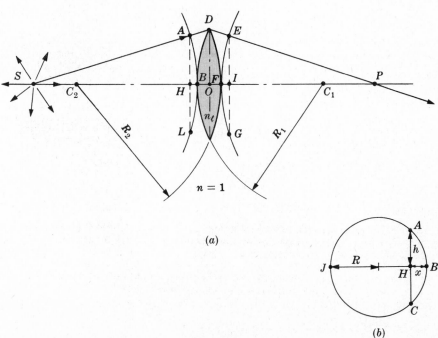

(a)

(b)

Fig. 4-15

If x is small,

$$h^2 \approx 2Rx \quad \text{or} \quad x \approx h^2/2R$$

Applying this relationship to each sagitta x ($\overline{BO}$ and $\overline{OF}$ are small since the lens is assumed thin), we find from (1):

$$\frac{(\overline{AH})^2}{2(\overline{SB})} + \frac{(\overline{AH})^2}{2(\overline{FP})} = (n_\ell - 1)\left[\frac{(\overline{DO})^2}{2(\overline{BC_1})} + \frac{(\overline{DO})^2}{2(\overline{FC_2})}\right] \tag{2}$$

For paraxial rays, $\overline{AH} \approx \overline{DO}$; and since $\overline{SB} = s_o$, $\overline{FP} = s_i$, $\overline{BC_1} = R_1$ and $\overline{FC_2} = -R_2$, we have from (2):

$$\frac{1}{s_o} + \frac{1}{s_i} = (n_\ell - 1)\left(\frac{1}{R_1} - \frac{1}{R_2}\right)$$

4.11. A point source S is located on the axis of, and 30 cm from, a plano-convex thin lens. Suppose that the glass lens is immersed in air ($n_{\ell m} = 1.5$) and that it has a radius of 5 cm. Determine the location of the image (a) when the flat surface is toward S and (b) when the curved surface is toward S.

(a) Since $R_1 = \infty$, $R_2 = -5$ cm,

$$\frac{1}{30} + \frac{1}{s_i} = (1.5 - 1)\left(\frac{1}{\infty} - \frac{1}{-5}\right)$$

and $s_i = 15$ cm. The image is real and to the right of the lens.

(b) With $R_1 = 5$ cm, $R_2 = \infty$, we have

$$\frac{1}{30} + \frac{1}{s_i} = (1.5 - 1)\left(\frac{1}{5} - \frac{1}{\infty}\right)$$

Hence $s_i = 15$ cm, as in (a).

4.12. What must be the focal length of a positive thin lens if the object and the image distances are to be 90 cm and 45 cm respectively?

From the Gaussian lens formula we get

$$\frac{1}{90} + \frac{1}{45} = \frac{1}{f}$$

and so $f = 30$ cm. Note that a positive or converging lens has a positive focal length while a negative or diverging lens has a negative focal length.

4.13. Compute the focal length of the bi-concave thin lens depicted in Fig. 4-16, if it is made of flint glass ($n_\ell = 1.66$) and immersed in water ($n_m = 1.33$).

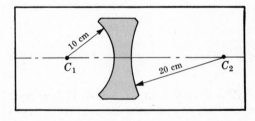

With light incident from the left, $R_1 = -10$ cm, $R_2 = +20$ cm and

$$\frac{1}{f} = \left(\frac{1.66}{1.33} - 1\right)\left(\frac{1}{-10} - \frac{1}{20}\right)$$

It follows that $f = -26.9$ cm, and the lens is negative as expected.

Fig. 4-16

4.14. Compute the focal length of the bi-concave thin lens depicted in Fig. 4-16, assuming it to be made of fluorite ($n_\ell = 1.43$) immersed in carbon disulfide ($n_m = 1.63$).

As in Problem 4.13, $R_1 = -10$ cm and $R_2 = +20$ cm, but now

$$\frac{1}{f} = \left(\frac{1.43}{1.63} - 1\right)\left(\frac{1}{-10} - \frac{1}{20}\right)$$

and $f = +54.3$ cm. The lens, surrounded by a higher-index medium, is now converging, i.e. rays bend toward the central axis.

4.15. Design an *aspherical* lens, not necessarily thin, which will convert spherical wavefronts from an axial point source into a collimated beam (plane waves).

Going back to Fig. 4-5, page 53, slice off the ellipsoidal end and grind in a hemispherical surface as in Fig. 4-17. If the point S is both the center of the sphere and a focus of the ellipsoid, the waves will be undeflected at the first interface and emerge planar.

Similarly, the solution to Problem 4.4 suggests using a hyperbolic plano-convex lens as shown in Fig. 4-18. Here S is at a focus of the hyperboloidal surface.

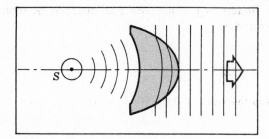

Fig. 4-17

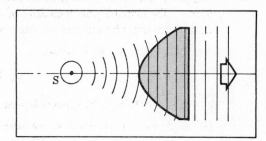

Fig. 4-18

4.16. A bi-convex thin lens of index 1.5 is known to have a focal length of 50 cm in air. When immersed in a transparent liquid the focal length is measured to be 250 cm. Determine the refractive index, n, of the liquid medium.

Since the focal length of a thin lens is given by

$$\frac{1}{f} = (n_{\ell m} - 1)\left(\frac{1}{R_1} - \frac{1}{R_2}\right)$$

the ratio of the focal lengths in the different media

$$\frac{1/50}{1/250} = \frac{1.5 - 1}{(1.5/n) - 1}$$

is independent of the radii. Hence

$$5\left(\frac{1.5}{n} - 1\right) = 0.5$$

$$7.5 = 5.5\,n$$

$$n = 1.36$$

4.17. Point S and its real image point P in Fig. 4-13 are said to be *conjugate*: a source at P would be imaged at S, and vice versa. It can be shown that the shortest distance between conjugate points for a thin positive lens is $4f$ and this occurs when $s_o = s_i$. Make a plot of $s_o + s_i$ versus s_o, where the latter is varied in multiples of some fixed f.

First make a table of values using the Gaussian lens formula in the form $s_i = s_o f/(s_o - f)$.

s_o	$6f$	$5f$	$4f$	$3f$	$2.5\,f$	$2f$	$1.5\,f$	$1.25\,f$	$1f$
s_i	$1.20\,f$	$1.25\,f$	$1.33\,f$	$1.50\,f$	$1.66\,f$	$2f$	$3f$	$5f$	∞
$s_o + s_i$	$7.20\,f$	$6.25\,f$	$5.33\,f$	$4.50\,f$	$4.16\,f$	$4f$	$4.50\,f$	$6.25\,f$	∞

Note that for values of $s_o < f$, $s_i < 0$; i.e. the image is virtual. The tabulation is graphed in Fig. 4-19.

4.18. What is the ratio of the focal length of a thin plano-convex lens to the focal length of a thin bi-convex lens, presuming that the indices are the same and all the spherical surfaces have the same curvature?

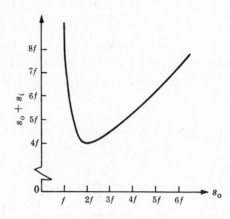

Letting f_{pc} be the focal length of the plano-convex lens and f_{bc} that of the bi-convex, we have

$$\frac{1}{f_{pc}} = (n_{\ell m} - 1)\left(\frac{1}{R} - \frac{1}{\infty}\right) = \frac{1}{R}(n_{\ell m} - 1)$$

and $$\frac{1}{f_{bc}} = (n_{\ell m} - 1)\left(\frac{1}{R} - \frac{1}{-R}\right) = \frac{2}{R}(n_{\ell m} - 1)$$

Thus, $f_{pc}/f_{bc} = 2$.

Fig. 4-19

4.5 SIMPLE THIN LENS IMAGERY

We saw in Fig. 4-14 that an axial parallel bundle of rays will be focused to a single point by a positive lens. Indeed, any parallel ray bundle will be brought to a focus on a surface σ passing through F_i, as in Fig. 4-20. Within the paraxial approximation, σ is planar and called the *second* or *back focal plane*. Similarly, there is a *first* or *front focal plane* perpendicular to the optical axis at F_o.

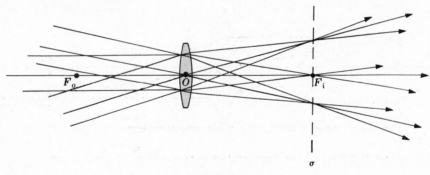

Fig. 4-20

We now examine — first graphically and then analytically — the image of a finite object formed by a thin lens.

There are three rays emerging from each object point which are especially simple to follow through the lens (Fig. 4-21). A ray (#1) traversing O, the center of the lens, will be undeviated, i.e. a straight line. A ray (#2) entering the lens parallel to the optical axis will pass through F_i. Similarly, a ray (#3) passing through F_o will emerge parallel to the axis. Accordingly, one need only construct any two of these rays for each point on the object; their corresponding point of intersection locates the image.

Of use in the analytical approach is yet another version of the thin lens equation. This *Newtonian* formulation reads:

$$x_i x_o = f^2$$

The various distances involved are illustrated in Fig. 4-22 and the concomitant sign con-

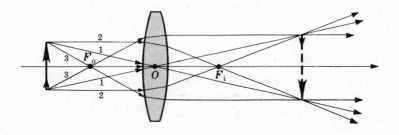

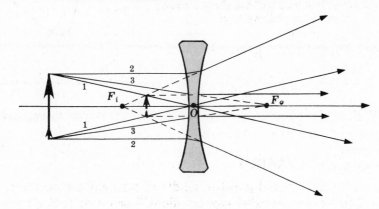

Fig. 4-21

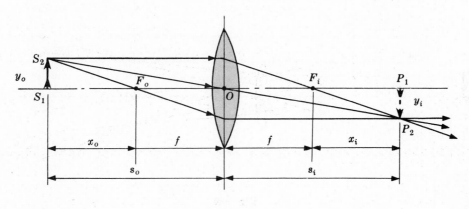

Fig. 4-22

vention is as shown in Table 4-1, page 55.   Observe that since the lenses dealt with here have negligible thicknesses, one can measure distance from a center plane through point O. Moreover, the rays can be drawn as if they underwent a single refraction at that plane rather than at both surfaces.

The size of the image (y_i) is of obvious importance and so we define the *lateral* or *transverse magnification* (M_T) as

$$M_T \equiv \frac{y_i}{y_o}$$

From the similar triangles $S_1 S_2 O$ and $P_1 P_2 O$, it follows that

$$M_T = -\frac{s_i}{s_o}$$

or in Newtonian form

$$M_T = -\frac{x_i}{f} = -\frac{f}{x_o}$$

When the image is right-side up, y_i is positive, M_T is positive and the image is said to be *erect*.   Table 4-2 summarizes the physical significance of the signs of the various quantities, while Table 4-3 lists image properties for both convex and concave thin lenses.   Keep in mind that a real image can be projected directly on a screen, while a virtual image (the kind you see in a planar mirror) cannot be.

Table 4-2.   The Physical Significance of the Signs of Thin Lens and Spherical Interface Parameters

Quantity	Sign	
	+	−
s_o	real object	virtual object
s_i	real image	virtual image
f	converging lens	diverging lens
y_o	erect object	inverted object
y_i	erect image	inverted image
M_T	erect image	inverted image

Table 4-3.   Thin Lens Image Characteristics for Real Objects

Convex				
OBJECT	IMAGE			
Location	Type	Location	Orientation	Relative Size
$\infty > s_o > 2f$	real	$f < s_i < 2f$	inverted	minified
$s_o = 2f$	real	$s_i = 2f$	inverted	same size
$f < s_o < 2f$	real	$\infty > s_i > 2f$	inverted	magnified
$s_o = f$		$\pm\infty$		
$s_o < f$	virtual	$|s_i| > s_o$	erect	magnified

Concave				
OBJECT	IMAGE			
Location	Type	Location	Orientation	Relative Size
anywhere	virtual	$|s_i| < |f|$	erect	minified

SOLVED PROBLEMS

4.19.   Compute the object and image distances for a thin bi-convex lens if the image is to be projected life-sized directly onto a screen.   The lens has equal radii of 60 cm and $n_{\ell m} = 1.5$.

The focal length is simply

$$\frac{1}{f} = (n_{\ell m} - 1)\left(\frac{1}{R} - \frac{1}{-R}\right) = 0.5\left(\frac{2}{R}\right)$$

or $f = R = 60$ cm. (This result is one to remember; it gives a good feel for the relationship between f and R even if it is a special case.)   Table 4-3 makes it apparent that a real life-sized image occurs when $s_o = s_i = 2f = 120$ cm.   Note that $M_T = -1$, i.e. the image is inverted.

4.20.   A 3-cm high wine-bottle cork is sitting 75 cm from a thin positive lens of 25-cm focal length.   Describe the resulting image completely, using the Gaussian formulation. Check your answers with Table 4-3.

Given that $s_o = 75$ cm and $f = 25$ cm, it follows from the thin lens equation that

$$\frac{1}{75} + \frac{1}{s_i} = \frac{1}{25}$$

and therefore $s_i = 37.5$ cm.   This is positive, which means that the image is real and located beyond the lens.   The size of the image can be determined from the expression

$$M_T = -\frac{s_i}{s_o} = \frac{y_i}{y_o}$$

Hence $$M_T = -\frac{37.5}{75.0} = \frac{y_i}{3}$$

The magnification is $-1/2$ and so $y_i = -1.5$ cm.   The image is minified and inverted ($M_T < 0$).

4.21.   Redo Problem 4.20, this time using the Newtonian formulation exclusively.

Since $s_o = 75 = f + x_o$ and $f = 25$ cm, $x_o = +50$ cm (see Table 4-1).   Accordingly,

$$x_i x_o = f^2 \quad\text{and}\quad x_i = \frac{(25)^2}{50} = +12.5 \text{ cm}$$

But, of course, $s_i = x_i + f = 12.5 + 25 = 37.5$ cm, as in Problem 4.20.   As for the size of the image,

$$M_T = -\frac{x_i}{f} = -\frac{f}{x_o}$$

and either of these forms will do.   Hence

$$M_T = -\frac{12.5}{25} = -0.5 \quad\text{or}\quad M_T = -\frac{25}{50} = -0.5$$

4.22.   It is required that a real image twice the size of the object be formed by a thin plano-convex lens.   If the lens has a radius of curvature of 50 cm and a refractive index of $n_{\ell m} = 1.5$, determine the locations of the object and image with respect to the lens (*a*) by use of the Newtonian expression, (*b*) by use of the Gaussian expression.

(*a*)   From Table 4-3 it is evident that a real magnified image would occur when $f < s_o < 2f$ and that it would be inverted and located such that $\infty > s_i > 2f$.   In this case

$$\frac{1}{f} = (n_{\ell m} - 1)\left(\frac{1}{R_1} - \frac{1}{R_2}\right)$$

yields $$\frac{1}{f} = \frac{0.5}{50}$$

that is, $f = 100$ cm.   Thus, since $M_T = -2$,

$$\frac{-x_i}{f} = \frac{-f}{x_o} = -2$$

With $f = 100$ cm, $x_i = +200$ cm and $x_o = 50$ cm, and therefore $s_o = 150$ cm, $s_i = 300$ cm.

(*b*)   The Gaussian formula

$$\frac{1}{s_o} + \frac{1}{s_i} = \frac{1}{f}$$

together with the fact that

$$M_T = -2 = -\frac{s_i}{s_o}$$

gives

$$\frac{1}{s_o} + \frac{1}{2s_o} = \frac{1}{100}$$

and, as in (a), $s_o = 150$ cm.

4.23. Suppose that an object positioned 10 inches to the left of a positive lens is imaged 30 inches to the right of the lens. Where will the image appear if the object is now moved so that it is 2.5 inches from the lens? Completely describe the image in both instances.

The focal length of the lens can be gotten from the thin lens equation:

$$\frac{1}{10} + \frac{1}{30} = \frac{1}{f}$$

or $f = +7.5$ inches. Furthermore

$$M_T = -\frac{s_i}{s_o} = -3$$

and so this image is real, inverted and magnified (as expected from Table 4-3 when $f < s_o < 2f$). In the second case, $s_o = 2.5$ inches and

$$\frac{1}{2.5} + \frac{1}{s_i} = \frac{1}{7.5}$$

yielding $s_i = -3.75$ inches. This time the image is virtual, erect ($M_T = +1.5$), magnified and located 3.75 inches in front of the lens.

4.24. Imagine that you'd like to look through a lens at your pet parakeet and see it standing right-side up but shrunk to one-third its normal height. Designating the focal length as f, determine the kind of lens needed, as well as the object and image distances in terms of f. Construct a ray diagram.

If the image is to be both minified and erect, the lens will have to be diverging, according to Table 4-3. Since $M_T = +1/3$,

$$\frac{1}{3} = -\frac{x_i}{f} = -\frac{f}{x_o}$$

and so $x_i = -f/3$ and $x_o = -3f$, where you should keep in mind that $f < 0$. Both x_i and x_o are therefore positive quantities although F_o is to the right of the lens and F_i is to the left of it. Since they are positive, x_o is measured to the left of F_o and x_i to the right of F_i. Consequently, $s_o = -3f - (-f) = -2f$, while $s_i = f - f/3 = 2f/3$. All this can be seen in Fig. 4-23.

Alternatively,

$$M_T = \frac{1}{3} = -\frac{s_i}{s_o}$$

and therefore

$$\frac{1}{s_i} + \frac{1}{(-3s_i)} = \frac{1}{f} \qquad \text{or} \qquad s_i = \frac{2f}{3}$$

4.25. A thin bi-convex glass ($n_g = 1.5$) lens has radii of curvature of 30 cm and 60 cm. If it is to cast a half-sized image of a ceiling lamp on a paper screen, what must be the lens-lamp and lens-screen distances? Construct an appropriate ray diagram.

First, to determine the focal length we use

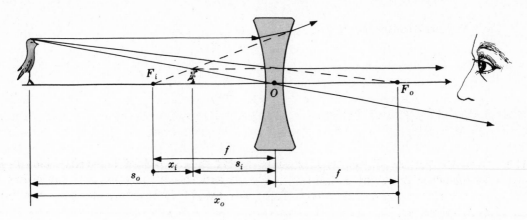

Fig. 4-23

$$\frac{1}{f} \;=\; (n_{\ell m}-1)\!\left(\frac{1}{R_1}-\frac{1}{R_2}\right) \;=\; 0.5\!\left(\frac{1}{30}-\frac{1}{-60}\right)$$

and find that $f = 40$ cm. The magnitude of the transverse magnification is given as 1/2, but if the image is to be real it must be inverted, hence

$$M_T \;=\; -\frac{1}{2} \;=\; -\frac{s_i}{s_o}$$

Substituting back into the thin lens equation, we have

$$\frac{1}{s_o}+\frac{1}{s_o/2} \;=\; \frac{1}{40}$$

and $s_o = 120$ cm, while $s_i = 60$ cm. A ray diagram is shown in Fig. 4-24.

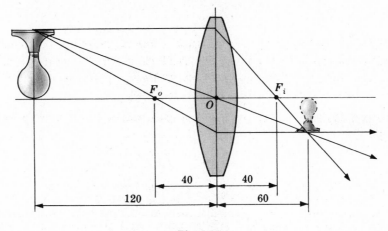

Fig. 4-24

4.26. A thin positive lens of focal length f is to cast a real image N-times larger than the object. Show that the lens-screen distance is equal to $(N+1)f$.

We know from Table 4-3 that the image must be inverted, i.e.

$$M_T \;=\; -N \;=\; -\frac{s_i}{s_o}$$

The lens equation then yields

$$\frac{1}{s_i}+\frac{1}{s_i/N} \;=\; \frac{1}{f}$$

or $s_i = (N+1)f$. Compare this with the results of Problems 4.19, 4.22 and 4.25.

4.6 COMPOUND THIN LENSES

We now examine the formation of images arising from combinations of simple thin lenses. The first approach is a graphical one and for that refer to Fig. 4-25. Observe that both f_1 and f_2 in this case are greater than d. The rays labeled 1 and 2 intersect at P_2 to form a minified, inverted, real image. Ray 2 is obvious enough since it passes through the two foci F_{o1} and F_{i2}. Ray 1 is a bit more problematic — it goes through O_2 but at an unknown angle. Note, however, that if lens L_2 were removed ray 1 would be unaffected. Thus imagine that L_2 vanished, as in Fig. 4-26. All rays entering lens L_1 from point S_2 will intersect at P_2'. We need only locate P_2' with any two convenient rays and then construct ray 1 running backwards from P_2' through O_2 to S_2.

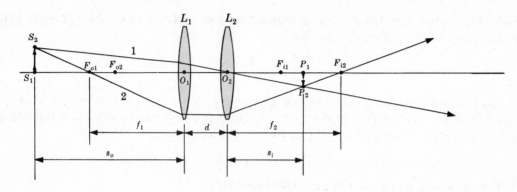

Fig. 4-25

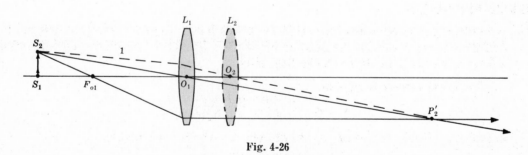

Fig. 4-26

To arrive at an analytical statement, we merely presume that the image formed by the first lens serves as the object for the second lens and so on through the system. In the case of any *two thin lenses* one finds that

$$s_i = \frac{f_2 d - [f_1 f_2 s_o/(s_o - f_1)]}{d - f_2 - [f_1 s_o/(s_o - f_1)]}$$

where s_o and s_i are the object and image distances, as measured in Fig. 4-25. If the magnifications of the individual lenses are M_{T_1} and M_{T_2}, it should be evident that the total magnification is

$$M_T = M_{T_1} M_{T_2}$$

In other words, the first lens produces an intermediate image of magnification M_{T_1}, which in turn is magnified by an amount M_{T_2} by the second lens. Explicitly,

$$M_T = \frac{f_1 s_i}{d(s_o - f_1) - s_o f_1}$$

If we let s_i go to ∞, s_o takes on the value designated by the abbreviation f.f.l. (*front focal length*), namely

$$\text{f.f.l.} \;=\; \frac{f_1(d-f_2)}{d-(f_1+f_2)}$$

Similarly, when s_o goes to infinity the corresponding image distance is called the *back focal length*, or b.f.l., where

$$\text{b.f.l.} \;=\; \frac{f_2(d-f_1)}{d-(f_1+f_2)}$$

Restating matters, if collimated light enters the compound lens from the left, it will be focused to the right of the last lens element at a distance b.f.l. If it enters from the right, it will be focused a distance f.f.l. in front of the first lens element.

Note that when the lenses are in contact $(d=0)$, b.f.l. = f.f.l.; the common value is called the *effective focal length* f, where

$$\frac{1}{f} \;=\; \frac{1}{f_1} + \frac{1}{f_2}$$

It has become usual practice to define a quantity called the *dioptric power* $\mathscr{D}$ of a lens; it is just the reciprocal of the focal length. When f is measured in meters, $\mathscr{D}$ has the units of m^{-1} or *diopters*. For the case of two thin lenses in contact

$$\mathscr{D} \;=\; \mathscr{D}_1 + \mathscr{D}_2$$

yields the combined power of the individual elements.

More on compound lenses will be found in Section 4.8.

SOLVED PROBLEMS

4.27. A compound lens consists of two thin bi-convex lenses L_1 and L_2 of focal lengths 10 cm and 20 cm, separated by a distance of 80 cm. Describe the image corresponding to a 5-cm tall object 15 cm from the first lens.

The image distance is given by

$$s_i \;=\; \frac{f_2 d - [f_1 f_2 s_o/(s_o - f_1)]}{d - f_2 - [f_1 s_o/(s_o - f_1)]}$$

wherein $f_1 = 10$ cm, $f_2 = 20$ cm, $d = 80$ cm and $s_o = 15$ cm. Accordingly,

$$s_i \;=\; \frac{(20)(80) - [(10)(20)(15)/(15-10)]}{80 - 20 - [(10)(15)/(15-10)]} \;=\; \frac{100}{3} \;=\; 33.3 \text{ cm}$$

The image is real and located 33.3 cm beyond the last lens. Its lateral magnification is

$$M_T \;=\; \frac{f_i s_i}{d(s_o - f_1) - s_0 f_1} \;=\; \frac{(10)(33.3)}{80(15-10) - (15)(10)} \;=\; 1.3$$

i.e. the image is magnified slightly and erect.

4.28. Construct a ray diagram to scale for Problem 4.27. Calculate the location of the intermediate image and see that it checks with your drawing.

The intermediate image, i.e. the one formed by the first lens, can be located using the thin lens equation

$$\frac{1}{15} + \frac{1}{s_{i1}} \;=\; \frac{1}{10}$$

Thus $s_{i1} = 30$ cm, which is 30 cm to the right of L_1.

Rays 1 and 2 in Fig. 4-27 easily locate the intermediate image. Any two rays from the tip of the intermediate image in turn will fix the position of the final image; rays 2 and 3 are the most convenient.

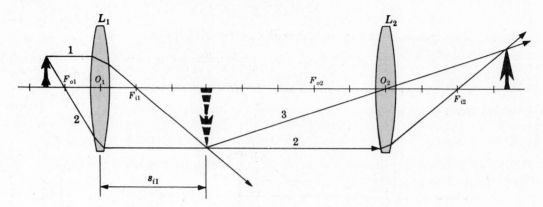

Fig. 4-27

4.29. Imagine a compound lens consisting of a thin positive lens followed at an interval of 20 cm by a thin negative lens. If these have focal lengths of +40 cm and −40 cm respectively, determine the f.f.l. and b.f.l.

Here $f_1 = 40$ cm, $f_2 = -40$ cm and $d = 20$ cm. Hence

$$\text{f.f.l.} \;=\; \frac{40[20 - (-40)]}{20 - [40 + (-40)]} \;=\; 120 \text{ cm}$$

$$\text{b.f.l.} \;=\; \frac{-40(20 - 40)}{20 - [40 + (-40)]} \;=\; 40 \text{ cm}$$

Thus the image of a very distant object like the sun would be formed 40 cm behind the second lens.

4.30. A bi-concave lens of focal length −60 mm is mounted in a cardboard cylinder 120 mm in front of a plano-convex lens of radius 60 mm and index 1.5. Determine completely the image which would result from a 3-mm ant located 180 mm in front of the device.

The focal length of the positive lens is arrived at from

$$\frac{1}{f} \;=\; (1.5 - 1)\!\left(\frac{1}{60} - \frac{1}{\infty}\right)$$

that is, $f_2 = +120$ mm. Hence, with $f_1 = -60$ mm, $d = 120$ mm and $s_o = 180$ mm, we find

$$s_i \;=\; \frac{(120)(120) - \{(-60)(120)(180)/[180 - (-60)]\}}{120 - 120 - \{(-60)(180)/[180 - (-60)]\}}$$

$$=\; \frac{12 + [6(180)/(240)]}{9/240} \;=\; 440 \text{ mm}$$

The image is real and to the right of the last lens. As for the lateral magnification,

$$M_T \;=\; \frac{(-60)(440)}{120[180 - (-60)] - (180)(-60)} \;=\; -\frac{2}{3}$$

and the image is inverted and minified; the ant appears 2 mm long.

4.31. A homemade microscope has a thin positive front lens L_1 of 2-cm focal length, 10 cm behind which is another positive lens L_2, with a 5-cm focal length. (a) Locate the image of an object 3 cm from the front lens and compute the magnification. (b) Construct a ray diagram for a single axial object point.

(a) The image will be found a distance s_i measured from L_2, where $f_1 = 2$ cm, $f_2 = 5$ cm, $d = 10$ cm and $s_o = 3$ cm. Accordingly,

$$s_i = \frac{(5)(10) - [(2)(5)(3)/(3-2)]}{10 - 5 - [(2)(3)/(3-2)]} = -20$$

The image is in front of L_2, that is, it's virtual. The lateral magnification is then

$$M_T = \frac{(2)(-20)}{10(3-2) - (3)(2)} = -10$$

and the image is inverted and 10 times larger than the object.

(b) See Fig. 4-28.

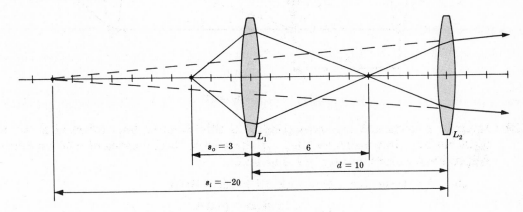

Fig. 4-28

4.32. A positive meniscus (covexo-concave) thin lens ($n = 1.5$) whose radii of curvature are 5 cm and 10 cm is positioned in contact with a plano-concave thin lens ($n = 1.6$) of radius 6 cm. What is the effective focal length of the lens system? Compute its power as well.

The first step is to calculate the two focal lengths, namely

$$\frac{1}{f_1} = (1.5 - 1)\left(\frac{1}{5} - \frac{1}{10}\right) \qquad \frac{1}{f_2} = (1.6 - 1)\left(\frac{1}{\infty} - \frac{1}{6}\right)$$

These lead to $f_1 = 20$ cm and $f_2 = -10$ cm. The combined focal length f is simply

$$\frac{1}{f} = \frac{1}{f_1} + \frac{1}{f_2} = \frac{1}{20} + \frac{1}{-10}$$

or $f = -20$ cm and $\mathscr{D} = -5$ dptr.

4.33. A negative meniscus thin lens of radii 60 cm and 30 cm and having an index of 1.5 is held horizontally with its concave side facing upwards. The concavity is then filled with a transparent oil of index 1.6. Determine the dioptric power of the compound thin lens assuming it to be immersed in air. Describe the image of an object 100 cm in front of the lens.

Inasmuch as $\mathscr{D}_1 = 1/f_1$, and here $R_1 = 0.6$ m, $R_2 = 0.3$ m and $n_{\ell m} = 1.5$,

$$\mathscr{D}_1 = (1.5 - 1)\left(\frac{1}{0.6} - \frac{1}{0.3}\right) = -0.83 \text{ dptr}$$

This is the power of the glass meniscus lens. As for the oil lens,

$$\mathscr{D}_2 = (1.6 - 1)\left(\frac{1}{0.3} - \frac{1}{\infty}\right) = +2 \text{ dptr}$$

The combined power of these two elements in contact is

$$\mathscr{D} = \mathscr{D}_1 + \mathscr{D}_2 = +1.17 \text{ dptr}$$

The effective focal length is then $\mathscr{D}^{-1} = f$ and the expression

$$\frac{1}{s_o} + \frac{1}{s_i} = \frac{1}{f}$$

will provide the needed image distance. Hence

$$\frac{1}{s_i} = 1.17 - 1$$

$s_i = 5.88$ m and $M_T = 5.88$, so that the image is real, erect and magnified.

4.7 THICK LENSES

Imagine that we position a point source S on the axis of a thick lens so that the emerging rays are parallel, as in Fig. 4-29(a). Evidently the distance from S to the vertex V_1 corresponds to what we have called the front focal length (f.f.l.). Similarly, an incident parallel bundle of rays will converge to a point a distance beyond V_2 equal to the back focal length (b.f.l.), as in Fig. 4-29(b).

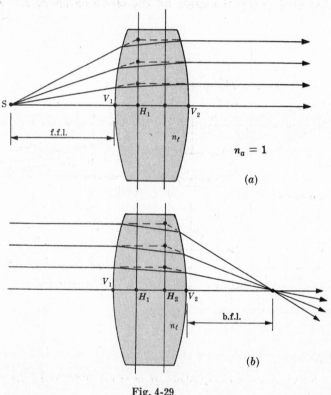

Fig. 4-29

If the incoming and outgoing rays are extended (as shown by the dotted lines) each pair will intersect on a surface. In the paraxial approximation these surfaces reduce to planes known as the *first* and *second principal planes*; their points of intersection with the central axis, H_1 and H_2, being the *first* and *second principal points*, respectively. As a rule of thumb, for glass lenses in air the distance $\overline{H_1 H_2}$ is roughly equal to one-third the thickness ($d = \overline{V_1 V_2}$) of the lens. Note that principal planes need not lie within the lens itself.

The simplest and most common case is that of a lens of index n_ℓ immersed in air, $n_a \approx 1$. The Gaussian lens formula

$$\frac{1}{s_o} + \frac{1}{s_i} = \frac{1}{f}$$

is again applicable provided that the object and image distances are measured from H_1 and H_2, respectively. The focal length, which is also referenced from the principal planes, is now given by

$$\frac{1}{f} = (n_\ell - 1)\left[\frac{1}{R_1} - \frac{1}{R_2} + \frac{(n_\ell - 1)d}{n_\ell R_1 R_2}\right]$$

which, of course, reduces to the thin lens expression when the thickness is made negligible $(d \to 0)$. The principal planes, in turn, can be located with respect to the vertices using the equations

$$\overline{V_1 H_1} \equiv h_1 = -\frac{f(n_\ell - 1)d}{R_2 n_\ell}$$

$$\overline{V_2 H_2} \equiv h_2 = -\frac{f(n_\ell - 1)d}{R_1 n_\ell}$$

Both h_1 and h_2 will be *positive* when the principal planes are to the *right* of their respective vertices, V_1 and V_2. The relationships between the various distances are illustrated in Fig. 4-30. Notice that a ray headed for any point on the first principal plane will leave the lens as if it originated at a point, the same height above or below the axis, on the second principal plane.

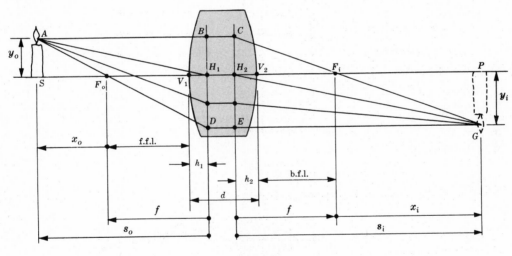

Fig. 4-30

SOLVED PROBLEMS

4.34. Show that the Newtonian lens equation once again obtains in the case of thick lenses.

Using the distances as defined in Fig. 4-30, we have $s_o = x_o + f$ and $s_i = x_i + f$. The Gaussian equation can, therefore, be rewritten as

$$f = \frac{s_o s_i}{s_o + s_i} = \frac{(x_o + f)(x_i + f)}{(x_o + f) + (x_i + f)}$$

yielding $x_i x_o = f^2$.

4.35. Derive an expression for the transverse magnification of a thick lens.

By definition the magnification is

$$M_T = \frac{y_i}{y_o}$$

In Fig. 4-30 triangles ASF_o and F_oH_1D are similar, as are triangles GPF_i and F_iH_2C.   Thus

$$\frac{y_o}{x_o} = \frac{\overline{DH_1}}{f} \qquad \frac{-y_i}{x_i} = \frac{\overline{CH_2}}{f}$$

Remember that y_i measured below the axis is a negative quantity.   Since $\overline{DH_1} = -y_i$ and $\overline{CH_2} = y_o$,

$$\frac{y_o}{x_o} = \frac{-y_i}{f} \qquad \frac{-y_i}{x_i} = \frac{y_o}{f}$$

Each of these relations yields an expression for the ratio y_i/y_o.   Accordingly,

$$M_T = \frac{-f}{x_o} = \frac{-x_i}{f}$$

which is identical to the thin lens formulation.

4.36. What kind of glass lens immersed in air will have a focal length which is independent of its thickness?

Since

$$\frac{1}{f} = (n_\ell - 1)\left[\frac{1}{R_1} - \frac{1}{R_2} + \frac{(n_\ell - 1)d}{n_\ell R_1 R_2}\right]$$

the dependence on d vanishes when either $R_1 = \infty$ or $R_2 = \infty$.   Hence, either a plano-convex or a plano-concave lens fills the requirement.

4.37. Show that one of the principal planes will always be tangent to the curved surface if the thick lens is either plano-concave or plano-convex.

Envision a parallel bundle of rays entering the planar surface perpendicularly.   All of the bending of the rays occurs at the second face, whether it is concave or convex.   The points of intersection of the rays thus all lie on the second surface of the lens.   Only in the paraxial approximation does this curved surface reduce to the principal plane, which is then tangent at the vertex.   Alternatively, in this instance, $R_1 = \infty$ and so

$$h_2 = -\frac{f(n_\ell - 1)d}{(\infty)n_\ell} = 0$$

independently of the value of R_2.

4.38. Suppose that an object is located at the first principal plane of a thick meniscus lens.   Determine the location and magnification of the image.

An object at the first principal plane is at a distance f from $\overline{F_o}$; that is, $x_o = -f$.   Since

$$x_i x_o = f^2$$

this means that $x_i = -f$.   In other words, the image is located on the second principal plane.   The magnification ($M_T = -f/x_o$) is clearly 1.   This is why the principal planes are also spoken of as *unit planes*.

4.39. Figure 4-31 depicts the principal and focal planes of a thick lens.   Graphically determine the conjugate image point corresponding to the object point S.

First, draw any ray from S to the first principal plane ($\overline{SA}$ in Fig. 4-32).   We know the ray emerges from point B; but at what angle?   To answer that, draw a ray passing through F_o parallel to $\overline{SA}$.   This one arrives at C, emerges at D and intersects the back focal plane at E.   Since $\overline{SA}$ and $\overline{F_oC}$ are parallel they must converge to the same point on the focal plane; viz., E.   Extending the line $\overline{BE}$ locates the image point at P.

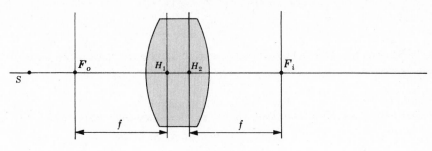

Fig. 4-31

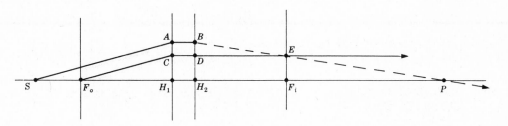

Fig. 4-32

4.40. (a) Write an expression for the focal length of a gypsy's crystal ball in terms of its radius R and refractive index n. (b) Determine the location of its principal points. (c) Where will sunlight be focused by a 4-inch diameter ball of index 1.5?

(a) Since $R_1 = R$, $R_2 = -R$ and $d = 2R$, substitution into the focal length equation yields

$$\frac{1}{f} = (n-1)\left[\frac{1}{R} + \frac{1}{R} + \frac{(n-1)2R}{-nR^2}\right]$$

or

$$f = \frac{nR}{2(n-1)}$$

(b) The principal points are to be found at

$$h_1 = -\left[\frac{nR}{2(n-1)}\right]\frac{(n-1)(2R)}{(-R)n} = +R$$

$$h_2 = -\left[\frac{nR}{2(n-1)}\right]\frac{(n-1)(2R)}{Rn} = -R$$

That is, H_1 is a distance R to the right of V_1 while H_2 is R to the left of V_2. Obviously the principal planes coincide and pass through the center of the sphere.

(c) For a 4-inch sphere, we have from (a):

$$f = \left(\frac{3}{2}\right)\frac{2}{2(1/2)} = 3 \text{ inches}$$

and the focal point appears 1 inch from the vertex.

4.41. A 10-cm diameter glass ($n_g = 1.5$) sphere is to be used to cast a real image of a window onto a screen. If the window is 3 meters from the center of the sphere, where must the screen be put? What will the magnification be?

From Problem 4.40

$$f = \left(\frac{3}{2}\right)\frac{5}{2(1/2)} = \frac{15}{2} \text{ cm}$$

The Gaussian equation then becomes

$$\frac{1}{300} + \frac{1}{s_i} = \frac{2}{15}$$

and $s_i = 7.69$ cm measured from the second principal plane (i.e. from the center of the sphere). We saw in Problem 4.35 that the magnification is the same as for a thin lens:

$$M_T = -\frac{s_i}{s_o} = -\frac{7.69}{300} = -0.026$$

The image is real, minified and inverted.

4.42. A double convex lens has radii of 5 cm and 20 cm, a thickness of 2 cm and an index of 3/2. (*a*) Locate both the principal and focal points and compute the image distance for an object 16.4 cm in front of V_1. (*b*) Determine the values of the f.f.l. and b.f.l.

(*a*) Substituting into the focal length expression leads to

$$\frac{1}{f} = \frac{1}{2}\left[\frac{1}{5} - \frac{1}{-20} + \frac{(1/2)2}{(3/2)5(-20)}\right]$$

or $f = 8.2$ cm. Also,

$$h_1 = -\frac{8.2(1/2)2}{(-20)3/2} = 0.3 \text{ cm}$$

$$h_2 = -\frac{8.2(1/2)2}{(10)3/2} = -0.5 \text{ cm}$$

The Gaussian lens equation yields

$$\frac{1}{16.4} + \frac{1}{s_i} = \frac{1}{8.2}$$

or $s_i = 16.4$ cm.

(*b*) From the values found in (*a*):

$$\text{b.f.l.} = f + h_2 = 7.7 \text{ cm}$$

$$\text{f.f.l.} = f - h_1 = 7.9 \text{ cm}$$

4.8 LENS COMBINATIONS

Two or more thick lenses can be combined to form a compound lens as typified in Fig. 4-33. Here the effective focal length f of the composite system is given by

$$\frac{1}{f} = \frac{1}{f_1} + \frac{1}{f_2} - \frac{d}{f_1 f_2}$$

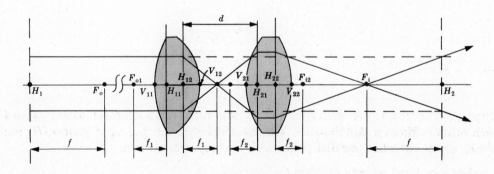

Fig. 4-33

in terms of the individual focal lengths (f_1 and f_2) and the lens separation d. Each constituent lens has its own pair of principal points (H_{11}, H_{12} and H_{21}, H_{22}), as does the compound lens itself (H_1 and H_2). Thus, if a beam of collimated light were to enter the compound lens, it would be brought to a focus at F_i a distance f from H_2. The expressions

$$\overline{H_{11}H_1} = \frac{fd}{f_2} \qquad\qquad \overline{H_{22}H_2} = \frac{-fd}{f_1}$$

specify the positions of H_1 and H_2.

For a system of thin lenses the individual principal planes coalesce and d simply becomes the center-to-center lens separation as in Fig. 4-25, page 68. Such a system behaves as a thick lens whose focal length and principal points are given by the above expressions.

In effect the procedure given above combines two lenses (thick or thin) into one. Thus, if you had five or six lenses in a centered system, you could go down the line replacing them two at a time until you had one equivalent lens representing the entire system.

SOLVED PROBLEMS

4.43. Two identical bi-convex thick lenses are placed in line with a separation of 25.7 mm, as in Fig. 4-33. Each lens has radii of 60 mm and 40 mm, a thickness of 20 mm and an index of 1.5. Calculate the focal length of each and locate the points H_{11}, H_{12}, H_{21} and H_{22}. Determine the effective focal length of the system immersed in air.

The focal length of each lens is computable as follows:

$$\frac{1}{f} = (n_\ell - 1)\left[\frac{1}{R_1} - \frac{1}{R_2} + \frac{(n_\ell - 1)d}{n_\ell R_1 R_2}\right]$$

$$= (1.5 - 1)\left[\frac{1}{60} - \frac{1}{-40} + \frac{(1.5 - 1)20}{(1.5)(60)(-40)}\right]$$

$$= \frac{1}{2}\left[\frac{6}{360} + \frac{9}{360} - \frac{10}{3600}\right] = \frac{7}{360}$$

and finally $f_1 = f_2 = 51.4$ mm. The principal points of the individual lenses are positioned at

$$h_1 = -\frac{(51.4)(1/2)20}{(-40)(3/2)} = \frac{25.7}{3} = 8.6 \text{ mm}$$

$$h_2 = -\frac{(51.4)(1/2)20}{60(3/2)} = -\frac{51.4}{9} = -5.7 \text{ mm}$$

Since the lenses are identical, these values fix the positions of the principal planes with respect to the vertices for both lenses. The compound lens has a focal length of

$$\frac{1}{f} = \frac{1}{f_1} + \frac{1}{f_2} - \frac{d}{f_1 f_2}$$

$$= \frac{1}{51.4} + \frac{1}{51.4} - \frac{25.7}{(51.4)(51.4)} = \frac{3}{102.8}$$

and $f = 34.3$ mm.

4.44. Suppose that the lenses in Problem 4.43 are arranged with their flatter sides facing each other. Make a sketch of the system and locate its principal planes (H_1 and H_2). Trace a ray entering parallel to the axis through the system.

The only things we need calculate are

$$\overline{H_{11}H_1} = \frac{fd}{f_2} = \frac{(34.3)(25.7)}{51.4} = 17.2 \text{ mm}$$

$$\overline{H_{22}H_2} \;=\; -\frac{fd}{f_1} \;=\; -\frac{(34.3)(25.7)}{51.4} \;=\; -17.3 \text{ mm}$$

Remember that positive values of these quantities are measured to the right, as in Fig. 4-34. Two rays are traced through the lens; the upper ray deflects at the principal planes of the constituent lenses, while the lower is drawn using H_1 and H_2 only. The two schemes are equivalent.

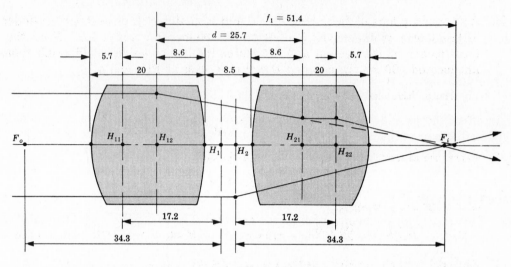

Fig. 4-34

4.45. Imagine a compound lens consisting of a thin positive lens followed at an interval of 20 cm by a thin negative lens. If these have focal lengths of +40 cm and −40 cm, respectively, determine the value of f, f.f.l. and b.f.l. (take a look at Problem 4.29).

The equation

$$\frac{1}{f} \;=\; \frac{1}{f_1} + \frac{1}{f_2} - \frac{d}{f_1 f_2}$$

now applies, wherein $f_1 = 40$ cm, $f_2 = -40$ cm and $d = 20$ cm. Hence

$$\frac{1}{f} \;=\; \frac{1}{40} + \frac{1}{-40} - \frac{20}{(40)(-40)}$$

and so $f = 80$ cm. The principal plane for each lens passes through its center, from which are measured H_1 and H_2. Thus

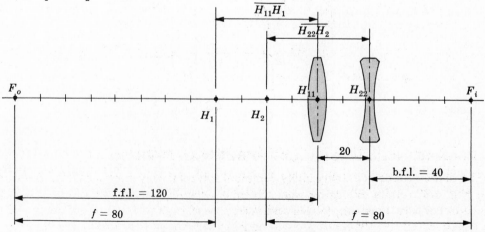

Fig. 4-35

$$\overline{H_{11}H_1} = \frac{(80)(20)}{-40} = -40 \text{ cm} \qquad \overline{H_{22}H_2} = -\frac{(80)(20)}{40} = -40 \text{ cm}$$

The situation is shown in Fig. 4-35. Clearly, since f.f.l. and b.f.l. are measured from the centers of the thin lenses to the focal points (F_o and F_i), their values are just 120 cm and 40 cm, respectively.

4.46. A bi-concave lens of focal length −60 mm is mounted in a cardboard cylinder 120 mm in front of a plano-convex lens of radius 60 mm and index 1.5. Find the effective focal length of the system and determine the image which would result from a 3-mm ant located 180 mm in front of the device (look at Problem 4.30).

The positive lens has a focal length

$$f_2 = (1.5-1)\left(\frac{1}{60} - \frac{1}{\infty}\right) = 120 \text{ mm}$$

Hence, the effective focal length of the combination is

$$\frac{1}{f} = \frac{1}{-60} + \frac{1}{120} - \frac{120}{(-60)(120)}$$

or $f = 120$ mm. Moreover,

$$\overline{H_{11}H_1} = \frac{(120)(120)}{120} = 120 \text{ mm} \qquad \overline{H_{22}H_2} = -\frac{(120)(120)}{-60} = 240 \text{ mm}$$

Measured from H_1, the object distance is $s_o = 300$ mm and so

$$\frac{1}{300} + \frac{1}{s_i} = \frac{1}{120} \qquad \text{or} \qquad s_i = 200 \text{ mm}$$

The image is 200 mm to the right of H_2 as in Fig. 4-36. The magnification is then

$$M_T = -\frac{s_i}{s_o} = -\frac{200}{300} = -0.66$$

and the ant is inverted and minified.

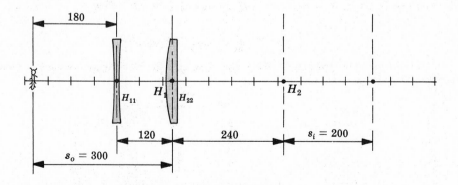

Fig. 4-36

4.9 PLANAR, ASPHERICAL AND SPHERICAL MIRRORS

The plane mirror is an extremely common and relatively simple device. A source point S, as in Fig. 4-37, emits diverging rays which bounce off the mirror and continue to diverge. An eye or camera lens can collect and focus these rays to form a real image of S, but the image generated by the mirror itself at P is virtual; it lies behind the mirror, cannot be projected and the rays appear to diverge from it.

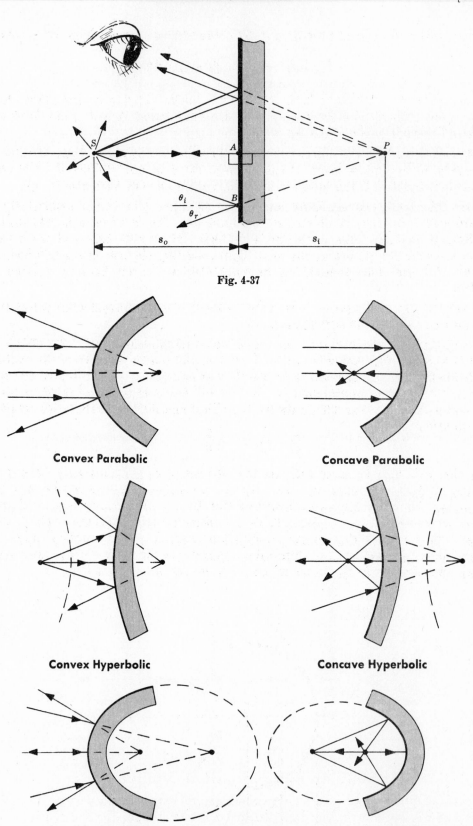

Fig. 4-37

Convex Parabolic   **Concave Parabolic**

Convex Hyperbolic   **Concave Hyperbolic**

Convex Elliptical   **Concave Elliptical**

Fig. 4-38. **Aspherical Mirrors**

In Fig. 4-37 right triangles ASB and APB are congruent, since side $\overline{AB}$ is common and

$$\angle ASB = \theta_i = \theta_r = \angle APB$$

Therefore, $|s_o| = |s_i|$. Unlike a lens, this virtual image appears on the right side of the interface. Accordingly, we shall adopt the convention that s_o *and* s_i *are both measured negative to the right of the reflecting interface.*

Each source point in the object space corresponds to a point, an equal distance behind the interface, in the image space. Consequently, for a planar mirror, M_T (the transverse magnification) equals +1; the image is life-size, virtual and erect (right-side up).

Curved mirrors are conveniently categorized as either spherical or aspherical. Several aspherical configurations are illustrated in Fig. 4-38. The fact that a paraboloidal mirror will reflect an incident plane wave into a perfectly converging spherical wave (see Problem 3.14) accounts for its use as the main light-collecting element in the 200-inch Palomar telescope. For the same reason, the antenna dish at Jodrell Bank is a huge 250-foot paraboloid.

The rays in Fig. 4-38 appear to converge toward or diverge from axial points which are the geometrical foci of the curved surfaces.

A comparison of the parabolic and spherical configurations (Fig. 4-39) shows that the two are almost indistinguishable in the vicinity of the central axis when the radius of the sphere is made equal to twice the focal length of the parabola. Accordingly, we can expect, at least in the paraxial approximation, that F will serve as the focal point of a spherical mirror centered on C. For such a device the object and image distances are related by the *mirror equation*:

$$\frac{1}{s_o} + \frac{1}{s_i} = -\frac{2}{R} = \frac{1}{f}$$

Observe that this has the same form as the lens equation, provided we adhere to the sign convention in Table 4-4, page 82. As a result, a concave spherical mirror has much the same imaging characteristics as a converging thin lens, while a convex spherical mirror behaves like a diverging lens. Indeed, Table 4-3 (page 64) is applicable to spherical mirrors or lenses. This implies that under appropriate conditions the spherical mirror has attributes of both the parabolic and elliptical configurations: like the former it can form images of distant objects and like the latter it can form images of nearby objects.

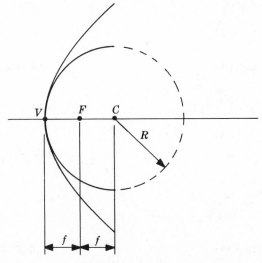

Fig. 4-39

Table 4-4.   Sign Convention for Spherical Reflecting Surfaces

s_o, f	+ left of V
s_i	+ left of V
R	+ when C is right of V
y_o, y_i	+ above optical axis

Table 4-5.   Physical Significance of the Signs of Spherical Mirror Parameters

Quantity	Sign	
	+	−
s_o	real object	virtual object
s_i	real image	virtual image
f	concave mirror	convex mirror
y_o	erect object	inverted object
y_i	erect image	inverted image
M_T	erect image	inverted image
R	convex mirror	concave mirror

SOLVED PROBLEMS

4.47. A pencil is held so that it is tilted away from a plane mirror. Construct a ray diagram locating the image.

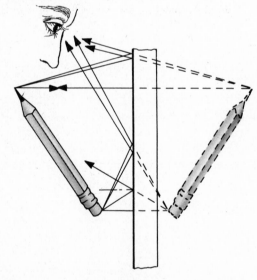

Since each point on the object results in an image point an equal distance behind the mirror (as in Fig. 4-37), we need only locate the ends of the pencil's image. Any two rays from an object point will determine the corresponding image point, but perhaps the simplest choice is one perpendicular ray. Figure 4-40 should be self-explanatory.

4.48. What is the length of the smallest vertical planar mirror in which you can see your entire body, and how should it be positioned? (A classic problem.)

Fig. 4-40

Whatever the geometry, the mirror plane Σ will be halfway between object and image $(s_o = s_i)$. If your toe is to be seen, a ray from it must enter your eye as in Fig. 4-41. We don't know the height of point H, but $\angle DHC$ must equal $\angle CHB$. This means that triangles BHC and DHC are congruent and so $\overline{GH} = \overline{HI} = \overline{BD}/2$. Similarly, if you are to see the top of your head, $\overline{EF} = \overline{FG} = \overline{AB}/2$. Thus a mirror of length $\overline{FH}$ should do the job, where

$$\overline{FH} = \overline{FG} + \overline{GH} = \frac{\overline{AB}}{2} + \frac{\overline{BD}}{2} = \frac{\overline{AD}}{2}$$

In other words, a mirror half your height, with its top edge lowered by half the distance between your eye and the top of your head, will serve the purpose.

4.49. Two front-surfaced plane mirrors at right angles to each other are set upon a table in

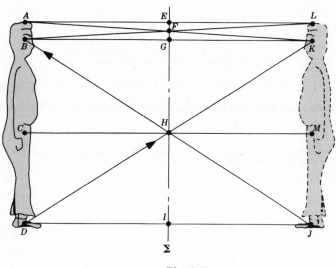

Fig. 4-41

front of a little green frog.   How many images of itself will the frog see?

Three (see Fig. 4-42). Two images each result from reflections off only one of the two **mirrors**; the third image arises when light is reflected from both mirrors.

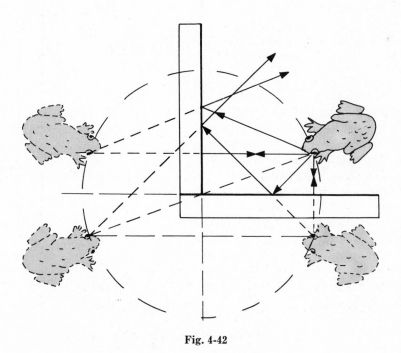

Fig. 4-42

4.50.   Show that the spherical mirror equation is applicable to a planar reflecting surface.

The mirror equation is

$$\frac{1}{s_o} + \frac{1}{s_i} = -\frac{2}{R} = \frac{1}{f}$$

For a planar surface the radius of curvature becomes infinite, hence

$$\frac{1}{s_o} + \frac{1}{s_i} = 0$$

or $s_o = -s_i$, as required. (Since the object distance is positive, s_i must be a negative number; the image is to the right of the interface.)

4.51. Envision a ray in a plane perpendicular to the two mirrors of Fig. 4-43. Prove that the ray will be deviated through an angle 2θ regardless of its incident angle.

The angle of deviation, call it γ, is an exterior angle of triangle ADC and therefore equal to the sum of the opposite interior angles; i.e., $\gamma = 2\alpha + 2\beta$. In triangle ABC we have

$$\angle CAB + \angle ACB + \theta = 180°$$

which leads to $\theta = 180° - (90° - \beta) - (90° - \alpha) = \alpha + \beta$. But $\gamma = 2(\alpha + \beta)$ and so $\gamma = 2\theta$.

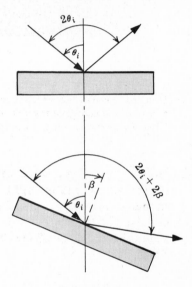

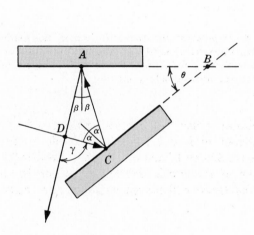

Fig. 4-43   Fig. 4-44

4.52. It is quite common to find a small planar mirror attached to the suspension system of such devices as torsion pendulums and galvanometers. Show that if the mirror rotates through an angle β, the beam will be deflected by an additional angle of 2β.

The setup is that of Fig. 4-44. When the mirror is rotated, the new incident angle is $\theta_i + \beta$, which also equals the angle of reflection. The total deflection is then $2\theta_i + 2\beta$, as compared with $2\theta_i$ before rotation.

4.53. Figure 4-45 depicts an ellipsoidal reflector whose foci are at F_1 and F_2. The positive thin lens has a focal length f and a tungsten filament is positioned at F_1. Trace the progress of rays emitted from the filament.

Most rays emanating from F_1 strike the ellipsoid and are reflected toward the second focus F_2, as in Fig. 4-38, page 80. The rays pass through F_2 and move on, much as if the point source were there rather than at F_1. Since F_2 is the object focus of the lens, these rays will emerge from the device as a collimated beam parallel to the central axis. Of course some rays will escape directly without reflection, while others will be multiply reflected before reaching the lens.

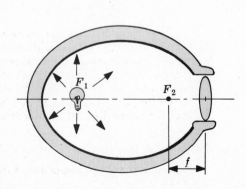

Fig. 4-45

4.54. The *Gregorian reflecting telescope* is a centered system consisting of a large parabolic primary mirror which collects incident light, bringing it to bear on a small concave ellipsoidal secondary mirror. The rays reflect off the secondary and converge back through a hole in the primary. Draw a ray diagram and discuss the locations of the various foci.

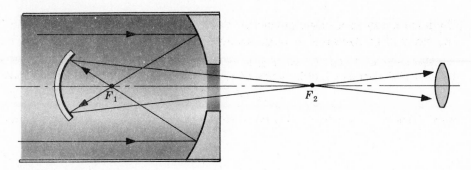

Fig. 4-46

The point F_1 in Fig. 4-46 is clearly the focal point of the parabolic mirror, the mathematical focus of the paraboloid. Figure 4-38, page 80, indicates that light diverging from one focus of an ellipsoid will converge toward the other focus. Hence F_1 and F_2 are the foci of the ellipsoidal mirror. Thus F_1 is the common focus of both the primary and secondary mirrors.

4.55. In the Kitt Peak solar telescope a planar mirror 80 inches across tracks the sun, reflecting collimated light down a 500-foot shaft to a 60-inch parabolic mirror. This primary mirror, in turn, focuses the beam 300 feet back up the shaft where the image can be photographed. If the diameter of the sun is 864,000 miles and its distance from the earth is 93,000,000 miles, how large will its image be at the focus of the telescope?

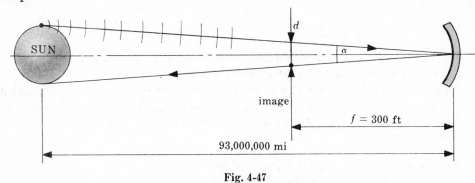

Fig. 4-47

Each point on the sun emits a spherical wave which increases in radius until it arrives and fills the aperture of a distant telescope with an almost planar wave. The nearly parallel bundle of rays is focused essentially to a point image a distance f from the mirror. Thus, point by point, parallel bundles of rays entering at slightly different directions build up a complete inverted image of the sun. Of course, only the axial point on the sun will be perfectly imaged by a parabolic mirror, but the subtented angle (α) is small and so there will be very little deterioration in the image over the entire disc. It follows from Fig. 4-47 that

$$\alpha = \frac{864,000 \text{ mi}}{93,000,000 \text{ mi}} = 0.0093 \text{ rad}$$

The diameter of the image disc is evidently given by

$$d = f\alpha = 300(0.0093) = 2.8 \text{ ft}$$

4.56. A concave spherical mirror has a radius of magnitude $|R|$ and is centered at C. A real erect object $|R|/6$ tall is located a distance $1.5\,|R|$ from the mirror's vertex. Draw a ray diagram showing the formation of the image.

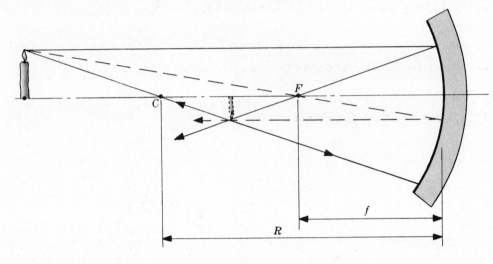

Fig. 4-48

It will take at least two rays to locate the image of the topmost object point. The simplest rays to use are those passing through C and the focal point F, where $f = -R/2$. A ray through C propagates down a radius of the sphere and is reflected back along itself. A ray entering parallel to the central axis will be reflected through F. The image of the top of the object is positioned at the intersection of these two rays. A ray from the bottom of the object along the axis passes through C and returns on itself. Thus the image of the object's base resides on the axis just above the point of intersection of the rays through C and F.

Another convenient ray is the one going through F before striking the mirror (see Fig. 4-48).

4.57. Compute the magnification and image location for Problem 4.56.

Because the radius R is actually a negative quantity here, we write it in terms of its absolute value as $R = -|R|$. The mirror equation

$$\frac{1}{s_o} + \frac{1}{s_i} = -\frac{2}{R}$$

becomes

$$\frac{1}{1.5\,|R|} + \frac{1}{s_i} = \frac{2}{|R|}$$

or $s_i = 3|R|/4$. The image is real and to the left of the vertex (see Tables 4-4 and 4-5, page 82). As for the magnification,

$$M_T = -\frac{s_i}{s_o} = -\frac{3|R|/4}{3|R|/2} = -\frac{1}{2}$$

The image is inverted and half-sized. Take another look at Fig. 4-48 and compare these results with Table 4-3 ($\infty > s_o > 2f$) on page 64.

4.58. A one-inch tall candle is set three inches in front of a concave spherical mirror having a one-foot radius. Describe the resulting image.

The mirror equation

$$\frac{1}{s_o} + \frac{1}{s_i} = -\frac{2}{R}$$

yields $$\frac{1}{3} + \frac{1}{s_i} = -\frac{2}{-12}$$

or $s_i = -6$ inches. The image is virtual because s_i is negative. (Refer to Table 4-3, page 64; the concave mirror behaves like a converging lens, hence, since $f = -R/2 = +6$ inches and $s_o = +3$ inches, we see immediately that $|s_i| > s_o$ and the image should be virtual, erect and magnified.) Proceeding,

$$M_T = -\frac{s_i}{s_o} = -\frac{-6}{2} = +2$$

so that the image is erect and twice the size of the object.

4.59. Draw a ray diagram for Problem 4.58.

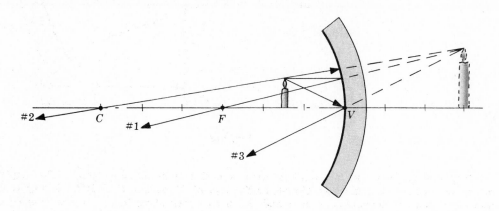

Fig. 4-49

The first thing to draw is a ray (#1) from the top of the candle parallel to the axis. It reflects off the mirror and returns through F. We should know from Table 4-3, page 64, that an object this close in ($s_o < f$) will form a virtual image, but let us just see how it turns out. A ray (#2) arriving along a radius will strike the mirror, return on itself and pass through C. It is clear that #1 and #2 will never intersect on the left side of V, but they do appear to diverge from a point 6 inches behind the mirror. Another ray (#3) is easily drawn; it is the one incident at V. (A ray along the line from F to the top of the object might also be used. It would reflect back parallel to the axis.)

4.60. A concave spherical mirror of 20-cm radius is to be used to project an image of a candle onto a wall 110 cm away. Where will the candle have to be placed and what will the image look like?

The object distance is to the left of V if the image is to be real, hence $s_i = +110$ cm and

$$\frac{1}{s_o} + \frac{1}{110} = -\frac{2}{-20}$$

so that $s_o = +11$ cm. This is slightly greater than $f = 10$ cm and less than $2f$, in agreement with Table 4-3. Moreover,

$$M_T = -\frac{s_i}{s_o} = -\frac{110}{11} = -10$$

which means that the image is inverted and magnified 10 times.

4.61. Design a spherical mirror which will form an erect half-sized image of an object if that object is 100 cm from the vertex. Where will the image be located?

We can determine the image distance from the magnification as follows:

$$M_T = -\frac{s_i}{s_o} = +\frac{1}{2}$$

or

$$-\frac{s_i}{100} = \frac{1}{2}$$

whence $s_i = -50$ cm. The mirror equation now yields the radius:

$$\frac{1}{100} + \frac{1}{-50} = -\frac{2}{R} \quad \text{or} \quad R = +200 \text{ cm}$$

The mirror is convex (see Table 4-5, page 82) and the image is virtual (see Table 4-3, page 64). Note that only a convex mirror generates an erect minified image.

Supplementary Problems

ASPHERICAL REFRACTING SURFACES

4.62. Return to the Cartesian ovoid depicted in Fig. 4-2, page 52, and construct a set of coordinate **axes** with the origin at the vertex V. Locate the x-axis along the SP line and the y-axis **perpendicular** to it. Now derive an equation for the ovoid in terms of s_o, s_i, n_1, n_2, x and y.

Ans. $n_2 s_i - n_1 s_o = [(x - s_o)^2 + y^2]^{1/2} n_1 + [(x - s_i)^2 + y^2]^{1/2} n_2$

4.63. Suppose we have a horizontal and vertical x-y coordinate system whose origin is at the **vertex** of the curved interface between two media (see Fig. 4-2). If plane waves entering from **the left** are to be focused at P, show that the boundary curve is given by

$$x^2 + \frac{n_2^2}{n_2^2 - n_1^2} y^2 - \frac{2 s_i n_2 (n_2 - n_1)}{n_2^2 - n_1^2} x = 0$$

By completing the square, prove that the interface is an ellipsoid of revolution when $n_2 > n_1$. Determine the semimajor and semiminor axes, a and b, as well as the eccentricity e.

Ans. $a = \dfrac{n_2 s_i}{n_2 + n_1}$, $\quad b = \left(\dfrac{n_2 - n_1}{n_2 + n_1}\right)^{1/2} s_i$, $\quad e = \dfrac{n_1}{n_2}$ and $\quad \dfrac{(x - a)^2}{a^2} + \dfrac{y^2}{b^2} = 1$

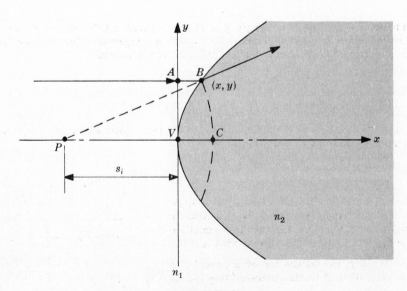

Fig. 4-50

4.64. A parallel axial bundle of rays entering from the left in Fig. 4-50 will appear to diverge from P provided $n_1 > n_2$. Remembering that here s_i is a negative quantity, show that the interface curve has the same equation as that in Problem 4.63. Prove that the configuration is a hyperboloid of revolution. *Hint*: Use the fact that $n_1 \overline{AB} = n_2 \overline{VC}$, where C is on the arc swept out by $\overline{PB}$. Point P is the first focus of the hyperboloid of two sheets.

SPHERICAL REFRACTING SURFACES

4.65. A diamond ($n_d = 2.42$) rod with one end ground into a convex hemisphere contains a small black flaw. If the radius of curvature is 20 cm and the flaw lies on the central axis 20 cm from the vertex, where will its image appear when the rod is imbedded in water ($n_w = 1.33$)?

Ans. Object at C; therefore $s_i = -20$ cm, regardless of surrounding medium.

4.66. A long glass rod ($n_g = 1.5$) is 10 cm in diameter and is immersed in air. It has a convex hemispherical surface as its left end, and a concave hyperboloidal surface as its right end. The hyperboloid has an eccentricity of 1.5 and its vertex is 5 cm to the right of its first focus F_1. Where must an axial point source be located if the rod is to form a virtual image of it at F_1?

Ans. For the spherical surface $f_o = 10$ cm and therefore $s_o = 10$ cm to the left.

4.67. A borosilicate crown glass sphere ($n_g = 1.5$) of radius 4 cm is surrounded by ethyl alcohol ($n_e = 1.36$). An ant drifting in the alcohol is 6 cm from the sphere's center; describe its image.

Ans. $s_i = -2.32$ cm (virtual image to the left of vertex)

4.68. A convex interface separates two media of refractive indices 1 and 2. An axial point source in the air a distance 40 cm from the vertex is imaged in the second medium 80 cm from the vertex. Determine the radius of curvature of the interface.

Ans. $R = +20$ cm

THE THIN LENS EQUATION

4.69. The radii of curvature of a double convex thin glass lens ($n = 1.5$) are in the ratio of 2 to 1. Write an expression for R, the smaller of the two radii, in terms of the focal length.

Ans. $R = 3f/4$

4.70. A thin positive lens of focal length f is placed between a point source S and a screen, which are themselves separated by a distance L. Write an expression for the two locations of the lens (measured from S) which will yield real images on the screen.

Ans. $s_o = \dfrac{1}{2}\left[L \pm \sqrt{L(L - 4f)}\,\right]$

4.71. An equiconvex thin lens of flint glass ($n_\ell = 1.65$) has a focal length of 62 cm when immersed in air. Determine its radii of curvature.

Ans. $R_1 = 80.6$ cm, $R_2 = -80.6$ cm

4.72. Figure 4-51 depicts a bundle of converging rays entering a diverging thin lens. Describe what's happening and then use the thin lens equation to verify your conclusions. (Note that $s_o < 0$.)

Ans. Point A locates the top of a virtual object, while B is the corresponding image point. Since $|s_o| < f$, the image is real, erect and magnified, and $s_i > |s_o|$.

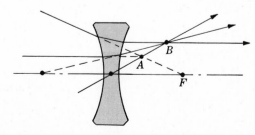

Fig. 4-51

SIMPLE THIN LENS IMAGERY

4.73. A positive thin lens is used to project the enlarged image of a slide onto a wall 10 m away. If the slide is 20×30 mm, and if its image is to be 2×3 m, what must be the focal length of the lens and the distance from it to the slide?

   Ans. $s_o = 0.1$ m, $M_T = -100$, $f = 0.099$ m

4.74. A thin positive lens generates an 8-cm tall erect image of a 5-cm tall object located 90 cm from the lens. Compute the focal length of the lens and locate the image.

   Ans. $s_i = -144$ cm, $f = 240$ cm

4.75. A simple camera consists of a thin positive lens which casts a real image on the film plane. Suppose that the lens has a 50-mm focal length. How far from a 1-m tall object must the camera be if the image is to appear 25 mm high? How far will the lens be from the film plane?

   Ans. $s_o = 2.04$ m, $s_i = 51.3$ mm

4.76. Envision a thin lens for which the object and image are separated by a distance L. Show that

$$L = \frac{-f(M_T - 1)^2}{M_T}$$

COMPOUND THIN LENSES

4.77. Telephoto camera lenses most often resemble the Galilean telescope, i.e. they consist of a positive lens L_1 followed by a negative lens L_2. If the focal length of L_1 is 20 cm, that of L_2 is -40 cm and the separation is 10 cm, determine the f.f.l. and the b.f.l.

   Ans. f.f.l. $= 33.33$ cm, i.e. the object focal point is to the left of L_1; b.f.l. $= 13.33$ cm, i.e. the image focal point is to the right of L_2

4.78. Three thin lenses of focal lengths $f_1 = 10$ cm, $f_2 = 20$ cm and $f_3 = -40$ cm are in contact, forming a single unit. If an object is located 16 cm in front of the lens, describe the resulting image.

   Ans. $f = 8$ cm, $s_i = +16$ cm

4.79. Two thin positive lenses are in contact, forming a compound lens of focal length 30 cm. If the power of one of the component lenses is twice that of the other, what are their two focal lengths?

   Ans. 45 cm, 90 cm

4.80. An object sits on a table 12 cm from a positive thin lens of focal length 9 cm, which in turn is 21 cm in front of a negative thin lens of focal length -18 cm. Locate the image formed by the system.

   Ans. $s_i = +90$ cm (image is 90 cm to the right of the negative lens)

THICK LENSES

4.81. Envision a thick lens with a refractive index of 2, whose radii of curvature are equal and negative. If the centers of curvature are separated by a distance d and if the lens is surrounded by air, describe its properties.

   Ans. The lens is positive since $f = 2R^2/d$; $h_1 = h_2 = -R$ $(R < 0)$, so that the principal planes are off to the right and separated by d.

4.82. A thick glass double convex lens $(n_g = 1.5)$ has radii of 2 cm and 4 cm and a thickness of 2 cm. Locate the principal and focal points with respect to the vertices V_1 and V_2.

Ans. $f = 3$ cm, $h_1 = +0.5$ cm, $h_2 = -1.0$ cm

4.83.   A hemispherical converging lens of radius $+12$ cm and refractive index of 2.0 is 36 cm from an axial point object. Locate the principal and focal planes and describe the image.

Ans. $f = 12$ cm; $h_1 = 0$ (first principal plane at V_1); $h_2 = -6$ cm (i.e. to the left of V_2); $s_i = 18$ cm to the right of H_2, image is real

4.84.   A thick lens with a refractive index of 2 obeys the special condition that both its surfaces have a common center of curvature outside of the lens. Describe its properties if the thickness is designated as d.

Ans. $f = -2|R|(|R| + d)/d$ (negative lens); $h_1 = -|R|$, $h_2 = -(|R| + d)$. Note that H_1 and H_2 coincide with the center of curvature.

4.85.   What is the focal length in air of a spherical droplet of benzene ($n_b = 1.501$) having a radius of 2 mm? Describe the image resulting from a 0.5-mm tall object 5.3 cm from the center of the droplet.

Ans. $f = 3$ mm; $s_i = 3.18$ mm from the sphere's center, $M_T = -0.06$, $y_i = -0.03$ mm (image is real, minified and inverted)

LENS COMBINATIONS

4.86.   The *Huygens ocular* is a combination of two thin plano-convex lenses. The first is known as the field lens and the second, nearest the observer's eye, is the eye lens. Suppose that the field lens has a focal length of $3f_1$ and the eye lens has a focal length of f_1. Both have their curved surfaces to the left and they are separated by $2f_1$. A bundle of rays converging toward the first focal plane of the ocular emerges as a parallel beam. Locate that focal plane.

Ans. $f = 3f_1/2$, $\overline{H_{11}H_1} = 3f_1$. The first focal plane resides between the lenses, $f_1/2$ to the left of the eye lens.

4.87.   Two thin positive lenses of focal lengths 40 cm and 60 cm are separated by 20 cm. Where must an object be positioned if its image is to reside on a screen 45 cm behind the second lens?

Ans. $f = 30$ cm, $\overline{H_{11}H_1} = 10$ cm, $\overline{H_{22}H_2} = -15$ cm; object 50 cm left of first lens or 60 cm from H_1

4.88.   Imagine that you have three thin lenses, two converging and one diverging, of focal lengths $f_1 = 4$ cm, $f_2 = -8$ cm and f_3. The first two lenses are separated by 6 cm and the last two by 1.4 cm. What must be the focal length f_3 of the last lens if the system is to be *afocal*, i.e. rays that enter parallel emerge parallel?

Ans.   Combining the first two lenses yields $f_{12} = +3.2$ cm and $\overline{H_{22}H_2} = -4.8$ cm. The third lens is a distance $d = 6.2$ cm from this combination, and for the resultant power to be zero, $f_3 = 3.0$ cm.

4.89.   A *Ramsden ocular* consists of two thin plano-convex lenses, each of focal length f_1, separated by $2f_1/3$, with the curved surfaces facing each other. Locate the object plane (Σ_o) in front of the ocular such that light diverging from any point on Σ_o emerges as a collimated beam. In practice, an objective lens would form a real image on Σ_o (where there might also be a pair of cross hairs), which is then converted into parallel light by the ocular so that the eye can view it in a relaxed (unaccommodated) fashion.

Ans. Σ_o must be a distance away equal to the effective focal length $f = 3f_1/4$.

4.90.   A thin converging lens having a power of 3.33 diopters is $\frac{1}{4}$ m in front of a thin diverging lens of -20 diopters. What is the focal length of the combination?

Ans.   The system is afocal; when parallel rays enter they emerge parallel.

PLANAR, ASPHERICAL AND SPHERICAL MIRRORS

4.91. An object positioned 300 cm from a spherical concave mirror generates a real image 150 cm from the mirror's vertex. To where must the object be moved if the new image is to reside in the object's original position?

Ans. $s_o = 150$ cm, $f = 100$ cm

4.92. Suppose that the compound lens of Problem 4.80 is placed so that its negative back lens is 60 cm from the vertex of a convex spherical mirror having a 15-cm radius. Locate the image formed by the mirror of an object 12 cm from the first lens.

Ans. $s_i = -10$ cm; image is virtual, inverted and 10 cm to the right of the mirror's vertex

4.93. A cone of rays converges toward an axial point S a distance d *behind* a convex mirror of focal length $f > d$. In other words, S is a virtual object and $d = |s_o| < f$. Use the mirror equation to arrive at a description of the resulting image. (Remember that $s_o < 0$.)

Ans. The image is real, erect, magnified and farther from the mirror than the object is ($|s_o| < s_i$).

4.94. A 1-cm high object is positioned 12 cm in front of a spherical concave mirror having a radius of curvature of 8 cm. Completely describe the resulting image.

Ans. $s_i = 6$ cm, $M_T = -1/2$ (image inverted, real and $\frac{1}{2}$ cm tall)

4.95. An object 4 cm high sits 200 cm in front of a convex mirror having a focal length of -400 cm. Describe the image.

Ans. $s_i = -133.3$ cm, $M_T = +0.66$ (image virtual, erect and minified)

4.96. A 3-cm tall object is located 180 cm from a spherical convex mirror having a radius of curvature of 90 cm. Describe the resultant image.

Ans. $s_i = -36$ cm, $M_T = +1/5$ (image virtual, erect and $\frac{3}{5}$ cm tall)

Chapter 5

Polarization

5.1 INTRODUCTION

Light is a transverse electromagnetic wave and thus far we have considered only cases where the electric field vector resided in a fixed plane. This plane is referred to as the *plane of vibration* and the light is said to be *plane polarized*. This chapter deals, for the most part, with the superposition of two orthogonal plane polarized light waves of the same frequency. The resultant electric field need not reside in a fixed plane; indeed, the field vector might even rotate in time. The amplitudes and relative phase of the interacting waves will determine the *state of polarization* of the composite disturbance. By comparison, the interaction of coplanar waves is the usual domain of interference theory (Chapter 6).

5.2 PLANE POLARIZATION

Consider two perpendicular harmonic optical fields given by

$$\mathbf{E}_x(z, t) = \hat{\mathbf{i}} E_{0x} \cos(kz - \omega t)$$

$$\mathbf{E}_y(z, t) = \hat{\mathbf{j}} E_{0y} \cos(kz - \omega t + \varepsilon)$$

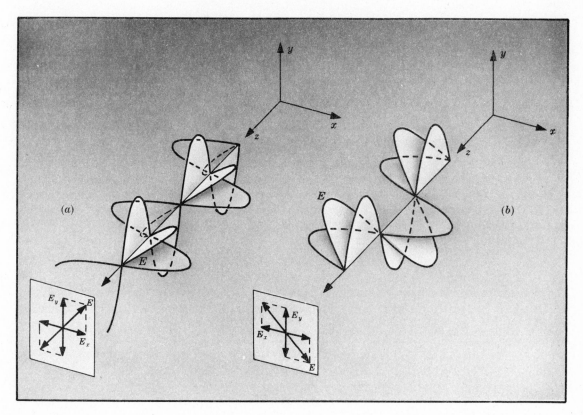

Fig. 5-1

93

The waves move in the positive z-direction and have a relative phase ε. The plane of vibration of $\mathbf{E}_x(z, t)$ corresponds to the xz-plane, while $\mathbf{E}_y(z, t)$ resides in the yz-plane. The resulting disturbance

$$\mathbf{E}(z, t) = \mathbf{E}_x(z, t) + \mathbf{E}_y(z, t)$$

varies with ε. In the specific instance when $\varepsilon = 0$ or an integral multiple of $\pm 2\pi$ the component fields are in phase and

$$\mathbf{E}(z, t) = (\hat{\mathbf{i}} E_{0x} + \hat{\mathbf{j}} E_{0y}) \cos (kz - \omega t)$$

The amplitude, $\hat{\mathbf{i}} E_{0x} + \hat{\mathbf{j}} E_{0y}$, is constant and so the resultant wave itself is *plane* or *linearly polarized*, as shown in Fig. 5-1(a). Similarly, when ε is an odd integral multiple of $\pm \pi$ the component fields are out of phase and

$$\mathbf{E}(z, t) = (\hat{\mathbf{i}} E_{0x} - \hat{\mathbf{j}} E_{0y}) \cos (kz - \omega t)$$

Again the resultant has a constant amplitude and the wave is linearly polarized, as depicted in Fig. 5-1(b).

An optical disturbance which is plane polarized is often simply referred to as $\mathcal{P}$-*state light*.

SOLVED PROBLEMS

5.1. The waves

$$\mathbf{E}(z, t) = (\hat{\mathbf{i}} E_{0x} + \hat{\mathbf{j}} E_{0y}) \cos (kz - \omega t)$$
$$\mathbf{E}'(z, t) = (\hat{\mathbf{i}} E'_{0x} - \hat{\mathbf{j}} E'_{0y}) \cos (kz - \omega t)$$

both represent $\mathcal{P}$-state light. Show that in general such waves are not orthogonal. Under what circumstances will their planes of vibration be normal to each other?

Letting $\mathbf{E}_0$ and $\mathbf{E}'_0$ be the amplitude vectors of $\mathbf{E}$ and $\mathbf{E}'$ respectively, their dot product is

$$\mathbf{E}_0 \cdot \mathbf{E}'_0 = E_0 E'_0 \cos \theta$$

where θ is the angle between $\mathbf{E}_0$ and $\mathbf{E}'_0$. But

$$\mathbf{E}_0 \cdot \mathbf{E}'_0 = (\hat{\mathbf{i}} E_{0x} + \hat{\mathbf{j}} E_{0y}) \cdot (\hat{\mathbf{i}} E'_{0x} - \hat{\mathbf{j}} E'_{0y}) = E_{0x} E'_{0x} - E_{0y} E'_{0y}$$

Thus
$$\cos \theta = \frac{E_{0x} E'_{0x} - E_{0y} E'_{0y}}{E_0 E'_0}$$

and θ is generally not $90°$, since the right side of this expression is generally nonzero. If, however, $E_{0x} E'_{0x} = E_{0y} E'_{0y}$, the right side vanishes and the waves are orthogonal. Perhaps the simplest case of this arises when $E_{0x} = E_{0y}$ and $E'_{0x} = E'_{0y}$.

5.2. Write an expression for a linearly polarized wave of angular frequency ω propagating in the positive z-direction with its plane of vibration at $30°$ to the zx-plane.

Suppose that the scalar amplitude of the wave is E_0. Then its x- and y-components are

$$E_{0x} = E_0 \cos 30° = 0.866 E_0$$
$$E_{0y} = E_0 \sin 30° = 0.5 E_0$$

Hence
$$\mathbf{E}(z, t) = (0.866 E_0 \hat{\mathbf{i}} + 0.5 E_0 \hat{\mathbf{j}}) \cos (kz - \omega t + \alpha)$$

where the unknown constant α depends on the initial conditions.

5.3. Write an expression for a plane polarized disturbance of angular frequency ω propagating in the positive z-direction such that the **E**-field makes an angle of $120°$ with the positive x-direction at $t = 0$ and $z = 0$. Verify that this wave is orthogonal to the wave of Problem 5.2.

With E_0 as the scalar amplitude,

$$E_{0x} = E_0 \cos 120° = -E_0 \cos 60° = -0.5\,E_0$$
$$E_{0y} = E_0 \sin 120° = E_0 \sin 60° = 0.866\,E_0$$

Thus
$$\mathbf{E}(z, t) = (-0.5\,E_0\,\hat{\mathbf{i}} + 0.866\,E_0\,\hat{\mathbf{j}}) \cos(kz - \omega t)$$

To verify that this wave is orthogonal to the wave of Problem 5.2, form the dot product of the amplitudes, i.e.

$$(0.866\,E_0\,\hat{\mathbf{i}} + 0.5\,E_0\,\hat{\mathbf{j}}) \cdot (-0.5\,E_0\,\hat{\mathbf{i}} + 0.866\,E_0\,\hat{\mathbf{j}})$$

Inasmuch as this is zero, the planes of vibration are normal.

5.4.   Describe the wave given by the expression

$$\mathbf{E} = \hat{\mathbf{j}}\,E_0 \sin\!\left(\frac{2\pi x}{\lambda} - \omega t\right)$$
$$- \hat{\mathbf{k}}\,E_0 \sin\!\left(\frac{2\pi x}{\lambda} - \omega t\right)$$

where $\hat{\mathbf{j}}$ and $\hat{\mathbf{k}}$ are unit basis vectors in Cartesian coordinates.

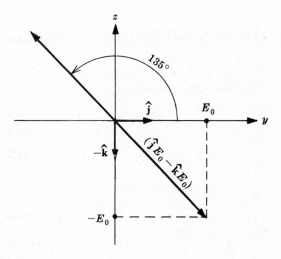

The wave can be reformulated as

$$\mathbf{E} = (\hat{\mathbf{j}}\,E_0 - \hat{\mathbf{k}}\,E_0) \sin\!\left(\frac{2\pi x}{\lambda} - \omega t\right)$$

It travels in the positive x-direction with a constant amplitude of $\hat{\mathbf{j}}\,E_0 - \hat{\mathbf{k}}\,E_0$ and is therefore linearly polarized. The plane of polarization, as shown in Fig. 5-2, is tilted at 135° to xy-plane. Note that the scalar amplitude of $\mathbf{E}$ is $\sqrt{2}\,E_0$.

Fig. 5-2

5.5.   Describe the wave $\mathbf{E}(y, t)$ which results from the superposition of the disturbances

$$\mathbf{E}_x(y, t) = \hat{\mathbf{i}}\,E_0 \cos k(y - vt) \qquad \mathbf{E}_z(y, t) = -\hat{\mathbf{k}}\,E_0 \cos k(y - vt)$$

Make a sketch of $\mathbf{E}(0, t)$ at $t = 0$, $t = \tau/4$, $t = \tau/2$, $t = 3\tau/4$ and $t = \tau$ (where, of course, τ is the period).

Inasmuch as $\cos(\theta + \pi) = -\cos\theta$, we can rewrite $\mathbf{E}_z$ as

$$\mathbf{E}_z(y, t) = \hat{\mathbf{k}}\,E_0 \cos[k(y - vt) + \pi]$$

wherein the relative phase, ε, is just π. Hence the resultant is linearly polarized. The phase of $\mathbf{E}_x$ can be written as $ky - (2\pi t/\tau)$, while that of $\mathbf{E}_z$ is $ky - (2\pi t/\tau) + \pi$. Thus, at $y = 0$,

$$\mathbf{E}(0, 0) = \hat{\mathbf{i}}\,E_0 + \hat{\mathbf{k}}\,E_0 \cos\pi$$
$$\mathbf{E}(0, \tau/4) = \hat{\mathbf{i}}\,E_0 \cos(-\pi/2) + \hat{\mathbf{k}}\,E_0 \cos(\pi/2) = 0$$
$$\mathbf{E}(0, \tau/2) = \hat{\mathbf{i}}\,E_0 \cos(-\pi) + \hat{\mathbf{k}}\,E_0 \cos 0$$
$$\mathbf{E}(0, 3\tau/4) = \hat{\mathbf{i}}\,E_0 \cos(-3\pi/2) + \hat{\mathbf{k}}\,E_0 \cos(-\pi/2) = 0$$
$$\mathbf{E}(0, \tau) = \hat{\mathbf{i}}\,E_0 \cos(-2\pi) + \hat{\mathbf{k}}\,E_0 \cos(-\pi)$$

Figure 5-3, page 96, depicts the corresponding disturbances.

5.6. Two linearly polarized waves having the forms

$$\mathbf{E}_1(z, t) = \hat{\mathbf{i}} E_{0x} \cos(\omega t - kz) + \hat{\mathbf{j}} E_{0y} \cos(\omega t - kz)$$

$$\mathbf{E}_2(z, t) = \hat{\mathbf{i}} E'_{0x} \cos(\omega t - kz) + \hat{\mathbf{j}} E'_{0y} \cos(\omega t - kz)$$

overlap in space. Show that the resultant is also linearly polarized.

The resultant $\mathbf{E} = \mathbf{E}_1 + \mathbf{E}_2$ is given by

$$\mathbf{E} = \hat{\mathbf{i}}(E_{0x} + E'_{0x}) \cos(\omega t - kz)$$
$$+ \hat{\mathbf{j}}(E_{0y} + E'_{0y}) \cos(\omega t - kz)$$
$$= [\hat{\mathbf{i}}(E_{0x} + E'_{0x}) + \hat{\mathbf{j}}(E_{0y} + E'_{0y})] \cos(\omega t - kz)$$

It is seen that the vector amplitude of $\mathbf{E}$ is independent of z and t, i.e. it's constant. Accordingly, $\mathbf{E}$ is linearly polarized.

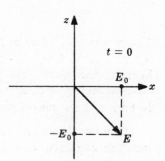

5.7. Write an expression for a linearly polarized harmonic plane wave of scalar amplitude E_0, propagating along a line in the xy-plane at 45° to the x-axis and having the xy-plane as its plane of vibration.

The vector amplitude $\mathbf{E}_0$ makes an angle of 135° with the x-axis. If it is to have a scalar value of E_0, then

$$\mathbf{E}_0 = -\frac{E_0}{\sqrt{2}}\hat{\mathbf{i}} + \frac{E_0}{\sqrt{2}}\hat{\mathbf{j}}$$

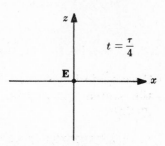

A harmonic plane wave has the general form

$$\mathbf{E} = \mathbf{E}_0 \cos(\mathbf{k} \cdot \mathbf{r} - \omega t)$$

In the present case the propagation vector is

$$\mathbf{k} = \frac{2\pi}{\lambda}\left(\frac{1}{\sqrt{2}}\hat{\mathbf{i}} + \frac{1}{\sqrt{2}}\hat{\mathbf{j}}\right)$$

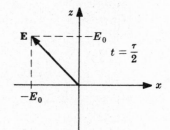

where $|\mathbf{k}| = k = 2\pi/\lambda$, and the position vector is

$$\mathbf{r} = x\hat{\mathbf{i}} + y\hat{\mathbf{j}} + z\hat{\mathbf{k}}$$

(this $\hat{\mathbf{k}}$ is a unit vector along the z-axis). Substituting into the wave function yields

$$\mathbf{E} = \frac{1}{\sqrt{2}}(-E_0\hat{\mathbf{i}} + E_0\hat{\mathbf{j}}) \cos\left[\frac{\sqrt{2}\pi}{\lambda}(x + y) - \omega t\right]$$

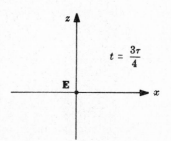

5.3 CIRCULAR POLARIZATION

Suppose now that the two orthogonal $\mathcal{P}$-states of Section 5.2 have a relative phase $\varepsilon = -\pi/2 + 2m\pi$ ($m = 0, \pm 1, \pm 2, \ldots$), i.e., $\varepsilon = -\pi/2, +3\pi/2, -5\pi/2, +7\pi/2, \ldots$. Then, if their scalar amplitudes are equal, that is, $E_{0x} = E_{0y} = E_0$, the two disturbances are expressible as

$$\mathbf{E}_x(z, t) = \hat{\mathbf{i}} E_0 \cos(kz - \omega t)$$

$$\mathbf{E}_y(z, t) = \hat{\mathbf{j}} E_0 \sin(kz - \omega t)$$

(the specific values of ε simply shift the cosine function to a sine function). The resultant wave $\mathbf{E} = \mathbf{E}_x + \mathbf{E}_y$ is

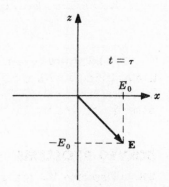

Fig. 5-3

$$\mathbf{E} = \hat{\mathbf{i}} E_0 \cos(kz - \omega t) + \hat{\mathbf{j}} E_0 \sin(kz - \omega t)$$
$$= E_0[\hat{\mathbf{i}} \cos(kz - \omega t) + \hat{\mathbf{j}} \sin(kz - \omega t)]$$

The magnitude of $\mathbf{E}$ is E_0 and is constant, but the direction of $\mathbf{E}$ is a function of z and t. As in Fig. 5-4(a), the electric field vector rotates clockwise (looking toward the source). Because the amplitude is constant, the endpoint of $\mathbf{E}$ sweeps out a circle (to be precise, a circular helix) with a frequency equal to that of the constituent waves. Such a field is said to be *right circularly polarized*, corresponding to an $\mathcal{R}$-*state*.

In much the same way, when $\varepsilon = \pi/2 - 2m\pi$ $(m = 0, \pm 1, \pm 2, \ldots)$, the cosine is shifted into the negative sine, giving

$$\mathbf{E}_x(z, t) = \hat{\mathbf{i}} E_0 \cos(kz - \omega t)$$
$$\mathbf{E}_y(z, t) = -\hat{\mathbf{j}} E_0 \sin(kz - \omega t)$$
$$\mathbf{E}(z, t) = E_0[\hat{\mathbf{i}} \cos(kz - \omega t) - \hat{\mathbf{j}} \sin(kz - \omega t)]$$

Once again $\mathbf{E}$ has a constant magnitude, but now it rotates counterclockwise (looking toward the source), as in Fig. 5-4(b). The field is *left circularly polarized*, corresponding to an $\mathcal{L}$-*state*.

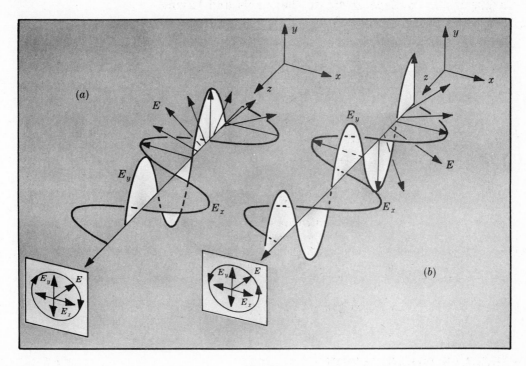

Fig. 5-4

$\mathcal{R}$- and $\mathcal{L}$-states are of particular significance in the quantum description, where they are associated with the spin angular momentum of the photons. All polarization states can be synthesized out of $\mathcal{R}$- and $\mathcal{L}$-states (see Problems 5.11 and 5.21), a feature which is a necessity in the photon model.

SOLVED PROBLEMS

5.8.   Describe the difference between the $\mathcal{R}$-state wave

$$\mathbf{E} = E_0[\hat{\mathbf{i}} \cos(kz - \omega t) + \hat{\mathbf{j}} \sin(kz - \omega t)]$$

and a wave of the form

$$\mathbf{E}' = E_0[\hat{\mathbf{i}}\sin{(kz-\omega t)} + \hat{\mathbf{j}}\cos{(kz-\omega t)}]$$

The wave $\mathbf{E}'(z, t)$ has a constant magnitude of E_0 and is circularly polarized. The two disturbances can easily be compared by examining their behavior at some fixed point in space, say $z = 0$. At $t = 0, \tau/4, \tau/2, 3\tau/4$ and τ, $\mathbf{E}(0, t)$ has values of $E_0\hat{\mathbf{i}}, -E_0\hat{\mathbf{j}}, -E_0\hat{\mathbf{i}}, E_0\hat{\mathbf{j}}$ and $E_0\hat{\mathbf{i}}$, respectively. In contrast, at these same values of t, $\mathbf{E}'(0, t)$ is equal to $E_0\hat{\mathbf{j}}, -E_0\hat{\mathbf{i}}, -E_0\hat{\mathbf{j}}, E_0\hat{\mathbf{i}}$ and $E_0\hat{\mathbf{j}}$, respectively. Thus $\mathbf{E}'(z, t)$ is an $\mathcal{L}$-state which is along the positive y-axis at $z = 0$ and $t = 0$.

5.9. Determine the state of polarization of the wave

$$\mathbf{E}(z, t) = E_0[\hat{\mathbf{i}}\sin{(kz-\omega t)} - \hat{\mathbf{j}}\cos{(kz-\omega t)}]$$

The magnitude of $\mathbf{E}$, i.e. $(\mathbf{E}\cdot\mathbf{E})^{1/2}$, is again constant at E_0, so the wave is circular. Fixing z at 0 we examine $\mathbf{E}(0, t)$ at $t = 0, \tau/4, \tau/2, 3\tau/4$ and τ, and find that $\mathbf{E}(0, t)$ has values of $-E_0\hat{\mathbf{j}}, -E_0\hat{\mathbf{i}}$, $E_0\hat{\mathbf{j}}, E_0\hat{\mathbf{i}}$ and $-E_0\hat{\mathbf{j}}$, respectively. The wave is evidently right circularly polarized, since the $\mathbf{E}$-field rotates clockwise in time.

5.10. Write an expression for a right circularly polarized wave propagating in the positive z-direction such that its $\mathbf{E}$-field points in the negative x-direction at $z = 0$ and $t = 0$.

As we saw in Problem 5.8,

$$\mathbf{E} = E_0[\hat{\mathbf{i}}\cos{(kz-\omega t)} + \hat{\mathbf{j}}\sin{(kz-\omega t)}]$$

is an $\mathcal{R}$-state which is directed along the positive x-axis at $z = 0$ and $t = 0$. That suggests that the wave we are looking for has the form

$$\mathbf{E}_\mathcal{R} = E_0[-\hat{\mathbf{i}}\cos{(kz-\omega t)} - \hat{\mathbf{j}}\sin{(kz-\omega t)}]$$

As a check, examine it at $z = 0$ and $t = 0, \tau/4, \tau/2, 3\tau/4$ and τ. At these values $\mathbf{E}_\mathcal{R}(0, t)$ equals $-\hat{\mathbf{i}}E_0, \hat{\mathbf{j}}E_0, \hat{\mathbf{i}}E_0, -\hat{\mathbf{j}}E_0$ and $-\hat{\mathbf{i}}E_0$, respectively. It is right circularly polarized and does have an initial negative x-component.

5.11. Show that the superposition of an $\mathcal{R}$- and an $\mathcal{L}$-state yields a $\mathcal{P}$-state provided that the scalar amplitudes of the constituent waves are equal.

Writing the two circular waves as

$$\mathbf{E}_\mathcal{R} = E_{0\mathcal{R}}[\hat{\mathbf{i}}\cos{(kz-\omega t)} + \hat{\mathbf{j}}\sin{(kz-\omega t)}]$$

$$\mathbf{E}_\mathcal{L} = E_{0\mathcal{L}}[\hat{\mathbf{i}}\cos{(kz-\omega t)} - \hat{\mathbf{j}}\sin{(kz-\omega t)}]$$

their sum becomes

$$\mathbf{E} = (E_{0\mathcal{R}}+E_{0\mathcal{L}})\hat{\mathbf{i}}\cos{(kz-\omega t)} + (E_{0\mathcal{R}}-E_{0\mathcal{L}})\hat{\mathbf{j}}\sin{(kz-\omega t)}$$

Note that at $z = 0$ and $t = 0$, $\mathbf{E} = (E_{0\mathcal{R}}+E_{0\mathcal{L}})\hat{\mathbf{i}}$, while at $z = 0$ and $t = \tau/4$, $\mathbf{E} = (E_{0\mathcal{R}}-E_{0\mathcal{L}})\hat{\mathbf{j}}$. Since both the magnitude and direction of $\mathbf{E}$ vary with z and t, the resultant is neither linearly nor circularly polarized. However, if $E_{0\mathcal{R}} = E_{0\mathcal{L}} = E_0$, then

$$\mathbf{E} = 2E_0\hat{\mathbf{i}}\cos{(kz-\omega t)}$$

which is a $\mathcal{P}$-state.

5.12. Write expressions for an $\mathcal{R}$- and an $\mathcal{L}$-state which combine to yield the $\mathcal{P}$-state

$$\mathbf{E}_p = E_0\hat{\mathbf{i}}\sin{(kz-\omega t)}$$

Bearing in mind the preceding three problems, along with the requirement that the cosine terms cancel, we consider the functions

$$\mathbf{E}_{\mathcal{R}} = E_{0\mathcal{R}}[\hat{\mathbf{i}}\sin(kz - \omega t) - \hat{\mathbf{j}}\cos(kz - \omega t)]$$

$$\mathbf{E}_{\mathcal{L}} = E_{0\mathcal{L}}[\hat{\mathbf{i}}\sin(kz - \omega t) + \hat{\mathbf{j}}\cos(kz - \omega t)]$$

We know from Problem 5.11 that a $\mathcal{P}$-state will arise when $E_{0\mathcal{R}} = E_{0\mathcal{L}}$, and so

$$\mathbf{E}_P = \mathbf{E}_{\mathcal{R}} + \mathbf{E}_{\mathcal{L}} = 2E_{0\mathcal{R}}\hat{\mathbf{i}}\sin(kz - \omega t)$$

which is the required function provided that $E_{0\mathcal{R}} = E_0/2$.

5.13. Write expressions for an $\mathcal{R}$- and an $\mathcal{L}$-state which when superimposed will yield a $\mathcal{P}$-state propagating along the z-axis with the yz-plane as its plane of vibration.

The component waves evidently must travel in the z-direction. Moreover, we require that $\mathbf{E}$ be in the yz-plane for all z and t. In other words, $\mathbf{E}$ must have only a $\hat{\mathbf{j}}$-component. From Problem 5.11, we also require that $E_{0\mathcal{R}} = E_{0\mathcal{L}}$, which we set equal to E_0. If we add an $\mathcal{R}$-state initially along x [i.e. $\mathbf{E}_{\mathcal{R}}(0,0) = E_0\hat{\mathbf{i}}$] to an $\mathcal{L}$-state initially along $-x$ [i.e. $\mathbf{E}_{\mathcal{L}}(0,0) = -E_0\hat{\mathbf{i}}$] the resulting $\mathcal{P}$-state would begin on a downward cycle. Accordingly, using

$$\mathbf{E}_{\mathcal{L}}(z,t) = E_0[-\hat{\mathbf{i}}\cos(kz - \omega t) + \hat{\mathbf{j}}\sin(kz - \omega t)]$$

$$\mathbf{E}_{\mathcal{R}}(z,t) = E_0[\hat{\mathbf{i}}\cos(kz - \omega t) + \hat{\mathbf{j}}\sin(kz - \omega t)]$$

we get

$$\mathbf{E}_P(z,t) = 2E_0\hat{\mathbf{j}}\sin(kz - \omega t)$$

The opposite choice (i.e. the $\mathcal{R}$-state initially along $-x$ and the $\mathcal{L}$-state initially along x) would give a $\mathcal{P}$-state

$$\mathbf{E}_P(z,t) = -2E_0\hat{\mathbf{j}}\sin(kz - \omega t)$$

that began on an upward swing. This is obviously just the negative of the previous solution.

5.14. Describe the state of polarization of the wave

$$\mathbf{E} = \hat{\mathbf{i}}E_0\cos(\omega t - kz + \pi/2) + \hat{\mathbf{j}}E_0\cos(\omega t - kz)$$

Making use of the fact that $\cos(\alpha + \beta) = \cos\alpha\cos\beta - \sin\alpha\sin\beta$, the wave function can be recast as

$$\mathbf{E} = -\hat{\mathbf{i}}E_0\sin(\omega t - kz) + \hat{\mathbf{j}}E_0\cos(\omega t - kz)$$

At $z = 0$ and $t = 0$, $\tau/4$, $\tau/2$, $3\tau/4$ and τ, $\mathbf{E}(0,t)$ equals $\hat{\mathbf{j}}E_0$, $-\hat{\mathbf{i}}E_0$, $-\hat{\mathbf{j}}E_0$, $\hat{\mathbf{i}}E_0$ and $\hat{\mathbf{j}}E_0$, respectively. Since the field vector is constant in length at E_0 and rotates counterclockwise, the wave is left circularly polarized.

Another approach uses the fact that $\sin(-\alpha) = -\sin\alpha$, while $\cos(-\alpha) = \cos\alpha$. Accordingly,

$$\mathbf{E} = \hat{\mathbf{i}}E_0\sin(kz - \omega t) + \hat{\mathbf{j}}E_0\cos(kz - \omega t)$$

which was seen in Problem 5.8 to be an $\mathcal{L}$-state.

5.4 ELLIPTICAL POLARIZATION

Linear and circular light are rather special cases. Both require specific values of the relative phase ε, and the latter demands equal component amplitudes as well. A more general superposition of orthogonal $\mathcal{P}$-states yields *elliptical* or *$\mathcal{E}$-state* light. Here the endpoint of the field vector sweeps out an ellipse (or, more accurately, an elliptical helix) as $\mathbf{E}$ changes in magnitude and direction.

We again write the $\mathcal{P}$-state components, this time in scalar form, as

$$E_x = E_{0x}\cos(kz - \omega t) \qquad E_y = E_{0y}\cos(kz - \omega t + \varepsilon)$$

The expression for E_y can be expanded to separate ε from the phase, and then, after some manipulation to remove the explicit dependence on $kz - \omega t$, one arrives at

$$\left(\frac{E_y}{E_{0y}}\right)^2 + \left(\frac{E_x}{E_{0x}}\right)^2 - 2\left(\frac{E_x}{E_{0x}}\right)\left(\frac{E_y}{E_{0y}}\right)\cos\varepsilon = \sin^2\varepsilon$$

This is the equation of an ellipse tilted at an angle α to the E_x-axis, as shown in Fig. 5-5. The value of α can be computed from the equation

$$\tan 2\alpha = \frac{2E_{0x}E_{0y}}{E_{0x}^2 - E_{0y}^2}\cos\varepsilon$$

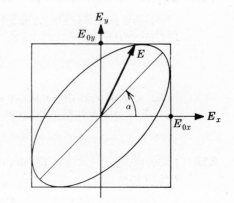

Fig. 5-5

SOLVED PROBLEMS

5.15. Verify that linear light is a special case of elliptical light.

We know (Section 5.2) that linearly polarized light occurs when $\varepsilon = (2m+1)\pi$ and when $\varepsilon = 2m\pi$, where $m = 0, \pm 1, \pm 2, \ldots$. In the former case, $\cos\varepsilon = -1$, $\sin\varepsilon = 0$ and

$$\left(\frac{E_y}{E_{0y}}\right)^2 + \left(\frac{E_x}{E_{0x}}\right)^2 + 2\left(\frac{E_x}{E_{0x}}\right)\left(\frac{E_y}{E_{0y}}\right) = 0$$

Factoring this yields

$$\left(\frac{E_y}{E_{0y}} + \frac{E_x}{E_{0x}}\right)\left(\frac{E_y}{E_{0y}} + \frac{E_x}{E_{0x}}\right) = 0 \quad\text{or}\quad E_y = -\frac{E_{0y}}{E_{0x}}E_x$$

This is the equation of a straight line (where the coordinates are E_x and E_y) passing through the point $E_y = 0$, $E_x = 0$ and having a slope of $-E_{0y}/E_{0x}$. Similarly, ε equal to zero or an even multiple of π yields $\cos\varepsilon = 1$, $\sin\varepsilon = 0$ and

$$\left(\frac{E_y}{E_{0y}}\right)^2 + \left(\frac{E_x}{E_{0x}}\right)^2 - 2\left(\frac{E_x}{E_{0x}}\right)\left(\frac{E_y}{E_{0y}}\right) = 0$$

or

$$E_y = \frac{E_{0y}}{E_{0x}}E_x$$

This is again a straight line, but now the slope is positive.

5.16. Show that the equation

$$\tan 2\alpha = \frac{2E_{0x}E_{0y}}{E_{0x}^2 - E_{0y}^2}\cos\varepsilon$$

obtains for linear light.

We can express the tangent as

$$\tan 2\alpha = \frac{2\tan\alpha}{1 - \tan^2\alpha}$$

If the ellipse of Fig. 5-5 degenerates to its axis in the first and third quadrants, then

$$\tan\alpha = \frac{E_{0y}}{E_{0x}}$$

and therefore

$$\tan 2\alpha = \frac{2(E_{0y}/E_{0x})}{1 - (E_{0y}/E_{0x})^2} = \frac{2E_{0x}E_{0y}}{E_{0x}^2 - E_{0y}^2}$$

This, of course, corresponds to the case where ε is zero or an even multiple of π. In the other case (ε an odd multiple of π) the ellipse degenerates to its axis in the second and fourth quadrants.

5.17. Verify that circular light is a special case of elliptical light.

Recall from Section 5.3 that circularly polarized light occurs when $\varepsilon = \pm\pi/2, \pm3\pi/2, \pm5\pi/2\ldots$. In that circumstance, $\cos\varepsilon = 0$, $\sin\varepsilon = \pm1$ and

$$\left(\frac{E_y}{E_{0y}}\right)^2 + \left(\frac{E_x}{E_{0x}}\right)^2 = 1$$

which describes an ellipse whose axes are the coordinate axes ($\alpha = 0$). When $E_{0x} = E_{0y} = E_0$, then $E_y^2 + E_x^2 = E_0^2$, which is the sought-after equation of a circle.

5.18. Determine the state of polarization of the wave whose orthogonal $\mathcal{P}$-state components are

$$\mathbf{E}_x(z, t) = \hat{\mathbf{i}}\, E_{0x} \cos{(kz - \omega t)}$$

$$\mathbf{E}_y(z, t) = \hat{\mathbf{j}}\, E_{0y} \cos{(kz - \omega t + \pi/2)}$$

The amplitudes E_{0x} and E_{0y} are not equal, so that even though $\varepsilon = \pi/2$, the resultant is not circular light. Inasmuch as $\cos\varepsilon = 0$, $\alpha = 0$ and we have an $\mathcal{E}$-state whose symmetry axes are the E_x- and E_y-axes. Examining the resultant wave $\mathbf{E}(z, t)$ at $z = 0$ as it unfolds in time we have $\mathbf{E}(0, 0) = \hat{\mathbf{i}}\, E_{0x}$, $\mathbf{E}(0, \tau/4) = \hat{\mathbf{j}}\, E_{0y}$, $\mathbf{E}(0, \tau/2) = -\hat{\mathbf{i}}\, E_{0x}$, $\mathbf{E}(0, 3\tau/4) = -\hat{\mathbf{j}}\, E_{0y}$, and $\mathbf{E}(0, \tau) = \hat{\mathbf{i}}\, E_{0x}$. The wave is left-handed and elliptically polarized.

Notice that because the time term is preceded by a minus sign a positive ε will cause $\cos{(kz - \omega t)}$ to reach any value prior to $\cos{(kz - \omega t + \varepsilon)}$. Accordingly, for $\varepsilon > 0$, E_x is said to *lead* E_y.

Figure 5-6 generalizes the result to other values of ε where $E_{0x} \neq E_{0y}$.

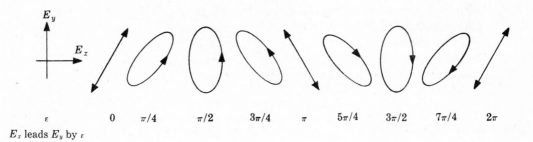

Fig. 5-6

5.19. Describe the state of polarization of the wave

$$\mathbf{E}(z, t) = \hat{\mathbf{i}}\, E_0 \cos{(kz - \omega t)} + \hat{\mathbf{j}}\, E_0 \cos{(kz - \omega t + \pi/4)}$$

giving its orientation as well.

The wave is neither linear nor circular, since $\varepsilon = \pi/4$, and so it must be elliptical. Watching it at $z = 0$, we have

$$\mathbf{E}(0, 0) = \hat{\mathbf{i}}\, E_0 + \hat{\mathbf{j}}\, E_0/\sqrt{2}$$

$$\mathbf{E}(0, \tau/8) = \hat{\mathbf{i}}\, E_0/\sqrt{2} + \hat{\mathbf{j}}\, E_0$$

$$\mathbf{E}(0, \tau/4) = 0 + \hat{\mathbf{j}}\, E_0/\sqrt{2}$$

$$\mathbf{E}(0, 3\tau/8) = -\hat{\mathbf{i}}E_0/\sqrt{2} + 0$$

$$\mathbf{E}(0, \tau/2) = -\hat{\mathbf{i}}E_0 - \hat{\mathbf{j}}E_0/\sqrt{2}$$

$$\mathbf{E}(0, 5\tau/8) = -\hat{\mathbf{i}}E_0/\sqrt{2} - \hat{\mathbf{j}}E_0$$

$$\mathbf{E}(0, 3\tau/4) = 0 - \hat{\mathbf{j}}E_0/\sqrt{2}$$

$$\mathbf{E}(0, 7\tau/8) = \hat{\mathbf{i}}E_0 + 0$$

$$\mathbf{E}(0, \tau) = \hat{\mathbf{i}}E_0 + \hat{\mathbf{j}}E_0/\sqrt{2}$$

The **E**-field rotates counterclockwise and as such is left-handed.   To find the tilt of the ellipse use

$$\tan 2\alpha = \frac{2E_0^2}{E_0^2 - E_0^2} \cos \tau = \infty \quad \text{or} \quad \alpha = 45°$$

5.20.   In Problem 5.19, we dealt with an $\mathcal{E}$-state tilted at 45°. When actually analyzing such light, it would be of practical interest to know the maximum and minimum field values.   Accordingly, determine both the semimajor and semiminor axes for the wave of Problem 5.19.

The semimajor axis occurs in time halfway between 0 and $\tau/8$, i.e. at $t = \tau/16$.   Consequently,

$$\mathbf{E}(0, \tau/16) = \hat{\mathbf{i}}E_0 \cos(-\pi/8) + \hat{\mathbf{j}}E_0 \cos(\pi/8)$$

and
$$(\mathbf{E}\cdot\mathbf{E})^{1/2} = \sqrt{2}E_0 \cos \pi/8 = 0.924\sqrt{2}E_0 = 1.31E_0$$

The semiminor axis occurs one quarter of a cycle later, i.e. at $t = \tau/16 + \tau/4 = 5\tau/16$.   Hence

$$\mathbf{E}(0, 5\tau/16) = \hat{\mathbf{i}}E_0 \cos(-5\pi/8) + \hat{\mathbf{j}}E_0 \cos(-3\pi/8)$$

and
$$(\mathbf{E}\cdot\mathbf{E})^{1/2} = \sqrt{2}E_0 \cos(3\pi/8) = 0.383\sqrt{2}E_0 = 0.542E_0$$

5.21.   Elliptical light can be synthesized via the superposition of an $\mathcal{R}$- and an $\mathcal{L}$-state.   Write expressions for the waves $\mathbf{E}_{\mathcal{R}}$ and $\mathbf{E}_{\mathcal{L}}$ propagating along the z-axis which, when combined, result in an $\mathcal{E}$-state rotating clockwise with its semimajor axis in the y-direction.

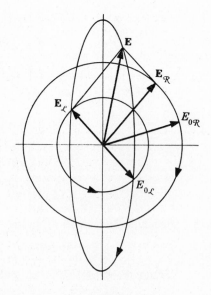

We evidently need an $\mathcal{R}$- and an $\mathcal{L}$-state which have only $\hat{\mathbf{j}}$-components at $z = 0$ and $t = 0$.   Furthermore, if $E_{0\mathcal{R}} > E_{0\mathcal{L}}$ the resultant will rotate clockwise along with the $\mathcal{R}$-state.   Hence

$$\mathbf{E}_{\mathcal{R}}(z, t) = E_{0\mathcal{R}}[-\hat{\mathbf{i}}\sin(kz - \omega t) + \hat{\mathbf{j}}\cos(kz - \omega t)]$$

$$\mathbf{E}_{\mathcal{L}}(z, t) = E_{0\mathcal{L}}[\hat{\mathbf{i}}\sin(kz - \omega t) + \hat{\mathbf{j}}\cos(kz - \omega t)]$$

so that

$$\mathbf{E}_{\mathcal{E}}(z, t) = (E_{0\mathcal{L}} - E_{0\mathcal{R}})\hat{\mathbf{i}}\sin(kz - \omega t)$$
$$+ (E_{0\mathcal{L}} + E_{0\mathcal{R}})\hat{\mathbf{j}}\cos(kz - \omega t)$$

The process is illustrated in Fig. 5-7.

Fig. 5-7

5.5 NATURAL AND PARTIALLY POLARIZED LIGHT

An ordinary light source consists of an exceedingly large number of randomly oriented

atomic emitters.   Each microscopic source radiates a polarized wave train extending in
time for about 10^{-8} s.   The superposition of such waves, having the same frequency but
enjoying no particular phase relationship, generates a resultant wave of a given polarization which is sustained for a time less than 10^{-8} s.   As new uncorrelated wave trains are
continuously emitted, the overall polarization state varies quite unpredictably.   Light of
this sort, where the state of polarization persists for a period too brief to perceive, is
referred to as *natural* or *unpolarized light*.   We can regard such a wave as if it were elliptically polarized, with both the shape and orientation varying rapidly and randomly.   Or,
equivalently, we can envisage the wave as the superposition of two orthogonal, *incoherent*
$\mathcal{P}$-states of equal amplitude.   In other words, ε changes rapidly and completely randomly.

   Usually light is neither totally polarized nor unpolarized but a mixture of the two types.
Thus the two orthogonal $\mathcal{P}$-states representing the wave will have unequal amplitudes or,
if you like, a non-randomly varying ε.   In such cases we say that the light is *partially
polarized*.   A measure of this condition is the *degree of polarization V*, defined as

$$V = \frac{I_p}{I_p + I_u}$$

Here I_p and I_u are the constituent flux densities of polarized and unpolarized light, respectively.   Clearly, $I_p + I_u$ is the total irradiance and V is then simply the fractional polarized
component.

SOLVED PROBLEMS

5.22.   A partially polarized beam is composed of 3 W/m² of polarized light and 7 W/m² of
natural light.   Determine the degree of polarization of the beam.

   By definition

$$V = \frac{I_p}{I_p + I_u}$$

where now $I_p = 3$ W/m²   and $I_u = 7$ W/m².   On substitution

$$V = \frac{3}{3 + 7} = 30\%$$

Notice that the limits of V are 0 (when $I_p = 0$) and 1 (when $I_u = 0$) and so partial polarization corresponds to $0 < V < 1$.

5.23.   Imagine that we have a detector which admits linear light polarized in a given direction and measures its irradiance.   Suppose a partially polarized linear wave impinges
on the device.   If the detector is then rotated about the propagation axis a maximum
value of the irradiance, I_{max}, will occur at some orientation and a minimum, I_{min}, perpendicular to it.   Derive an expression for the degree of polarization in terms of
I_{max} and I_{min}.

   Let us resolve the natural light into two incoherent, orthogonal $\mathcal{P}$-states.   Their directions are
arbitrary and so set one parallel to the linear light and the other perpendicular to it.   The component
field amplitudes of the unpolarized wave are equal, as are their flux densities.   Thus, if the total
unpolarized irradiance is I_u, each component has an irradiance of $I_u/2$.   The polarized irradiance is

$$I_p = I_{max} - I_{min}$$

where $I_{min} = I_u/2$.   Thus

$$V = \frac{I_p}{I_p + I_u} = \frac{I_{max} - I_{min}}{(I_{max} - I_{min}) + 2I_{min}} = \frac{I_{max} - I_{min}}{I_{max} + I_{min}}$$

As derived, this relationship obtains only for a mixture of linear and natural light.

5.24. When a beam of natural light impinges on a transparent dielectric, a portion of it is reflected while the rest is transmitted. Because $r_{||}$ and $r_\perp$, the amplitude reflection coefficients, are generally unequal, the refracted beam will usually be partially polarized. Write a formula for the degree of polarization of that beam in terms of the transmitted irradiance components parallel and perpendicular to the incident plane, i.e. $I_{t||}$ and $I_{t\perp}$. For what values of θ_i will $V = 0$?

> The incident beam can be resolved into two orthogonal, incoherent $\mathcal{P}$-states, one in the plane of incidence (with irradiance $I_{i||}$) and one perpendicular to it (with irradiance $I_{i\perp}$). Thus while the total incident flux density is $I_{i||} + I_{i\perp}$, the total transmitted flux density is $I_{t||} + I_{t\perp}$. From Section 3.4 we see that at $\theta_i = 0°$ and $90°$, $|r_{||}| = |r_\perp|$, the reflectances, $R_{||}$ and $R_\perp$, are equal and the transmitted light is unpolarized ($V = 0$). At any other angle $R_\perp > R_{||}$, which means that $I_{t||} > I_{t\perp}$. Since these two components are incoherent, the polarized irradiance is $I_p = I_{t||} - I_{t\perp}$ and

$$V = \frac{I_p}{I_p + I_u} = \frac{I_{t||} - I_{t\perp}}{I_{t||} + I_{t\perp}}$$

5.25. A beam of natural light is incident on an air-glass interface ($n_{ti} = 1.54$) at an angle of $57°$, such that $R_{||} = 0$ and $R_\perp = 0.165$. Determine the degree of polarization of the transmitted wave.

> Since $R_{||} = I_{r||}/I_{i||} = 0$, the corresponding transmittance

$$T_{||} = \frac{n_{ti}\cos\theta_t}{\cos\theta_i}\frac{I_{t||}}{I_{i||}} = 1$$

inasmuch as $R_{||} + T_{||} = 1$. Similarly, $T_\perp = 1 - R_\perp$ and

$$T_\perp = \frac{n_{ti}\cos\theta_t}{\cos\theta_i}\frac{I_{t\perp}}{I_{i\perp}} = 0.835$$

Because the incident light is natural, $I_{i\perp} = I_{i||}$ and we can write

$$I_{t||} = C \qquad I_{t\perp} = 0.835\,C$$

where

$$C = \frac{I_{i||}\cos\theta_i}{n_{ti}\cos\theta_t}$$

Hence, from Problem 5.24,

$$V = \frac{C - 0.835\,C}{C + 0.835\,C} = 8.9\%$$

5.6 DICHROISM AND POLAROID

In the very broadest sense *dichroism* corresponds to the selective absorption of one of the two orthogonal $\mathcal{P}$-states comprising an incident beam. The earliest usage applied to naturally occurring dichroic crystals such as *tourmaline*. Light whose **E**-field is parallel to the crystal's optic axis is transmitted with little absorption, while a field component normal to that axis is strongly absorbed.

Of more practical concern nowadays are the man-made dichroic devices, the simplest of which is the *wire grid polarizer*, depicted in Fig. 5-8. Here an unpolarized beam of, say, microwaves is shown impinging on a set of closely spaced, fine conducting wires. The constituent $\mathcal{P}$-state parallel to the wires drives electrons within them, thereby generating an alternating current. In addition to joule heating, which corresponds to the removal of energy from the vertical field component, the electrons reradiate a wave which tends to

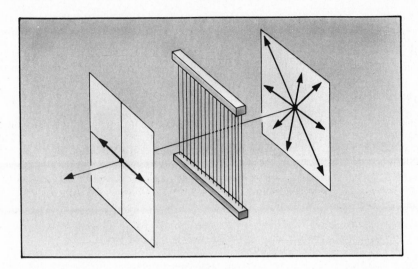

Fig. 5-8

further weaken this $\mathcal{P}$-state. By contrast, the horizontal field is affected little by the restricted motion of electrons transverse to the wires. Therefore the transmitted beam is strongly linearly polarized perpendicular to the wires.

Modern H-sheet polaroid is a molecular analog of the wire grid. Here a clear piece of polyvinyl alcohol has its long-chain molecules aligned in a particular direction by heating and stretching it. The sheet is then dyed with an iodine solution and the iodine, in turn, lines up along the straight polyvinyl alcohol molecules. Conduction electrons associated with the iodine can then circulate up and down the molecules as if they were microscopic wires. The result is a *linear polarizer*, that is, a device which passes only light whose **E**-field is parallel to a given direction (the *transmission axis*).

SOLVED PROBLEMS

5.26. Figure 5-9 shows two polaroid linear polarizers oriented with an angle θ between their transmission axes. Derive an expression for the irradiance of the emerging beam as a function of θ.

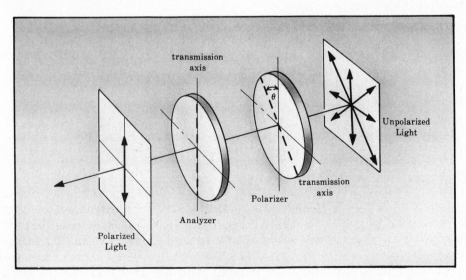

Fig. 5-9

Let the electric field amplitude emerging from the first polaroid (the *polarizer*) be $\mathbf{E}_0$. It is tilted at an angle θ to the vertical transmission axis of the second polaroid (the *analyzer*) and so has a component along that transmission axis of $E_0 \cos \theta$. From Problem 2.13 we know that I varies with the square of the field amplitude:

$$I(\theta) = \frac{c\varepsilon_0}{2}(E_0 \cos \theta)^2$$

This has a maximum value of $I(0) = c\varepsilon_0 E_0^2/2$ when $\theta = 0$. Hence

$$I(\theta) = I(0) \cos^2 \theta$$

which is known as *Malus's law*.

5.27. An unpolarized light beam of irradiance I_i is made to pass through a sequence of two perfect linear polarizers. What must be their relative orientation if the exiting beam is to have an irradiance of (*a*) $I_i/2$, (*b*) $I_i/4$?

(*a*) Imagine the incident beam to be resolved into two orthogonal, incoherent $\mathcal{P}$-states, each of flux density $I_i/2$. Since there are no losses for the components parallel to the transmission axes (perfect polarizers), one polarizer alone will pass $I_i/2$. Clearly, both polarizers aligned with their transmission axes parallel will also pass $I_i/2$.

(*b*) Malus's law is $I(\theta) = I(0) \cos^2 \theta$, wherein, from (*a*), $I(0) = I_i/2$. We require that $I(\theta) = I_i/4$; in other words,

$$\frac{I_i}{4} = \frac{I_i}{2} \cos^2 \theta$$

Hence, $\cos \theta = \sqrt{1/2} = 1/\sqrt{2}$ and $\theta = 45°$.

5.28. What must be the relative orientation of two perfect linear polarizers if under natural illumination the emerging beam is to be reduced to half its maximum transmitted value?

The maximum transmitted irradiance occurs at $\theta = 0$. Thus we want θ such that $I(\theta) = I(0)/2$. Malus's law becomes

$$\frac{I(0)}{2} = I(0) \cos^2 \theta$$

Hence, $\cos \theta = 1/\sqrt{2}$ and $\theta = 45°$.

5.29. Imagine two crossed linear polarizers with transmission axes vertical and horizontal. Now insert a third linear polarizer between them with its transmission axis at 45° to the vertical. Determine the emerging irradiance before and after insertion of the third polarizer in terms of I_i, the flux density of a beam of incident natural light.

Clearly, no light will emerge from a pair of perfect linear polarizers crossed at 90°.

With the middle polarizer in place, light exiting from the first polarizer is linearly polarized in the vertical direction and has an irradiance of $I_i/2$. From Malus's law, with $I(0) = I_i/2$, a flux density of

$$I(45°) = \frac{I_i}{2} \cos^2 45° = \frac{I_i}{4}$$

leaves the middle polarizer heading toward the last. The angle between these two is again 45° and so the final irradiance is given by

$$I = \frac{I_i}{4} \cos^2 45° = \frac{I_i}{8}$$

Light leaving the three polarizers is a horizontal $\mathcal{P}$-state.

5.30. Three perfect linear polarizers are stacked normal to a central axis along which is incident a beam of natural light of irradiance I_i. If the first and last polarizers are crossed and if the middle one rotates at a rate ω about the axis, write an expression for the irradiance I of the emerging beam as a function of ω.

The beam leaving the first polarizer is assumed to be vertically polarized with a flux density of $I_1 = I_i/2$. By Malus's law, the irradiance leaving the second polarizer (when it makes an angle θ with the first) is $I_2 = I_1 \cos^2 \theta$, where $\theta = \omega t$. The angle between the transmission axes of the second and third polarizers is $90° - \theta$. Neglecting any effects due to the finite value of c, the irradiance emerging from the third polarizer must be

$$I = I_2 \cos^2 (90° - \theta) = (I_1 \cos^2 \theta) \sin^2 \theta = \frac{I_i}{2} \sin^2 \theta \cos^2 \theta$$

Applying the identity $\cos 4\theta = 1 - 2 \sin^2 2\theta = 1 - 8 \sin^2 \theta \cos^2 \theta$, we obtain

$$I = \frac{I_i}{16}(1 - \cos 4\theta) = \frac{I_i}{16}(1 - \cos 4\omega t)$$

Interestingly enough, the emerging irradiance oscillates at four times the rotation rate.

5.31. An elliptically polarized light beam given by

$$\mathbf{E}(z, t) = \hat{\mathbf{i}} E_0 \sin (kz - \omega t) + \hat{\mathbf{j}} E_0 \sin (kz - \omega t + \pi/4)$$

passes normally through an ideal linear polarizer whose transmission axis is tilted at 45° in the xy-plane. Write an expression for the emerging beam and describe its state of polarization.

To find the transmitted component of $\mathbf{E}$ form a unit vector along the transmission axis of the polarizer, i.e.

$$\hat{\mathbf{a}} = \frac{1}{\sqrt{2}}(\hat{\mathbf{i}} + \hat{\mathbf{j}})$$

The dot product $\mathbf{E} \cdot \hat{\mathbf{a}}$ equals the required scalar component:

$$E'(z, t) = \frac{E_0}{\sqrt{2}}\left[\sin (kz - \omega t) + \sin \left(kz - \omega t + \frac{\pi}{4} \right)\right]$$

These two sine terms can be combined into a single sine function oscillating in the plane of z and the transmission axis; of course, the beam is linearly polarized. In fact, any two harmonic functions of the same frequency

$$A = A_0 \sin (\omega t + \alpha) \qquad B = B_0 \sin (\omega t + \beta)$$

combine to yield

$$C = C_0 \sin (\omega t + \gamma)$$

where

$$C_0^2 = A_0^2 + B_0^2 + 2A_0 B_0 \cos (\beta - \alpha)$$

$$\tan \gamma = \frac{A_0 \sin \alpha + B_0 \sin \beta}{A_0 \cos \alpha + B_0 \cos \beta}$$

In our case $A_0 = B_0 = 1$, $\alpha = 0$, $\beta = \pi/4$, and we shall replace ωt by $kz - \omega t$. Thus

$$C_0 = \left(2 + \frac{2}{\sqrt{2}} \right)^{1/2} \qquad \tan \gamma = \frac{1/\sqrt{2}}{1 + (1/\sqrt{2})} = \frac{1}{2.414}$$

and

$$E'(z, t) = \left(1 + \frac{1}{\sqrt{2}} \right)^{1/2} E_0 \sin (kz - \omega t + \gamma)$$

The emerging irradiance is therefore

$$I' = \frac{c\varepsilon_0}{2}\left(1 + \frac{1}{\sqrt{2}} \right)E_0^2 = \frac{c\varepsilon_0}{2}(1.707)E_0^2$$

We can check this result rather quickly by sketching the two sine functions. Evidently the resultant reaches a peak where the two cross, which by symmetry is at 67.5°. This maximum value of E' is its amplitude, equal to $2(\sin 67.5°)E_0/\sqrt{2}$ or $1.848\,E_0/\sqrt{2}$. Squaring this yields $1.707\,E_0^2$, in agreement with the above.

5.32. Two ideal linear polarizers with horizontal transmission axes are illuminated by natural light. If a third such polarizer is placed between them with its transmission axis at 30° to the horizontal, find the emerging irradiance in terms of its value I' before insertion.

Let the field amplitude leaving the first polarizer be E_0. The field emerging from the second one is then $E_0 \cos 30°$ or $E_0/2$. The field leaving the last polarizer is $(E_0/2)(\cos 30°) = E_0/4$. The corresponding irradiance is then

$$I = \frac{c\varepsilon_0}{2}\left(\frac{E_0}{4}\right)^2$$

The exiting irradiance without the middle polarizer is just

$$I' = \frac{c\varepsilon_0}{2}E_0^2$$

Hence, $I = I'/16$.

5.7 POLARIZATION BY REFLECTION

In Section 3.4 we found that

$$r_\perp \equiv \left(\frac{E_{0r}}{E_{0i}}\right)_\perp = \frac{n_i \cos \theta_i - n_t \cos \theta_t}{n_i \cos \theta_i + n_t \cos \theta_t}$$

$$r_\| \equiv \left(\frac{E_{0r}}{E_{0i}}\right)_\| = \frac{n_t \cos \theta_i - n_i \cos \theta_t}{n_i \cos \theta_t + n_t \cos \theta_i}$$

are the amplitude coefficients of reflection perpendicular and parallel to the plane of incidence. The corresponding irradiance ratios, which are called the reflectances, are simply $R_\perp = r_\perp^2$ and $R_\| = r_\|^2$. Thus

$$R_\| = \frac{I_{r\|}}{I_{i\|}} = \frac{\tan^2 (\theta_i - \theta_t)}{\tan^2 (\theta_i + \theta_t)}$$

$$R_\perp = \frac{I_{r\perp}}{I_{i\perp}} = \frac{\sin^2 (\theta_i - \theta_t)}{\sin^2 (\theta_i + \theta_t)}$$

Figure 5-10 is a plot of these reflectances for external reflection at various values of the incident angle θ_i for an air-glass interface. The figure also contains a graph of the corresponding reflectance of natural light which, as we saw in Problem 3.24, is given by

$$R_n = \frac{1}{2}(R_\| + R_\perp)$$

Observe that $R_\| = 0$ when its denominator equals ∞; that is, when $\theta_i + \theta_t = 90°$. The special value of the incident angle for

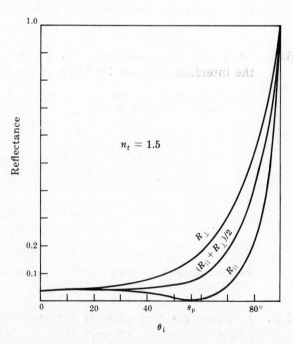

Fig. 5-10

which this is true ($\theta_i \equiv \theta_p$) is spoken of as the *polarization angle* because for that angle the reflected light is completely polarized normal to the plane of incidence.

SOLVED PROBLEMS

5.33. Show that the polarization angle, θ_p, can be determined from the formula

$$\tan \theta_p = \frac{n_t}{n_i}$$

This is known as *Brewster's law*.

Beginning with Snell's law,

$$n_i \sin \theta_i = n_t \sin \theta_t$$

we require that when $\theta_i = \theta_p$, $\theta_i + \theta_t = 90°$, i.e. $\theta_p + \theta_t = 90°$. Thus

$$n_i \sin \theta_p = n_t \sin (90° - \theta_p)$$

or

$$n_i \sin \theta_p = n_t \cos \theta_p$$

or

$$\tan \theta_p = \frac{n_t}{n_i}$$

5.34. Determine the polarization angle for external reflection at an air-glass interface ($n_{ti} = 1.5$). Describe the state of polarization of the reflected beam for unpolarized light incident at θ_p.

Brewster's law can be written as $\tan \theta_p = n_{ti}$. Here $n_{ti} = 1.5$ and so

$$\theta_p = \tan^{-1} 1.5 = 56.3°$$

A beam of natural light incident in air at $56.3°$ will be partially reflected and partially transmitted. Inasmuch as the incoming beam can be imagined to be composed of two orthogonal, incoherent $\mathcal{P}$-states we conclude that the reflected wave is a $\mathcal{P}$-state *parallel to the interface* ($R_{\parallel} = 0$). This suggests an easy way to determine the transmission axis of an unmarked piece of polaroid. When it passes the glare on a horizontal surface the transmission axis is horizontal.

5.35. Show that the polarization angles (θ_p and θ_p') for external and internal reflection at the interface between the same media are complements of each other.

Suppose that we have an interface between two media of indices n_1 and n_2, where $n_2 > n_1$. For external reflection ($n_t > n_i$)

$$\tan \theta_p = \frac{n_t}{n_i} = \frac{n_2}{n_1}$$

while for internal reflection ($n_t < n_i$)

$$\tan \theta_p' = \frac{n_t}{n_i} = \frac{n_1}{n_2}$$

Accordingly,

$$\frac{\sin \theta_p}{\cos \theta_p} = \frac{\cos \theta_p'}{\sin \theta_p'}$$

or

$$\sin \theta_p \sin \theta_p' - \cos \theta_p \cos \theta_p' = 0$$

This is equivalent to $\cos (\theta_p + \theta_p') = 0$ which means that $\theta_p + \theta_p' = 90°$.

5.36. A planar sheet of glass is immersed in water. Show that a beam of natural light which strikes the first surface at the polarization angle will, in part, be transmitted to the second surface, where it is again incident at the polarization angle.

Figure 5-11 depicts the situation in question wherein $\theta_{i1} = \theta_p$ and we are to prove that $\theta_{i2} = \theta_p'$. By Snell's law,

$$n_w \sin \theta_p \;=\; n_g \sin \theta_{t1}$$
$$=\; n_g \sin \theta_{i2}$$

It follows that

$$\sin \theta_{i2} \;=\; \frac{n_w}{n_g} \sin \theta_p$$

But

$$\tan \theta_p \;=\; \frac{n_g}{n_w}$$

and so

$$\sin \theta_{i2} \;=\; \cos \theta_p$$

From Problem 5.35, $\theta_p = 90° - \theta_p'$. Therefore

$$\sin \theta_{i2} \;=\; \sin \theta_p'$$

or

$$\theta_{i2} \;=\; \theta_p'$$

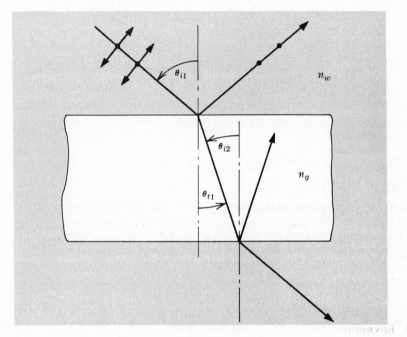

Fig. 5-11

5.37. As the sun rises over a still pond, an angle will be reached where its image seen on the water's surface ($n_w = 1.33$) will be completely linearly polarized in a plane parallel to the surface. Compute the appropriate incident angle. At what angle will the transmitted beam propagate through the water?

Brewster's law, $\tan \theta_p = n_t/n_i$, yields

$$\theta_p \;=\; \tan^{-1} \frac{1.33}{1} \;=\; 53.1°$$

This, then, is the angle the sun makes with the vertical. As in Problem 5.36 we find that

$$\theta_t \;=\; 90° - \theta_p \;=\; 36.9°$$

5.38. A beam of natural light is incident on an air-glass interface ($n_{ti} = 1.5$) at 30°. Determine the degree of polarization of the reflected beam.

Values of $r_{||}$ and $r_{\perp}$ can be computed from the Fresnel equations as was done in Problem 3.20. There we found that $r_{||} = 0.15$ and $r_{\perp} = -0.24$. Consequently, $R_{||} = r_{||}^2 = 2.25\%$, $R_{\perp} = r_{\perp}^2 = 5.76\%$. The degree of polarization

$$V = \frac{I_p}{I_p + I_u}$$

can be gotten by realizing that

$$I_p = R_{\perp} I_{i\perp} - R_{||} I_{i||}$$

since the orthogonal $\mathcal{P}$-states are incoherent. In much the same way it should be clear that the total reflected wave has an irradiance of

$$I_p + I_u = R_{\perp} I_{i\perp} + R_{||} I_{i||}$$

All of this means that

$$V = \frac{R_{\perp} I_{i\perp} - R_{||} I_{i||}}{R_{\perp} I_{i\perp} + R_{||} I_{i||}}$$

But we know, in addition, that $I_{i||} = I_{i\perp} = I_i/2$, and so

$$V = \frac{R_{\perp} - R_{||}}{R_{\perp} + R_{||}}$$

In this specific case

$$V = \frac{5.76 - 2.25}{5.76 + 2.25} = 43.8\%$$

5.8 BIREFRINGENCE

An optically isotropic material is one in which the index of refraction or, if you will, the phase velocity of a wave is the same in all directions. This obtains for cubic crystals like NaCl, as well as for noncrystalline substances such as unstressed glass and plastic, water and air.

Generally, however, crystals are anisotropic; the atomic binding forces on the electron clouds are different in different directions and as a result so are the refractive indices. We shall concern ourselves here only with *uniaxial birefringent* crystals (which encompass the trigonal, hexagonal and tetragonal systems). Such a crystal contains a single symmetry axis (actually a direction) known as the *optic axis* and displays *two* distinct *principal indices of refraction*. The latter correspond to light field oscillations parallel and perpendicular to the optic axis. Calcite, $CaCO_3$, is a good example. Here the carbonate (CO_3) groupings all lie in parallel planes which are normal to an axis of three-fold symmetry, the optic axis. The distribution of atoms is clearly anisotropic as is its response to light.

We can trace the progress of a wavefront through some medium by applying *Huygens' principle* which states that *in a homogeneous, isotropic substance every point on a wavefront can be envisioned as a source of secondary wavelets whose envelope at some later time corresponds to the primary wave at that time. The secondary wavelets are spherical, moving out in all directions with the same velocity and frequency as the primary wave.* In a material medium we can think of the primary wave stimulating atoms into re-emitting the secondary wavelets which, in turn, advance to the next layer of atoms.

To see how this applies to a uniaxial birefringent material, examine Fig. 5-12, page 112, which shows an edge view of a calcite plate cut so that the optic axis is in the plane of the drawing. In Fig. 5-12(a) a plane wave, linearly polarized normal to the page, impinges on the crystal. The electric field is everywhere perpendicular to the optic axis, the wavelets

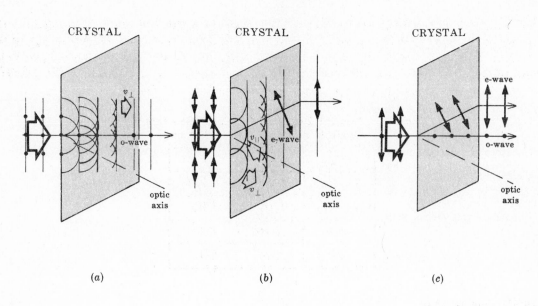

Fig. 5-12

spherical, and the disturbance passes through the plate (with a speed $v_\perp$) in the usual y—this is the *ordinary* or *o-wave*. In Fig. 5-12(*b*) the electric field of the incoming state is in the plane of the drawing and therefore has components parallel and perpenular to the optic axis, which propagate at speeds of $v_{||}$ and $v_\perp$, respectively. In calcite $> v_\perp$ and the wavelets can be thought of as elongating into ellipsoids of revolution about optic axis. The envelope of the ellipsoids is a planar wavefront moving upward across crystal as the *extraordinary* or *e-wave*. Notice that the **E**-field is not in the planar vefronts within the anisotropic medium. Figure 5-12(*c*) shows how the crystal splits incident unpolarized beam into its constituent $\mathcal{P}$-states, thereby forming two distinct erging beams.

The crystal actually has to be cut in a special y if the optic axis, the e- and the o-ray are be in a common plane (the o-ray always ides in the plane of incidence). Calcite will it naturally to form smooth *cleavage planes*. $CaCO_3$ crystal whose six faces are all cleavage nes is said to be a *cleavage form*, which here responds to a rhombohedron. A plane conning the optic axis is known as a *principal ne*. A particular principal plane which is pendicular to a pair of opposite faces of the avage form is a *principal section*, as depicted Fig. 5-13. The principal planes containing o- and e-rays coincide with the principal tion as in Fig. 5-12. Notice that the **E**-fields the o- and e-waves are normal and parallel the principal section, respectively.

It is customary to define the two principal lices of refraction for a uniaxial crystal as $\equiv c/v_\perp$ and $n_e \equiv c/v_{||}$. Values of these for

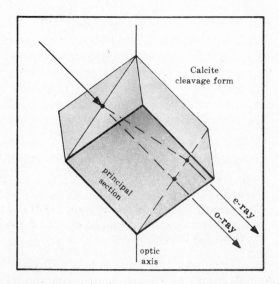

Fig. 5-13

various substances are displayed in Table 5-1. Keep in mind that while $v_\perp$ is always the speed of the o-wave, $v_{||}$ is the speed of the e-wave only when that wave travels at 90° to the optic axis. When propagating along the optic axis, the o- and e-waves both have fields normal to that axis and both advance with the same speed, $v_\perp$. Accordingly, the effective index for an e-wave moving in some intermediate direction lies between n_o and n_e.

Table 5-1. Principal Refractive Indices of Some Uniaxial Birefringent Crystals ($\lambda_0 = 589.3$ nm)

Crystal	n_o	n_e
Calcite	1.6584	1.4864
Ice	1.309	1.313
Quartz	1.5443	1.5534
Rutile	2.616	2.903

SOLVED PROBLEMS

5.39. The quantity $\Delta n = n_e - n_o$ is often referred to as the *birefringence*. A material is denoted as either *positive* or *negative* uniaxially birefringent depending on the sign of Δn. Assuming a point source to be imbedded in either such material, draw wavefronts for the o- and e-waves.

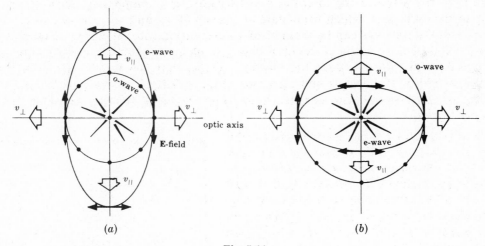

Fig. 5-14

We know that in calcite, which is negative uniaxial, $1.486\,v_{||} = 1.658\,v_\perp = c$, i.e. $v_{||} > v_\perp$. Moreover both waves, e and o, move at the same speed along the optic axis. The spherical o-wave and the ellipsoidal e-wave are therefore tangent at the optic axis, as in Fig. 5-14(a). The maximum velocity of the e-wave, i.e. $v_{||}$, occurs in a direction perpendicular to the optic axis. Figure 5-14(b) corresponds to a positive uniaxial crystal such as quartz. Here $n_e > n_o$, which means that $v_{||} < v_\perp$. This time $v_{||}$ is the minimum velocity of the e-wave and once more it occurs in a direction perpendicular to the optic axis.

5.40. Two calcite parallel plates are cut so that the optic axis in one plate is normal to the front face, while in the other it is parallel to the front face. Make an edge-view sketch of the o- and e-waves in both plates and discuss what is happening.

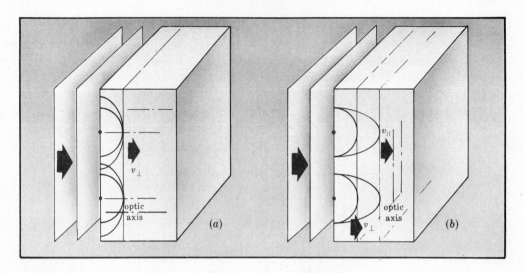

Fig. 5-15

In the plate of Fig. 5-15(a) the spherical and ellipsoidal wavelets both move through the crystal with a speed $v_\perp$. Thus the o- and e-wavefronts are coincident: in effect, only one wave traverses the plate. By contrast, in Fig. 5-15(b) we see the ellipsoidal e-fronts advancing at $v_{||}$ and the spherical o-fronts moving more slowly at $v_\perp$. Thus two separate waves traverse the crystal. At any point in space beyond the back surface both disturbances overlap to form a single resultant whose form depends on the relative phase difference introduced on traversing the plate.

5.41. Figure 5-16 depicts a *Wollaston polarizing beamsplitter,* consisting of two quartz segments cemented together with glycerine. Discuss how a beam of natural light incident on the prism is split into o- and e-waves.

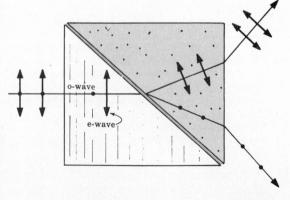

Fig. 5-16

Imagine the incident beam to consist of two orthogonal, incoherent $\mathcal{P}$-states, one parallel to the optic axis and one normal to it, as in Fig. 5-17. Because the beam strikes the first face of the prism perpendicularly, there is no refraction, although there is a phase difference as in Fig. 5-15(b). On crossing the diagonal interface the e-wave enters the second segment, where, its field being normal to the optic axis, it becomes an o-wave. Since $n_o = 1.54$ and $n_e = 1.55$, $n_e > n_o$ and Snell's law implies that the o-ray in the second segment bends *away* from the normal to the interface. Similarly, the o-wave in the first segment is transformed into an e-wave on traversing the diagonal interface, where it then bends *toward* the normal.

Our use of Snell's law relies on the fact that in the second segment of the prism the optic axis is normal to the plane of incidence. Consequently, both the o- and e-wavelets have circular cross sections in the second segment. Generally an e-wavelet has an elliptical cross section and, therefore, Snell's law does not apply.

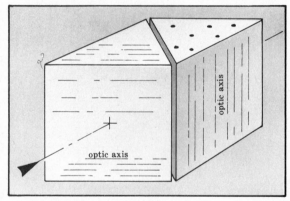

Fig. 5-17

5.42. A uniaxial birefringent crystal is cut to form a parallel plate with its optic axis parallel to the front face, as in Fig. 5-15(b). Such a device is known as a *retarder* or *retardation plate*. Assuming it to have a thickness d, write an expression for the relative phase difference, introduced by the plate, between the o- and e-waves traversing it.

The optical path lengths traversed by the e- and o-waves within the plate are $n_e d$ and $n_o d$, respectively. The optical path difference is then

$$\Lambda = |n_o - n_e| d$$

where the absolute value sign is merely to keep Λ positive. The number of wavelengths that the two waves are shifted by is simply

$$\frac{\Lambda}{\lambda_0} = \frac{1}{\lambda_0} |n_o - n_e| d$$

Each wavelength corresponds to an angular phase shift of 2π radians, and so the relative phase shift, $\Delta\varphi$, is given by

$$\Delta\varphi = \frac{2\pi}{\lambda_0} |n_o - n_e| d$$

In a negative uniaxial crystal $v_{||} > v_\perp$, and the direction of the optic axis of the retarder is denoted as the *fast* axis. The opposite is true in the case of a positive crystal, where the optic axis is the *slow* axis.

5.43. A collimated beam of sodium light ($\lambda_0 = 589.3$ nm) is incident normally on a parallel calcite plate whose optic axis is perpendicular to the beam. Determine the frequencies and wavelengths of the o- and e-waves within the calcite.

Provided that the response of the medium is linear, as it generally is for all but gigantic fields, the frequency is unchanged when a beam enters the medium. Accordingly,

$$\nu = \frac{c}{\lambda_0} = \frac{3 \times 10^8 \text{ m/s}}{589.3 \times 10^{-9} \text{ m}} = 5.1 \times 10^{14} \text{ Hz}$$

is the frequency of both waves, in and out of the plate. For the ordinary wave

$$\lambda_{\text{ord}} = \frac{\lambda_0}{n_o} = \frac{589.3 \times 10^{-9}}{1.66} = 355 \text{ nm}$$

and for the extraordinary wave

$$\lambda_{\text{ext}} = \frac{\lambda_0}{n_e} = \frac{589.3 \times 10^{-9}}{1.49} = 396 \text{ nm}$$

5.44. Compute the angle α between the o- and e-rays emerging from a calcite Wollaston prism whose wedge angle is 15°.

As in Problem 5.41, we can apply Snell's law at the diagonal interface. For the emerging e-ray we have

$$n_o \sin 15° = n_e \sin \theta_{e1}$$

or

$$\sin \theta_{e1} = \frac{1.66}{1.49} \sin 15° = 0.288$$

where θ_{e1} is, of course, between the local normal and the e-ray in the second segment. Similarly, for the o-ray,

$$n_e \sin 15° = n_o \sin \theta_{o1}$$

or

$$\sin \theta_{o1} = \frac{1.49}{1.66} \sin 15° = 0.232$$

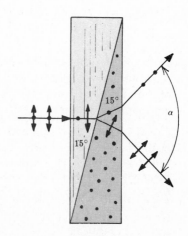

Fig. 5-18

Hence $\theta_{e1} = 16°46'$, $\theta_{o1} = 13°26'$ and the angle between the two rays within the prism is $\theta_{e1} - \theta_{o1} = 3°20'$. The e- and o-rays are incident on the back face at angles of $1°46'$ and $1°34'$, respectively. Thus they emerge at angles θ_{e2} and θ_{o2} with respect to the normal to the back face, where

$$n_o \sin 1°34' = \sin \theta_{o2} \qquad n_e \sin 1°46' = \sin \theta_{e2}$$

Thus, $\theta_{o2} = 2°34'$ and $\theta_{e2} = 2°39'$. Accordingly,

$$\alpha = (\theta_{e1} - \theta_{o1}) + (2°34' - 1°34') + (2°39' - 1°46') = 5°13'$$

5.45. Given incident 590-nm light, compute the minimum thickness which a quartz retarder must have if it is to be a *quarter-wave plate,* i.e. if the relative phase shift between the e- and o-waves is to be $\Delta\varphi = \pi/2$.

From Problem 5.42 we know that

$$\Delta\varphi = \frac{2\pi}{\lambda_0} |n_o - n_e| d$$

Since $\Delta\varphi$ is to equal an odd multiple of $\pi/2$,

$$(2m+1)\frac{\pi}{2} = \frac{2\pi}{\lambda_0} |n_o - n_e| d$$

with $m = 0, 1, 2, 3, \ldots$. The minimum thickness for such a quarter-wave plate occurs when $m = 0$, whereupon

$$\frac{\lambda_0}{4} = |n_o - n_e| d$$

Thus

$$d = \frac{590 \times 10^{-9}}{4|1.54 - 1.55|} = 1.48 \times 10^{-5} \text{ m} = 0.015 \text{ mm}$$

5.46. As indicated in Fig. 5-6, page 101, a linearly polarized beam whose plane of vibration is at $45°$ can be converted into circular light by causing E_x to either lead or lag E_y by $90°$. Accordingly, design a right circular polarizer, i.e. a device which will convert natural light into an $\mathcal{R}$-state.

The first step is to get the appropriate $\mathcal{P}$-state. Thus, start with a polaroid or other linear polarizer with its transmission axis at $45°$. This yields equal, in-phase, orthogonal field components. To get an $\mathcal{R}$-state, E_x must lag E_y by $\pi/2$ radians. That calls for a quarter-wave plate oriented with its fast axis parallel to E_y.

A common arrangement is a polaroid linear polarizer bonded to a quarter-wave plate made of polyvinyl alcohol.

Note that circular polarizers have distinct input and output faces.

5.47. A beam of quasi-monochromatic light linearly polarized in the y-direction is incident on a *half-wave* plate (i.e. $\Delta\varphi = \pi$). The plate is rotated so that its fast axis makes an angle of $30°$ with the y-axis. What are the amplitudes of the emerging x- and y- field components in terms of the amplitude, E_0, of the incident $\mathcal{P}$-state?

As we saw in Fig. 5-6, page 101, a phase shift of π radians will flip the $\mathcal{P}$-state as if mirrored in the axis of the retarder. As shown in Fig. 5-19, page 117, since the field is initially at $30°$, it flips over to $30°$ on the opposite side of the fast axis. Thus, the emerging $\mathcal{P}$-state is at $60°$ to the y-axis. The amplitudes of the x- and y-components are then

$$E_0 \sin 60° = \frac{\sqrt{3}\,E_0}{2} = 0.866\,E_0$$

$$E_0 \cos 60° = \frac{E_0}{2} = 0.500\,E_0$$

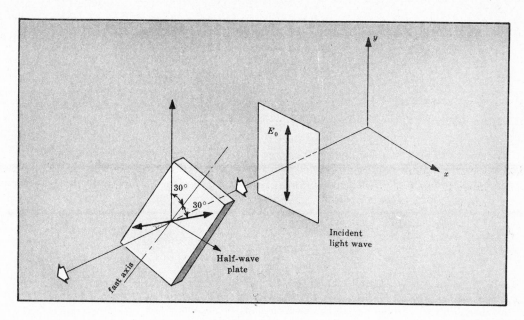

Fig. 5-19

Supplementary Problems

PLANE POLARIZATION

5.48.   Write an expression describing a harmonic $\mathcal{P}$-state, of amplitude E_0, propagating in the y-direc[t] with a speed v, and having a plane of vibration at 45° to the x-axis.

Ans. $\mathbf{E}(y, t) = \dfrac{E_0}{\sqrt{2}}(\hat{\mathbf{i}} + \hat{\mathbf{k}}) \cos \omega\left(\dfrac{y}{v} - t\right)$

(Any equivalent expression for the phase will do.)

5.49.   A beam of linear light polarized in the z-direction propagates in the x-direction through a quar[ter] wave plate whose fast axis is along the y-direction.   Assuming an incident amplitude of E_0, w[rite] an expression for the emerging harmonic wave.

Ans. $\mathbf{E}(x, t) = E_0\hat{\mathbf{k}} \cos (kx - \omega t)$

5.50.   Write an expression for a $\mathcal{P}$-state, of amplitude E_0, moving in the y-direction, whose plane of vib[ra-]tion corresponds to the xy-plane, and whose magnitude at $y = 0$ and $t = 0$ is zero.

Ans. $\mathbf{E}(y, t) = E_0\hat{\mathbf{i}} \sin (ky - \omega t)$

5.51.   Describe the main characteristics and state of polarization of the wave

$$\mathbf{E}(y, t) = -E_0\hat{\mathbf{k}} \cos (ky + \omega t) - \sqrt{3}\, E_0\hat{\mathbf{i}} \cos (ky + \omega t)$$

Ans.   A $\mathcal{P}$-state, at 60° to the yz-plane, of amplitude $2E_0$, propagating in the minus y-direction[.]

5.52.   Arrive at an equation representing a harmonic, linearly polarized wave propagating along the x-a[xis] of amplitude E_0, and having a plane of vibration tilted up from the xy-plane by 17.5°.

Ans. $\mathbf{E}(x, t) = E_0(0.9537\,\hat{\mathbf{j}} + 0.3007\,\hat{\mathbf{k}}) \cos (kx - \omega t)$

CIRCULAR POLARIZATION

5.53. Two $\mathcal{L}$-states having the same wavelength, of amplitudes $2E_0$ and E_0, and propagating in the **same** direction, overlap in space. Assuming the waves to be in phase, describe the resultant.

    Ans. An $\mathcal{L}$-state of amplitude $3E_0$.

5.54. Describe a means for determining whether a beam of light is right or left circular.

    Ans. If an $\mathcal{R}$-state is incident on the *output* side of a right circular polarizer it will emerge **as** a $\mathcal{P}$-state, while no light will emerge when the incident beam is an $\mathcal{L}$-state. The opposite is true for a beam impinging on the output side of a left circular polarizer. Take another look at Problem 5.46.

5.55. Write an equation for an $\mathcal{R}$-state wave moving along the positive z-axis, of amplitude E_0, for which $\mathbf{E}(0,0)$ is at $-45°$ measured from the x-axis.

    Ans. $\mathbf{E}(z,t) = E_0[\hat{\mathbf{i}} \cos(kz - \omega t - \pi/4) + \hat{\mathbf{j}} \sin(kz - \omega t - \pi/4)]$

5.56. Formulate a mathematical representation of a right circularly polarized wave, of amplitude E_0, moving in the positive z-direction, such that at $t = 0$ and $z = 0$,

$$\mathbf{E}(0,0) = \frac{E_0}{2}(\hat{\mathbf{i}} + \sqrt{3}\,\hat{\mathbf{j}})$$

    Ans. $\mathbf{E}(z,t) = E_0[\hat{\mathbf{i}} \cos(kz - \omega t - \pi/3) + \hat{\mathbf{j}} \sin(kz - \omega t + \pi/3)]$

5.57. Give an expression for an $\mathcal{L}$-state, of amplitude E_0, propagating in the positive z-direction, **for** which the $\mathbf{E}$-field at $z = 0$ and $t = 0$ is at $+30°$ to the x-axis.

    Ans. $\mathbf{E}(z,t) = E_0[\hat{\mathbf{i}} \cos(kz - \omega t - \pi/6) - \hat{\mathbf{j}} \sin(kz - \omega t - \pi/6)]$

ELLIPTICAL POLARIZATION

5.58. Write an expression for a right-handed, harmonic $\mathcal{E}$-state tilted at $45°$ to the y-axis and **propa**gating along the x-axis.

    Ans. $\mathbf{E}(x,t) = \hat{\mathbf{j}} E_0 \cos(kx - \omega t) + \hat{\mathbf{k}} E_0 \cos(kx - \omega t - \pi/4)$

5.59. Write an expression for a left-handed harmonic $\mathcal{E}$-state propagating along the z-axis with its major axis at $135°$ to the x-axis.

    Ans. $\mathbf{E}(z,t) = \hat{\mathbf{i}} E_0 \cos(kz - \omega t) + \hat{\mathbf{j}} E_0 \cos(kz - \omega t + 3\pi/4)$

5.60. Determine the state of polarization of the wave whose orthogonal $\mathcal{P}$-state components are

$$\mathbf{E}_x(z,t) = \hat{\mathbf{i}} E_0 \cos \omega\left(t - \frac{z}{v}\right)$$

$$\mathbf{E}_y(z,t) = \hat{\mathbf{j}} E_0 \cos\left[\omega\left(t - \frac{z}{v}\right) - \frac{5\pi}{4}\right]$$

    Ans. Because of the way the phase is written, E_x leads E_y by $5\pi/4$. The resultant is a **right**-handed ellipse tilted at $135°$ with respect to the x-axis.

5.61. Formulate an expression for an elliptically polarized harmonic wave propagating along the z-axis. It must be right-handed and have its major axis, which is to be twice its minor axis, along **the** x-direction.

    Ans. One possible form is $\mathbf{E}(z,t) = 2E_0 \hat{\mathbf{i}} \cos(\omega t - kz) - E_0 \hat{\mathbf{j}} \sin(\omega t - kz)$

5.62. Show analytically that elliptical light

$$\mathbf{E}_{\mathcal{E}}(z, t) \;=\; E_{0x}\hat{\mathbf{i}}\,\sin{(kz - \omega t)} \;+\; E_{0y}\hat{\mathbf{j}}\,\cos{(kz - \omega t)}$$

can be envisioned as the superposition of linear and circular light.

NATURAL AND PARTIALLY POLARIZED LIGHT

5.63. Is it accurate to maintain that a monochromatic wave is, by necessity, polarized?

Ans.   Yes, a perfectly monochromatic wave must be polarized.

5.64. How might one distinguish between partially linearly-polarized light (i.e. a mixture of $\mathcal{P}$-state and natural light) and partially elliptically-polarized light?

Ans.   With a linear polarizer determine the orientation corresponding to either the maximum or minimum irradiance. Align the axis of a quarter-wave plate in that direction and use the linear polarizer to examine the light emerging from it. If the light were a partial $\mathcal{P}$-state, insertion of the retarder would have no effect and the irradiance extrema would not be shifted. By contrast, if the light were a partial $\mathcal{E}$-state, both the maximum and minimum transmitted irradiance would be found, on rotating the polarizer, to have changed in orientation.

5.65. Imagine that you have a beam of light which might either be natural, circular, or a mixture of the two. How might you determine its actual nature?

Ans.   Insert a quarter-wave plate followed by a linear polarizer. If the light were circular, it would emerge from the plate linear and a zero irradiance minimum would appear on rotating the polarizer. Natural light would be unaffected by the plate and so no maximum or minimum would result. And partially circularly-polarized light would show a nonzero minimum as the polarizer was rotated.

5.66. A beam of partially linearly-polarized light propagating horizontally is being examined through a perfect linear polarizer. It is found that the maximum irradiance transmitted by the polarizer is 43 W/m² when oriented with its transmission axis at 30° to the right of vertical. The polarizer is rotated and it transmits an irradiance of 22 W/m² at an orientation of 60° left of vertical. Compute the degree of polarization of the beam.

Ans. $V = 32.3\%$

DICHROISM AND POLAROID

5.67. A planar, unpolarized light wave of irradiance I_i impinges, along the x-axis, on a perfect linear polarizer whose transmission axis is at 45° to both the y- and z-axes. Write an expression for the emerging wave function in terms of I_i, assuming it to have a wavelength λ.

Ans. $\mathbf{E} = \dfrac{E_0}{2}(\hat{\mathbf{j}} + \hat{\mathbf{k}})\,\sin\!\left(\dfrac{2\pi x}{\lambda} - \omega t\right)$ where $E_0 = \sqrt{\dfrac{I_i}{c\varepsilon_0}}$

5.68. Two polaroid linear polarizers are aligned with their transmission axes parallel. One of the polarizers is rotated to 30° and then to 60°. What is the ratio of the transmitted irradiances at these two positions?

Ans. $I(30°)/I(60°) = 3.00$

5.69. Three perfect linear polarizers are aligned in a row at angles, measured from the vertical, of 0°, 36° and 76°. What is the emerging irradiance in terms of I_i, the incident irradiance?

Ans. $0.1920\,I_i$

5.70. Ten perfect linear polarizers, arranged one behind the next, are oriented with 45° between the transmission axes of consecutive filters. Write an expression for the exiting irradiance in terms of the incident unpolarized irradiance, I_i. What would the emerging irradiance be for N such filters?

Ans. $9.77 \times 10^{-4} I_i$, $I_i(1/2)^N$

5.71. Four perfect linear polarizers are stacked such that the transmission axes of consecutive polarizers are separated by $30°$. What is the irradiance of the emerging beam in terms of the incident unpolarized irradiance, I_i? What would it be if the two middle polarizers were removed?

Ans. $0.2109 I_i$, zero

5.72. Imagine three perfect linear polarizers arranged one behind the other and denoted as 1, 2 and 3. Describe the emerging beam when 2 is $45°$ to the right of 1, while 3 is $45°$ to the left of 1. Compare this with Problem 5.29, wherein 1 is positioned between 2 and 3.

Ans.   No light emerges.

POLARIZATION BY REFLECTION

5.73. Unpolarized mercury light (546.072 nm) is made to fall on a glass plate at precisely $58°01'$, whereupon the reflected beam is found to be totally linearly polarized in the plane of the interface. Compute the index of refraction of the glass.

Ans. $n_g = 1.6014$

5.74. Determine the polarization angle for external reflection at the surface of a borosilicate crown glass plate ($n_g = 1.5170$) immersed in air. At what angle will the transmitted beam traverse the plate when light is incident at the polarization angle?

Ans. $\theta_p = 56°36'$, $\theta_t = 33°24'$

5.75. Figure 5-20 depicts a ray of light reflecting off two parallel dielectric plates at the polarization angle θ_p. Rotate the upper plate about the $\overline{O_1O_2}$ line through an angle θ so that the reflected ray comes out of the plane of the paper. Describe the irradiance of the emerging beam as a function of θ.

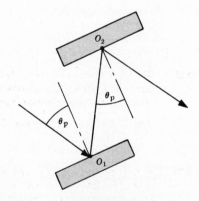

Ans.   The irradiance varies as $\cos^2 \theta$. This is Malus's experiment, which leads, as you might guess, to Malus's law.

5.76. Natural light is incident on an air-glass interface at the polarization angle. It is found that the reflectance of the perpendicular component, $R_\perp$, equals 0.15. Calculate the degree of polarization of both the reflected and transmitted light.

Ans. $V_r = 100\%$, $V_t = 8.1\%$

Fig. 5-20

5.77. A plate of extra dense flint glass ($n_g = 1.673$) is immersed in water ($n_w = 1.333$). Determine the polarization angles for both internal and external reflection at an interface.

Ans. $\theta_p = 51°27'$, $\theta_p' = 38°33'$

BIREFRINGENCE

5.78. A beam of right circular light propagating along the z-axis passes through a quarter-wave plate with a vertical (y-direction) fast axis. Describe the state of polarization of the emerging light.

Ans.   A $\mathcal{P}$-state at $135°$ to the horizontal (x-axis), i.e. in the second and fourth quadrants.

5.79. A beam of left circular light propagating along the z-axis passes through a quarter-wave plate with a vertical (y-direction) fast axis. Describe the state of polarization of the emerging light.

Ans.   A $\mathcal{P}$-state at 45° to the horizontal (x-axis).

5.80.   Figure 5-21 shows the formation of an o- and e-wave within a calcite principal section (see Fig. 5-12). Determine the angles α and β, where the former corresponds to the direction of the e-ray and the latter to the orientation of the optic axis with respect to the cleavage face.

Ans. $\beta = 45°24'$ and $\alpha = 6°14'$. The small value of α means that the ellipticity of the e-wave in Fig. 5-21 is much exaggerated.

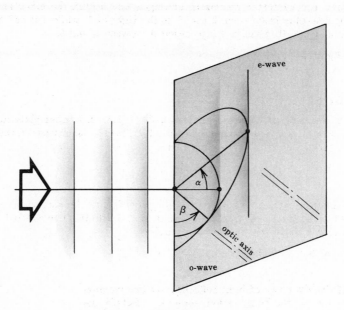

Fig. 5-21

5.81.   We have a source of left circular light ($\lambda_0 = 656$ nm) which we wish to convert to right circular by passing it through a quartz retarder ($n_e = 1.551$, $n_o = 1.542$). Compute the minimum thickness of the retarder and describe the necessary orientation.

Ans. $d = 3.64 \times 10^{-3}$ cm for a half-wave plate; the fast axis can be at any orientation normal to the beam

5.82.   A half-wave plate for red light ($\lambda_r = 780$ nm) is positioned between two crossed linear polarizers with its fast axis at 45° to the transmission axis of the polarizers. Describe the effect of such an arrangement on an incident beam of unpolarized red light.

Ans.   Red light will exit linearly polarized parallel to the analyzer's transmission axis.

5.83.   Neglecting the frequency dependence of the refractive indices in the Problem 5.82, what will happen to a beam of violet light ($\lambda_v = 390$ nm) traversing the arrangement?

Ans.   The retarder is a full-wave plate for λ_v and therefore no violet will emerge.

5.84.   Remove the analyzer in Problem 5.82 and determine which visible wavelengths, if any, will then be transmitted as circular light.

Ans.   yellow-green ($\lambda_{yg} = 520$ nm)

5.85.   A calcite retarder is positioned between two parallel linear polarizers. Determine the minimum thickness of the plate and its necessary orientation if no light ($\lambda_0 = 589.3$ nm) is to emerge from the arrangement when the incident beam is unpolarized.

Ans.   A half-wave plate at 45° with $d = 1.713 \times 10^{-4}$ cm.

5.86.   A crystalline negative uniaxial prism, as depicted in Fig. 5-22, generates minimum angles of deviation for o- and e-rays of 46° and 40°, respectively.   Determine the principal indices of refraction, n_o and n_e.

Ans. $n_e = 1.532$, $n_o = 1.597$

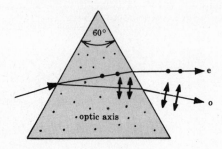

Fig. 5-22

Chapter 6

Interference and Coherence

6.1 INTRODUCTION

In Chapter 5 we dealt with the superposition of orthogonal $\mathcal{P}$-states. By contrast, interference theory treats, for the most part, with the superposition of coplanar electric fields.

As we saw in Chapter 1, wave phenomena are described by a second-order *linear* differential equation and so the *principle of superposition* obtains. Thus, at a point where two or more optical fields overlap, the resultant electric field intensity $\mathbf{E}$ is the vector sum of the constituent disturbances. Much like ripples on a pond, the fields in some regions will partially or completely cancel each other, while in still other locations the troughs and crests will be accentuated in the resultant. An irradiance distribution will often arise which differs from the simple algebraic sum of the irradiances of the contributing waves.

6.2 INTERFERENCE OF TWO WAVES

Imagine that we have two linearly polarized plane waves of the same wavelength, given by

$$\mathbf{E}_1(\mathbf{r}, t) = \mathbf{E}_{01} \cos(\mathbf{k}_1 \cdot \mathbf{r} - \omega t + \varepsilon_1)$$

$$\mathbf{E}_2(\mathbf{r}, t) = \mathbf{E}_{02} \cos(\mathbf{k}_2 \cdot \mathbf{r} - \omega t + \varepsilon_2)$$

which overlap at point P as in Fig. 6-1. Here, of course, $\mathbf{k}_1$, $\mathbf{k}_2$, ω, ε_1 and ε_2 are all constant. These waves may arise, for example, from two, very distant, point sources. The resultant field is simply

$$\mathbf{E} = \mathbf{E}_1 + \mathbf{E}_2$$

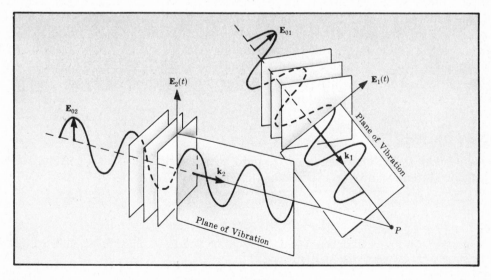

Fig. 6-1

But since at optical frequencies the fields oscillate at in excess of 10^{14} Hz the irradiance becomes the quantity of practical concern, it being directly measurable. Neglecting a constant factor, we can write the irradiance as just the time average of the total field:

$$I = \langle \mathbf{E}^2 \rangle$$

where $\mathbf{E}^2 \equiv \mathbf{E} \cdot \mathbf{E}$. Accordingly,

$$\mathbf{E}^2 = (\mathbf{E}_1 + \mathbf{E}_2) \cdot (\mathbf{E}_1 + \mathbf{E}_2) = \mathbf{E}_1^2 + \mathbf{E}_2^2 + 2\mathbf{E}_1 \cdot \mathbf{E}_2$$

Taking the average we obtain

$$I = I_1 + I_2 + I_{12}$$

where $I_1 = \langle \mathbf{E}_1^2 \rangle$, $I_2 = \langle \mathbf{E}_2^2 \rangle$ and $I_{12} = 2\langle \mathbf{E}_1 \cdot \mathbf{E}_2 \rangle$, the last being known as the *interference term*. It is by virtue of this last term that I differs from a simple sum of the irradiances of the component waves, i.e. $I_1 + I_2$.

The time average of some function $f(t)$ is given by

$$\langle f(t) \rangle = \frac{1}{T} \int_t^{t+T} f(t') \, dt'$$

Here the detection interval T is much greater than the period of oscillation of the waves, τ. Thus if we carried out the indicated calculation for the above plane waves (see Problem 2.13), the interference term would become

$$I_{12} = 2\langle \mathbf{E}_1 \cdot \mathbf{E}_2 \rangle = \mathbf{E}_{01} \cdot \mathbf{E}_{02} \cos \delta$$

where the *phase difference* δ is given by

$$\delta = (\mathbf{k}_1 \cdot \mathbf{r}) - (\mathbf{k}_2 \cdot \mathbf{r}) + \varepsilon_1 - \varepsilon_2$$

All of this means that as one moves from point to point in space, $\mathbf{r}$ varies, as does δ, and therefore I_{12} and I both vary as well.

The $(\mathbf{k}_1 \cdot \mathbf{r}) - (\mathbf{k}_2 \cdot \mathbf{r})$ contribution to the phase difference arises from a path-length difference sustained by the waves in going from the source points to P. The $\varepsilon_1 - \varepsilon_2$ contribution is due to an initial phase difference at the emitters, and if it is constant, as we have assumed, the sources are said to be *coherent*. We examine the more general condition of *partial coherence* in Section 6.6. For now, a simple rule will suffice: If the overlapping waves are coherent, their electric fields can combine with each other in a sustained fashion and will be added first and then squared to yield the irradiance. If the waves are incoherent, the individual fields, which are effectively independent, will be squared first and then these component irradiances added.

SOLVED PROBLEMS

6.1. Assume that the electric fields of the two waves in Fig. 6-1 are parallel. Derive a symbolic statement of I in terms of I_1, I_2 and δ.

For the superposition of coherent waves,

$$I = I_1 + I_2 + \mathbf{E}_{01} \cdot \mathbf{E}_{02} \cos \delta$$

but now the fields are parallel and $\mathbf{E}_{01} \cdot \mathbf{E}_{02} = E_{01}E_{02}$. Bearing in mind the results of Problem 2.13, the component irradiances can be written as

$$I_1 = \langle \mathbf{E}_1^2 \rangle = \frac{E_{01}^2}{2} \qquad I_2 = \langle \mathbf{E}_2^2 \rangle = \frac{E_{02}^2}{2}$$

Thus $\mathbf{E}_{01} \cdot \mathbf{E}_{02} = \sqrt{2I_1} \sqrt{2I_2}$ and

$$I = I_1 + I_2 + 2\sqrt{I_1 I_2} \cos \delta$$

To find the distribution of light in space one need only go to each point, determine δ there, and then compute I knowing I_1 and I_2.

6.2. Describe the resultant irradiance distribution which would occur in Problem 6.1 if the two waves were *incoherent*, i.e. if their phase angles were to vary randomly and rapidly as compared to the measuring time.

The interference term

$$I_{12} = 2\langle \mathbf{E}_1 \cdot \mathbf{E}_2 \rangle$$

can be written as

$$I_{12} = 2\langle \mathbf{E}_{01} \cdot \mathbf{E}_{02} \cos \left[\mathbf{k}_1 \cdot \mathbf{r} - \omega t + \varepsilon_1(t) \right] \cos \left[\mathbf{k}_2 \cdot \mathbf{r} - \omega t + \varepsilon_2(t) \right] \rangle$$

$$= 2E_{01}E_{02}\langle \cos (\mathbf{k}_1 \cdot \mathbf{r} - \omega t) \cos (\mathbf{k}_2 \cdot \mathbf{r} - \omega t) \cos \varepsilon_1 \cos \varepsilon_2$$

$$- \cos (\mathbf{k}_1 \cdot \mathbf{r} - \omega t) \sin (\mathbf{k}_2 \cdot \mathbf{r} - \omega t) \cos \varepsilon_1 \sin \varepsilon_2$$

$$- \sin (\mathbf{k}_1 \cdot \mathbf{r} - \omega t) \cos (\mathbf{k}_2 \cdot \mathbf{r} - \omega t) \sin \varepsilon_1 \cos \varepsilon_2$$

$$+ \sin (\mathbf{k}_1 \cdot \mathbf{r} - \omega t) \sin (\mathbf{k}_2 \cdot \mathbf{r} - \omega t) \sin \varepsilon_1 \sin \varepsilon_2 \rangle$$

Since $\varepsilon_1(t)$ and $\varepsilon_2(t)$ fluctuate irregularly and rapidly, as compared to the measuring interval, the average value of each term taken over that interval must be zero, hence $I_{12} = 0$. Alternatively, notice that the relative phase $\varepsilon_1 - \varepsilon_2 = \varepsilon$ is also rapidly varying and random, and if you let $\varepsilon_2 = 0$ and $\varepsilon_1 = \varepsilon$, then $I_{12} = 0$, as expected. In effect, the interference pattern changes so rapidly that what is observed is just

$$I = I_1 + I_2$$

This is why we can simply square the individual fields and then add irradiances when dealing with incoherent sources.

6.3. Suppose that two identical waves of natural light were made to overlap. Would interference result? (That is, would I_{12} be nonzero?)

Each of the unpolarized waves can be envisioned as composed of two orthogonal, mutually incoherent $\mathcal{P}$-states. These, in turn, can be labeled with respect to any convenient plane (e.g. the one containing $\mathbf{k}_1$ and $\mathbf{k}_2$) as $\mathbf{E}_1 = \mathbf{E}_{1\|} + \mathbf{E}_{1\perp}$ and $\mathbf{E}_2 = \mathbf{E}_{2\|} + \mathbf{E}_{2\perp}$. Since the waves are identical, $\mathbf{E}_{1\|}$ and $\mathbf{E}_{2\|}$ are coherent, as are $\mathbf{E}_{1\perp}$ and $\mathbf{E}_{2\perp}$. Consequently two independent, precisely overlapping interference patterns,

$$\langle (\mathbf{E}_{1\|} + \mathbf{E}_{2\|})^2 \rangle \qquad \text{and} \qquad \langle (\mathbf{E}_{1\perp} + \mathbf{E}_{2\perp})^2 \rangle$$

would result. The equation for the irradiance distribution therefore applies whether the waves are polarized or not, provided they are identical.

6.4. Reconsider the waves of Problem 6.1 and examine the conditions under which I assumes maximum and minimum values.

Inasmuch as

$$I = I_1 + I_2 + 2\sqrt{I_1 I_2} \cos \delta$$

$I_{\max}$ occurs when $\cos \delta = 1$, i.e. when $\delta = 0, \pm 2\pi, \pm 4\pi, \ldots$. Thus

$$I_{\max} = I_1 + I_2 + 2\sqrt{I_1 I_2}$$

The two waves are in phase, trough overlaps trough and crest overlaps crest, as in Fig. 6-2(a), page 126. Similarly, when $\cos \delta = -1$, i.e. when $\delta = \pm\pi, \pm 3\pi, \ldots$, $I = I_{\min}$ and

$$I_{\min} = I_1 + I_2 - 2\sqrt{I_1 I_2}$$

Here the troughs of one wave overlie the crests of the other, thereby tending to cancel each other, as in Fig. 6-2(b).

Whenever $I_1 + I_2 < I$ the situation is referred to as *constructive interference*, and whenever

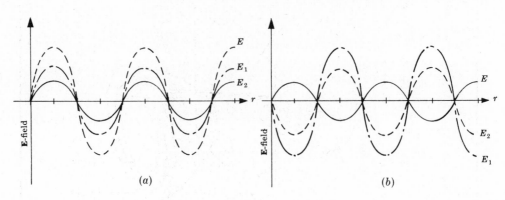

Fig. 6-2

$I_1 + I_2 > I$, it is spoken of as *destructive interference*.

6.5. Write an expression for the irradiance at point P in Fig. 6-1, assuming the field amplitudes to be both equal and parallel.

Since the fields are parallel, we know from Problem 6.1 that

$$I = I_1 + I_2 + 2\sqrt{I_1 I_2} \cos \delta$$

Furthermore, $E_{01} = E_{02}$, which means that $I_1 = I_2$, and we set both of these equal to I_0. This leads to

$$I = 2I_0(1 + \cos \delta) = 4I_0 \cos^2 \frac{\delta}{2}$$

The interference pattern varies as the cosine squared.

6.6. The *visibility* of the fringes in an interference pattern was defined by Michelson to be

$$\mathcal{V} \equiv \frac{I_{\max} - I_{\min}}{I_{\max} + I_{\min}}$$

(a) Derive an expression for the visibility of the pattern resulting from the two coherent waves in Fig. 6-1, assuming their fields to be parallel. (b) What is the visibility when the two field amplitudes are equal?

(a) Direct substitution of the expressions for $I_{\max}$ and $I_{\min}$ from Problem 6.4 leads to

$$\mathcal{V} = \frac{2\sqrt{I_1 I_2}}{I_1 + I_2}$$

(b) When $E_{01} = E_{02}$, $I_1 = I_2 = I_0$ and so

$$\mathcal{V} = \frac{2\sqrt{I_0^2}}{2I_0} = 1$$

Alternatively, from Problem 6.5, $I_{\min} = 0$ and again $\mathcal{V} = 1$.

6.7. Suppose that point P in Fig. 6-1 is moved nearer to the source points S_1 and S_2 so that the overlapping waves are spherical. Assume that the field amplitudes are equal and parallel at P and discuss the form of the interference pattern.

The spherical waves at P can be expressed by

$$\mathbf{E}_1(r_1, t) = \mathbf{E}_{01}(r_1) \cos (kr_1 - \omega t + \varepsilon_1)$$

$$\mathbf{E}_2(r_2, t) = \mathbf{E}_{02}(r_2) \cos (kr_2 - \omega t + \varepsilon_2)$$

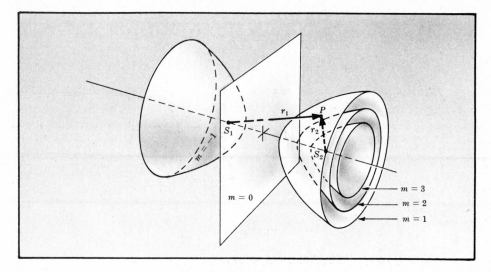

Fig. 6-3

with the appropriate geometry depicted in Fig. 6-3. Here

$$\delta \;=\; k(r_1 - r_2) + \varepsilon_1 - \varepsilon_2$$

and neglecting a constant factor, $\langle \mathbf{E}_1^2 \rangle = \langle \mathbf{E}_2^2 \rangle = I_0$. Thus, by Problem 6.5,

$$I \;=\; 4I_0 \cos^2 \frac{1}{2}\left[k(r_1 - r_2) + \varepsilon_1 - \varepsilon_2 \right]$$

and its maxima and minima occur when $\delta = 2\pi m$ and $\delta = \pi(2m+1)$, respectively, m being $0, \pm 1, \pm 2, \ldots$. Accordingly, maxima in I correspond to the situation where

$$r_1 - r_2 \;=\; \frac{2\pi m + \varepsilon_2 - \varepsilon_1}{k}$$

and minima where

$$r_1 - r_2 \;=\; \frac{\pi(2m+1) + \varepsilon_2 - \varepsilon_1}{k}$$

These are both equations of families of surfaces in space; namely, concentric hyperboloids of revolution. Figure 6-3 shows several such surfaces over which $I = I_{\max}$ in the case $\varepsilon_1 = \varepsilon_2$.

The region in which the two sources are immersed is filled with an interference pattern that would be evidenced by a series of bright and dark fringes on a viewing screen.

6.8. Describe the distant radiation pattern arising from two equal-strength point sources, S_1 and S_2, which are in phase ($\varepsilon_1 = \varepsilon_2$), where now their separation, a, equals $\lambda_0/2$.

In most cases of optical interference the separation between the sources is large, $a \gg \lambda_0$, but instances where $a \approx \lambda_0$ are of considerable interest as well. Examining Fig. 6-4, we see that the distances $\overline{S_1 P_1}$ and $\overline{S_2 P_1}$ are equal (and, although not drawn that way, are many wavelengths long).

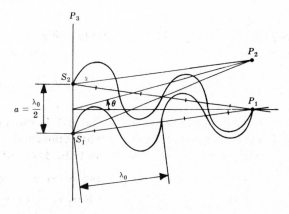

Fig. 6-4

The waves traverse equal optical path lengths and arrive at P_1 nearly parallel and in phase, interfering constructively. If I_0 is the irradiance of either source, the resultant at P_1 is $4I_0$.

At P_2, which is again far from the sources, the two fields will be assumed to be nearly equal in amplitude. The path difference $\overline{S_1P_2} - \overline{S_2P_2}$ is approximately $a \sin \theta$. Consequently, the difference in the number of waves spanning the two paths is $(a \sin \theta)/\lambda_0$. Since each whole wave corresponds to a phase shift of 2π radians, the phase difference between the disturbances arriving at P_2 is

$$\delta = \frac{2\pi}{\lambda_0} a \sin \theta$$

Hence, from Problem 6.7,

$$I(\theta) = 4I_0 \cos^2\left(\frac{\pi}{\lambda_0} a \sin \theta\right)$$

and for this particular situation wherein $a = \lambda_0/2$,

$$I(\theta) = 4I_0 \cos^2\left(\frac{\pi}{2} \sin \theta\right)$$

Notice that at P_1, $\theta = 0$ and $I(0)$ does indeed equal $4I_0$. At P_3 the disturbances would be $\lambda_0/2$ or π radians out of phase, completely canceling each other. In that case $\theta = 90°$, $\sin \theta = 1$, $\cos \pi/2 = 0$ and $I(90°) = 0$ as anticipated.

Figure 6-5 is a polar plot of I versus θ showing the two-lobe pattern consisting of radiation predominantly at $\theta = 0$ and $\theta = 180°$.

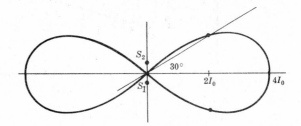

Fig. 6-5

6.3 WAVEFRONT-SPLITTING INTERFEROMETERS

As we saw in Section 5.5, an ordinary light source can be envisioned as emitting a large number of uncorrelated, nearly sinusoidal wave trains each lasting for roughly 10^{-8} s. The resultant light wave maintains a constant phase for a time somewhat less than this *coherence time*, Δt. The waves from two such sources could interfere but the pattern would be sustained for a time less than Δt. As the waves varied in phase rapidly and randomly the pattern would shift erratically, washing out over the comparatively long detection interval. Although two phase-locked lasers can generate a detectable interference pattern, the more usual approach is simply to split a single wave into two coherent pieces.

In the broadest sense, a device which generates interference fringes is referred to as an *interferometer*. The subclass of wavefront-splitting interferometers are all characterized by the fact that two separate segments of a wave are brought into superposition after having been made to traverse different paths.

Wavefront-splitting interferometers, and there are many of them, are perhaps best typified by the setup depicted in Fig. 6-6, which is known as *Young's experiment*. The sources S, S_1 and S_2 are either small holes or long narrow slits perpendicular to the plane of the drawing. The wave spreading out from S impinges on S_1 and S_2 which, in turn, serve as a pair of in-phase coherent emitters. For any distant point P not far from the central axis, the optical path length difference ($n \approx 1$) is $\overline{S_1P} - \overline{S_2P} = r_1 - r_2$. This is approximately equal to $\overline{S_1A} = a \sin \theta$ or, more simply, $a\theta$. Since $\tan \theta \approx \theta = y/s$, when y is small compared to s,

$$r_1 - r_2 = \frac{a}{s} y$$

The two waves arriving at P are in phase and interfere constructively whenever $r_1 - r_2 = m\lambda_0$, where $m = 0, \pm 1, \pm 2, \dots$. Thus, interference maxima occur whenever

$$y_m = \frac{sm\lambda_0}{a}$$

and the plane of observation is covered with alternately bright and dark parallel bands running perpendicular to the plane of the figure. Indeed, the region of space between Σ_a and Σ_o is filled with the interference pattern.

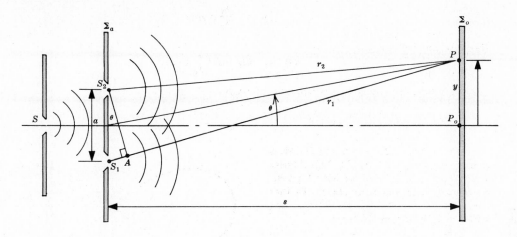

Fig. 6-6

SOLVED PROBLEMS

6.9. Write an expression for the irradiance distribution over the plane of observation in Young's experiment and plot I as a function of y.

Inasmuch as the waves are emitted at S_1 and S_2 in phase, the phase angle difference at P is simply

$$\delta = k_0(r_1 - r_2)$$

From Problem 6.5 it follows that

$$I = 4I_0 \cos^2 \frac{k_0(r_1 - r_2)}{2}$$

and since $r_1 - r_2 = ay/s$ and $k_0 = 2\pi/\lambda_0$, we have

$$I = 4I_0 \cos^2 \frac{ya\pi}{s\lambda_0}$$

The resulting pattern is plotted in Fig. 6-7.

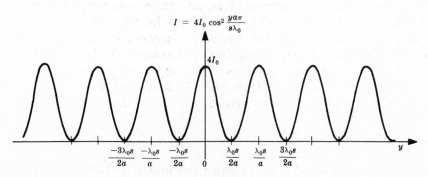

Fig. 6-7

6.10. Derive an expression for the separation Δy between alternate pairs of fringes (i.e. the distance between consecutive maxima or minima) in Young's experiment.

The mth and $(m+1)$th maxima are located at

$$y_m = \frac{sm\lambda_0}{a} \qquad\qquad y_{m+1} = \frac{s(m+1)\lambda_0}{a}$$

Accordingly, $$\Delta y = y_{m+1} - y_m = \frac{s\lambda_0}{a}$$

which can be easily verified from Fig. 6-7.

6.11. Use the plane wave formalism to verify that the two waves arriving at point P near the central axis in Young's experiment are out of phase by $k_0 ay/s$.

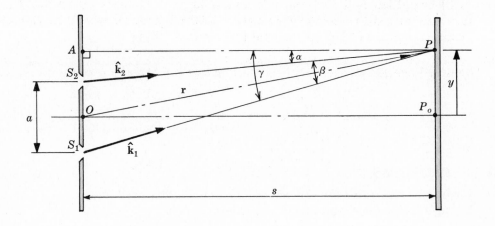

Fig. 6-8

Figure 6-8 shows the two unit propagation vectors $\hat{\mathbf{k}}_1 = \mathbf{k}_1/k_1$ and $\hat{\mathbf{k}}_2 = \mathbf{k}_2/k_2$, as well as the position vector $\mathbf{r}$ locating P relative to an origin on the central axis. The two disturbances arriving at P can be written as

$$\mathbf{E}_1(\mathbf{r}, t) = \mathbf{E}_0 \cos(\mathbf{k}_1 \cdot \mathbf{r} - \omega t)$$

$$\mathbf{E}_2(\mathbf{r}, t) = \mathbf{E}_0 \cos(\mathbf{k}_2 \cdot \mathbf{r} - \omega t)$$

assuming $\mathbf{E}_{01} = \mathbf{E}_{02} = \mathbf{E}_0$ and $\varepsilon_1 = \varepsilon_2$. Consequently,

$$\delta = (\mathbf{k}_1 \cdot \mathbf{r}) - (\mathbf{k}_2 \cdot \mathbf{r}) = (\mathbf{k}_1 - \mathbf{k}_2) \cdot \mathbf{r}$$

Now to find $\mathbf{k}_1 - \mathbf{k}_2$. Since $k_1 = k_2 = k_0$ and β is a small angle,

$$|\mathbf{k}_1 - \mathbf{k}_2| \approx k_0\beta$$

Moreover, $\tan\gamma \approx \gamma$ and $\tan\alpha \approx \alpha$; hence

$$\beta = \gamma - \alpha \approx \frac{\overline{S_1A}}{s} - \frac{\overline{S_2A}}{s} = \frac{a}{s}$$

Therefore $$\mathbf{k}_1 - \mathbf{k}_2 \approx \hat{\mathbf{j}}\frac{k_0 a}{s}$$

and $$\delta = (\mathbf{k}_1 - \mathbf{k}_2) \cdot \mathbf{r} = \frac{k_0 a}{s}\hat{\mathbf{j}} \cdot (\hat{\mathbf{i}}x + \hat{\mathbf{j}}y)$$

$$= \frac{k_0 ay}{s}$$

6.12. A quasi-monochromatic beam of wavelength λ_0 illuminates Young's experiment, generating a fringe pattern having a 5.6-mm separation between consecutive dark bands. If the distance between the plane containing the apertures, Σ_a, and the plane of observation, Σ_o, is 10 m, and if the two sources S_1 and S_2 are separated by 1.0 mm, what is the wavelength of the light?

It follows from Problem 6.10 that the fringe separation is $\Delta y = s\lambda_0/a$. Hence

$$\lambda_0 = \frac{a\,\Delta y}{s} = \frac{(1 \times 10^{-3})(5.6 \times 10^{-3})}{10} = 5.6 \times 10^{-7}\text{ m} = 560\text{ nm}$$

6.13. A glass chamber 25 mm long filled with air is positioned in front of one of the secondary sources in Young's experiment. The air is removed from the chamber and a test gas is entered in its stead. On comparing the fringe system corresponding to air with that of the test gas, it is found that the entire interference pattern on Σ_o was displaced by 21 bright bands toward the side containing the chamber. Given that the illumination was the red Fraunhofer C line ($\lambda_0 = 656.2816$ nm), for which the refractive index of air is $n_a = 1.000276$, determine the index of the gas (n_g).

The optical path length of the chambered region when occupied by air is $n_a(25 \times 10^{-3})$; when occupied by the gas it is $n_g(25 \times 10^{-3})$. Thus the optical path-length difference (O.P.D.) is given by

$$\text{O.P.D.} = (n_g - n_a)(25 \times 10^{-3})$$

But this must correspond to the shift of 21 wavelengths; that is, O.P.D. $= 21\lambda_0$. Consequently,

$$21\lambda_0 = (n_g - n_a)(25 \times 10^{-3})$$

or

$$n_g = \frac{21(656.2816 \times 10^{-9})}{25 \times 10^{-3}} + 1.000276$$

$$= 551.28 \times 10^{-6} + 1.000276 = 1.000827$$

6.14. *Fresnel's double mirror* consists of two specularly reflecting surfaces making a small angle α with respect to each other. The arrangement is shown in Fig. 6-9. Discuss how the system works and derive an expression for α in terms of R, λ_0, the fringe separation Δy and the distance d.

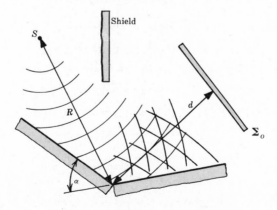

Fig. 6-9

Different portions of the primary wave from S are reflected off the two mirrors, thereafter overlapping and interfering in the region in front of Σ_o. We can imagine that these coherent waves originate at the two virtual sources shown in Fig. 6-10. Notice that the arrangement is then identi-

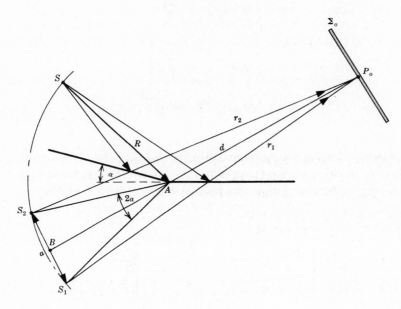

Fig. 6-10

cal to that of Young's experiment and so

$$\Delta y \;=\; \frac{s\lambda_0}{a}$$

where now $s = \overline{BP}_o$, the perpendicular distance from $\overline{S_1 S_2}$ to Σ_o. Here $s = \overline{BA} + d$ and, assuming $a \ll s$, $\overline{BA} \approx R$. Therefore we can write

$$\Delta y \;=\; \frac{(R+d)\lambda_0}{a}$$

The lines $\overline{S_1 A}$ and $\overline{S_2 A}$ can be taken as the images of the incident ray $\overline{SA}$ by the two mirrors. Then, by Problem 4.52, $\angle S_1 A S_2 = 2\alpha$, and so

$$a \;=\; 2R \sin \alpha \;\approx\; 2R\alpha$$

Hence
$$\Delta y \;=\; \frac{(R+d)\lambda_0}{2R\alpha} \quad \text{or} \quad \alpha \;=\; \frac{(R+d)\lambda_0}{2R\,\Delta y}$$

6.15. Compute the angle α between the reflecting surfaces of a Fresnel double mirror when the source is 1 m from the intersection of the mirrors, the screen Σ_o is 2 m from that intersection, $\lambda_0 = 500$ nm and the fringe separation is found to be 1 mm.

We are given that $R = 1$ m, $d = 2$ m, $\lambda_0 = 500 \times 10^{-9}$ m and $\Delta y = 10^{-3}$ m. Hence

$$\alpha \;=\; \frac{(R+d)\lambda_0}{2R\,\Delta y} \;=\; \frac{(1+2)(500 \times 10^{-9})}{2(1)10^{-3}} \;=\; 7.5 \times 10^{-4} \text{ rad}$$

or $0.043°$.

6.16. Obtain an expression for the irradiance distribution over the plane of observation for a Fresnel double mirror, in terms of $I(0)$ (the irradiance on the central axis), y measured from that axis, R, d and α.

We found in connection with Young's experiment (see Problem 6.9) that

$$I(y) \;=\; I(0) \cos^2 \frac{k_0(r_1 - r_2)}{2}$$

Moreover,

$$r_1 - r_2 \;=\; \frac{ya}{s}$$

and $a \approx 2R\alpha$, $s = R + d$.   Hence

$$r_1 - r_2 \;=\; \frac{2yR\alpha}{R + d}$$

and

$$I(y) \;=\; I(0)\,\cos^2 \frac{k_0 y R\alpha}{R + d}$$

6.17. Fresnel devised yet another wavefront-splitting system (Fig. 6-11) known as a *biprism*. Describe how it works and derive an equation for the fringe separation Δy in terms of R, d, α, n and λ_0, provided that the prisms are very thin ($\alpha \approx 1°$).

The minimum deviation δ_m of a prism of index n in air is given implicitly by

$$n \;=\; \frac{\sin\,[(\delta_m + \alpha)/2]}{\sin \alpha/2}$$

Since the biprism is thin, both δ_m and α are quite small and

$$n \;\approx\; \frac{(\delta_m + \alpha)/2}{\alpha/2}$$

$$=\; \frac{\delta_m + \alpha}{\alpha}$$

We can drop the subscript m, because the thin prisms function at or near minimum deviation, whereupon

$$(n - 1)\alpha \;=\; \delta$$

(Do not confuse this δ with the phase angle difference.) As can be seen in Fig. 6-12, the

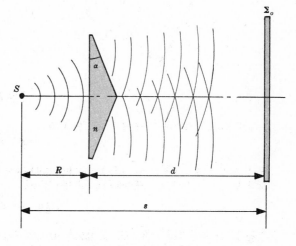

Fig. 6-11

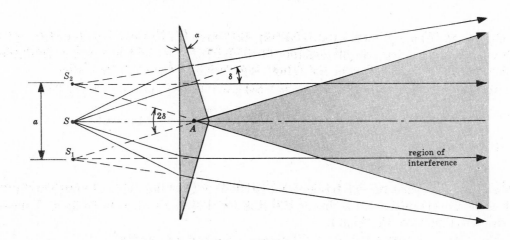

Fig. 6-12

primary wave is split in two by the biprism, in effect replacing the source S by two coherent virtual sources S_1 and S_2. Angle $S_1AS_2 = 2\delta$ and so $a \approx 2R\delta$. The formula for the fringe separation,

$$\Delta y = \frac{s\lambda_0}{a}$$

from Young's experiment again applies, and becomes

$$\Delta y = \frac{(R+d)\lambda_0}{2R\delta} = \frac{(R+d)\lambda_0}{2R(n-1)\alpha}$$

6.18. A collimated laser beam ($\lambda_0 = 632.8$ nm) impinges normally on a Fresnel biprism. Unlike Young's experiment, where the fringe separation Δy increased with increasing distance from the source, show that here it is independent of the location of Σ_o. Determine Δy when $n = 1.520$ and $\alpha = 1°30'$.

From Problem 6.17

$$\Delta y = \frac{(R+d)\lambda_0}{2R(n-1)\alpha}$$

but now $R \gg d$ and so $R + d \approx R$. Hence

$$\Delta y = \frac{\lambda_0}{2(n-1)\alpha}$$

For the given values,

$$\Delta y = \frac{632.8 \times 10^{-9}}{2(0.520)(0.0262)} = 0.023 \text{ mm}$$

Notice that even a value of α as small as $1°30'$ does not produce conveniently wide fringes.

6.19. Figure 6-13 depicts a segmented light wave; one portion of it travels directly to the plane of observation, while the other arrives there after undergoing specular reflection from a smooth surface. This wavefront-splitting arrangement is known as *Lloyd's mirror*. Discuss the manner in which it generates interference fringes, paying particular attention to the reflection process. Derive an expression for $I(y)$.

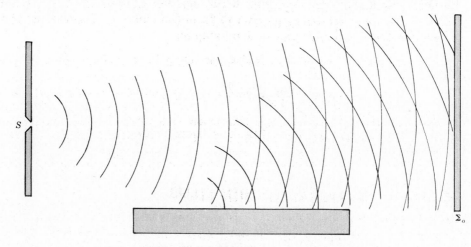

Fig. 6-13

The reflected beam appears to come from a virtual source S_1, as shown in Fig. 6-14. Once again we have a situation similar to Young's experiment in most respects. The crucial distinction is that the reflected beam undergoes a 180° phase shift at the interface. Recall the Fresnel equations of Section 3.4 which describe the reflection process via the amplitude coefficients $r_{||}$ and $r_{\perp}$. At glancing

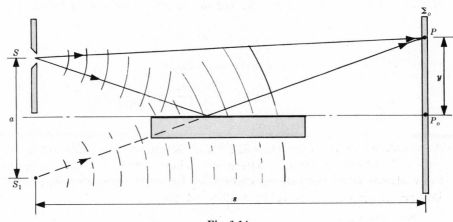

Fig. 6-14

incidence ($\theta_i \approx 90°$) these are both equal to -1, which means that the incident and reflected electric field vectors are antiparallel at the boundary. Thus, at any point P,

$$\delta = k_0(r_1 - r) \pm \pi$$

and

$$I = 4I_0 \cos^2 \frac{k_0(r_1 - r) \pm \pi}{2}$$

Alternatively,

$$I = 4I_0 \sin^2 \frac{k_0(r_1 - r)}{2}$$

and since, as in Young's setup, $r_1 - r = ay/s$, we get

$$I(y) = 4I_0 \sin^2 \frac{\pi a y}{\lambda_0 s}$$

Now there is a black band at $y = 0$ instead of a maximum.

6.20. Referring to Lloyd's mirror (Fig. 6-14), suppose that the source is 2 mm above the plane of the mirrored surface and 1 m from the plane of observation. If $\lambda_0 = 460$ nm, determine the location of the first maximum.

The separation between fringes is given, as before, by $\Delta y = s\lambda_0/a$. Here $a = 4$ mm, $s = 1$ m, and so

$$\Delta y = \frac{1(460 \times 10^{-9})}{4 \times 10^{-3}} = 1.15 \times 10^{-4}\,\text{m} = 0.115\,\text{mm}$$

The center of the dark central fringe falls at the level of the reflecting surface. The next dark fringe is 0.115 mm up from this surface and the first maximum is midway between these two at $y = 0.0575$ mm.

6.4 AMPLITUDE SPLITTING BY THIN FILMS

In all the examples of the previous section, the two interfering waves had just about the same electric field amplitude as the primary wave from which they were derived. We now show how to shear the entire primary wavefront into two segments, each with a diminished amplitude.

In Fig. 6-15 light from a quasi-monochromatic point source impinges on a thin transparent plate. Because the reflection coefficients are generally small, we shall limit the discussion to the two rays shown, each of which undergoes a single reflection, and omit the far weaker, multiply reflected rays. These waves, E_{1r} and E_{2r}, have nearly the same amplitudes and can be thought to have originated at two virtual coherent point sources. The optical path-length difference (O.P.D. or Λ) from S to P for these rays is given by

$$\Lambda = n_f(\overline{AB} + \overline{BC}) - n(\overline{AD})$$

or, since $\overline{AB} = \overline{BC} = d/\cos\theta_t$,

$$\Lambda = \frac{2n_f d}{\cos\theta_t} - n(\overline{AD})$$

By Snell's law,

$$\overline{AD} = \overline{AC}\sin\theta_i$$

$$= \overline{AC}\frac{n_f}{n}\sin\theta_t$$

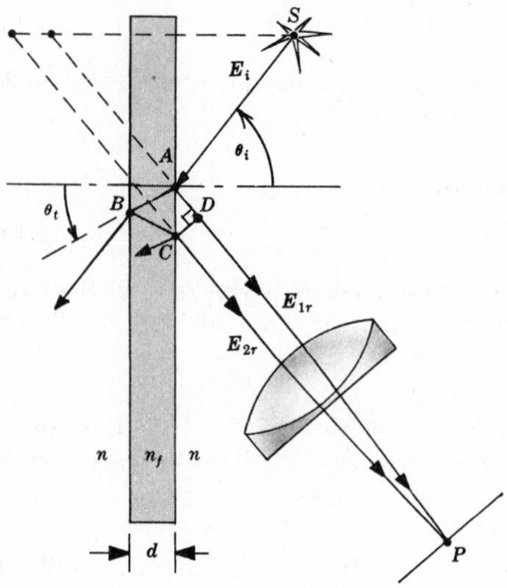

Fig. 6-15

Moreover, $\overline{AC} = 2d\tan\theta_t$. Consequently,

$$\Lambda = \frac{2n_f d}{\cos\theta_t}(1 - \sin^2\theta_t) = 2n_f d\cos\theta_t$$

There is an additional relative phase shift of π radians between the waves, since one is internally reflected while the other is externally reflected. Accordingly,

$$\delta = k_0\Lambda \pm \pi = \frac{4\pi n_f}{\lambda_0}d\cos\theta_t \pm \pi$$

or, in terms of θ_i,

$$\delta = \frac{4\pi d}{\lambda_0}(n_f^2 - n^2\sin^2\theta_i)^{1/2} \pm \pi$$

There is once again a double-beam interference pattern. Because δ is constant for all values of θ_i which are the same, these fringes are spoken of as *fringes of equal inclination*. Again, approximating the two fields to be equal, we have

$$I = 4I_0\cos^2\frac{\delta}{2}$$

and wherever θ_i is such that $\delta = (2m+1)\pi$ a minimum will exist. On the other hand, when $\delta = 2m\pi$, the irradiance will be a maximum. The number m ($m = 0, \pm 1, \pm 2, \ldots$) is called the *order number* of the dark or bright fringe. It should be noted that the fringe of order m need not correspond to the mth fringe as counted on the observing screen (see Problem 6.39).

As a consequence of the fact that δ is dependent on θ_i and not on the location of S, other source points would contribute to the pattern without obscuring it and, indeed, an extended or broad source could be used as well.

SOLVED PROBLEMS

6.21. Blue light ($\lambda_0 = 487.99$ nm) falls perpendicularly on a film of thickness 1.648×10^{-6} m which has a refractive index of 1.555 immersed in air. The beam is split as in Fig. 6-15. Compute the O.P.D. introduced in traversing the film. What is the phase angle difference between the two disturbances after they leave the film?

The O.P.D. is given by

$$\Lambda = 2n_f d \cos \theta_t$$

which can be expressed more usefully via Snell's law as

$$\Lambda = 2d(n_f^2 - n^2 \sin^2 \theta_i)^{1/2}$$

Because $n = 1$ and $\theta_i = 0$,

$$\Lambda = 2dn_f = 2(1.648 \times 10^{-6})(1.555) = 5.125 \times 10^{-6} \text{ m}$$

This corresponds to a difference in the two paths of $\Lambda/\lambda_0 = 10.5$ wavelengths, and each wavelength gives rise to a 2π phase difference. Clearly, then,

$$\delta = 21\pi \pm \pi$$

including the phase shift of π due to the reflection. This means that the two waves, one reflected from each of the film's surfaces, are back in phase (δ is an integral multiple of 2π).

6.22. A ray of green light ($\lambda_0 = 565.69$ nm) impinges at 30° on a thin planar film of index 1.500 immersed in air. (a) What is the smallest film thickness for which the point of reflection appears on a maximum fringe? (b) What would that point look like if the film were 1500 nm thick?

(a) An irradiance maximum will occur when

$$\delta = \frac{4\pi d}{\lambda_0}(n_f^2 - n^2 \sin^2 \theta_i)^{1/2} \pm \pi$$

is an integral multiple of 2π. This, in turn, will result for the smallest d when

$$\frac{d}{\lambda_0}(n_f^2 - n^2 \sin^2 \theta_i)^{1/2} = \frac{1}{4}$$

or

$$d = \frac{\lambda_0}{4}(1.5^2 - \sin^2 30°)^{-1/2} = 100 \text{ nm}$$

(b) If d were 1500 nm, we would have $\delta = 15\pi \pm \pi$, which again would correspond to a maximum.

6.23. Figure 6-16 is an arrangement for viewing a thin-film interference pattern at near-normal incidence. Describe the shape of the resulting bands of light on Σ_o (which are called *Haidinger fringes*).

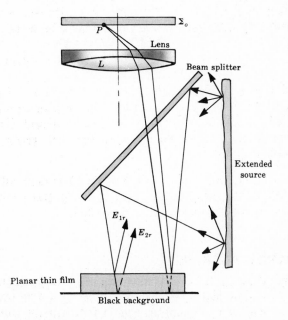

Fig. 6-16

Because these are fringes of equal inclination, all pairs of rays at a given angle to the normal will arrive at Σ_o with the same value of δ. The waves in each such pair are mutually coherent, but the various pairs arise from different source points and so are incoherent. The result is a large number of overlapping, identical, noninteracting contributions to I at any point on Σ_o. Figure 6-17 shows two such pairs of rays reaching P, each with the same δ. Of course, rays not in the plane of Σ_R can enter the lens at a particular θ_i. These will arrive with the same phase difference at some other point P'. Thus if there is a bright spot at P there

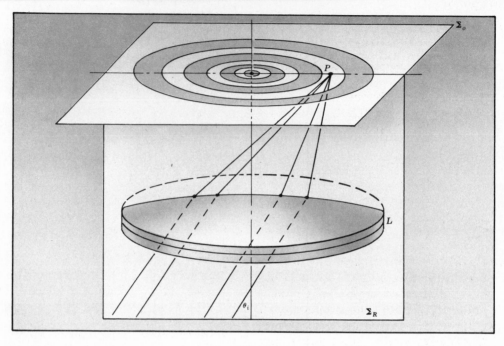

Fig. 6-17

will be an identical one at P'. In fact, if we rotate Σ_R about the central axis of the lens, the locus of P is a circular fringe. If L and Σ_o compose the viewer's eyeball, the pattern will be a series of concentric circles centered on the lens and moving along with it as the observer moves.

Whenever $\delta = 0, \pm 2\pi, \pm 4\pi, \ldots$, a maximum appears, i.e. whenever

$$2 n_f d \cos \theta_t \pm \frac{\lambda_0}{2} = m \lambda_0$$

where $m = 0, 1, 2, \ldots$. Thus a given circular fringe has a particular value of θ_t, and therefore of θ_i, as well as a specific *order* m.

Notice that the circular interference pattern appears on a plane normal to $\overline{S_1 S_2}$, as in Fig. 6-3.

6.24. Compute the smallest thickness which a transparent film of index 1.455 may have if it is to generate a minimum in reflected light under normal illumination at 500 nm surrounded by air.

A minimum occurs for δ equal to an odd multiple of π, and the smallest appropriate value of d corresponds to $\delta = \pi$. Thus, with $\theta_i = 0$,

$$\delta = \frac{4\pi d}{\lambda_0} n_f - \pi = \pi$$

whence
$$d = \frac{\lambda_0}{2 n_f} = \frac{500 \times 10^{-9}}{2(1.455)} = 171.8 \text{ nm}$$

All of this means that such a film will be a very poor reflector of 500-nm light and therefore a good transmitter at that wavelength.

6.25. Suppose that a thin transparent coating of index n_f is layered on a glass substrate of index n_s and the whole business is immersed in air of index n_0. It can be shown that the reflection coefficients at the two interfaces are equal when

$$n_f = \sqrt{n_0 n_s}$$

How can this be used to produce an *antireflection* coating, as for example, on a camera lens? Design a single-layer antireflection coating for glass in air at 500 nm.

Consider the case where $n_0 < n_f < n_s$. Here there would be no relative phase shift for the two waves since both are reflected externally ($n_i < n_t$). Consequently, a film $\lambda_f/4$ thick (where $\lambda_f = \lambda_0/n_f$) would introduce a phase shift of $180°$ in the wave that traverses it perpendicularly twice. Thus, complete destructive interference results for the waves of that particular wavelength reflected at the two surfaces. Bear in mind that since that wavelength is effectively not reflected at all it must be transmitted quite efficiently.

In the above situation $n_0 = 1$, $n_s = 1.5$ and

$$n_f = \sqrt{(1)(1.5)} = 1.225$$

Furthermore, since $\lambda_f = \lambda_0/n_f$,

$$d = \frac{\lambda_f}{4} = \frac{\lambda_0}{4n_f} = 102 \text{ nm}$$

What we need, then, is a single layer 102 nm thick of a transparent material of index 1.225.

6.26. Envision a thin transparent film in the form of a wedge, as in Fig. 6-18. Describe the resulting fringe pattern, locating both maxima and minima, and find the corresponding film thickness and fringe separation. Incidentally, these are known as *fringes of equal thickness*.

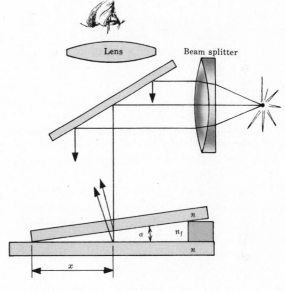

Fig. 6-18

The thickness at any point in the film, d, is approximated by $d = x\alpha$. Again the reflection of the wave at the top of the wedge is internal, while that at the bottom is external (assuming $n_f < n$); therefore there is a relative phase shift of π radians. Maxima in the irradiance occur when these two disturbances leave the wedge in phase, i.e. when

$$\delta = 2\pi m = \frac{4\pi n_f}{\lambda_0}d_m \pm \pi$$

where, under near-normal illumination ($\theta_i \approx \theta_t \approx 0$), $\cos\theta_t \approx 1$. Hence

$$(m \mp \tfrac{1}{2})\lambda_0 = 2n_f d_m$$

Choosing the plus sign (we could otherwise replace m by $m+1$), substituting $d_m = x_m\alpha$ and $\lambda_f = \lambda_0/n_f$, we obtain

$$x_m = \frac{m + \frac{1}{2}}{2\alpha}\lambda_f$$

The fringe separation Δx is then given by

$$\Delta x = x_{m+1} - x_m = \frac{\lambda_f}{2\alpha}$$

The thickness of the film at the mth maximum is simply

$$d_m = \left(m + \frac{1}{2}\right)\frac{\lambda_0}{2n_f}$$

It is clear that the interference pattern is a series of alternately bright and dark straight bands running parallel to the wedge's apex edge.

6.27. Two sheets of flat plate glass 25 cm long are separated at one end by a spacer $\frac{1}{4}$ mm thick, thereby forming a thin wedge-shaped air film. How many fringes per centimeter will be observed under normal illumination with red light ($\lambda_0 = 694.3$ nm) from a ruby laser?

We can determine α using $x = 25 \times 10^{-2}$, $d = 0.25 \times 10^{-3}$ and the fact that $\alpha = d/x$. Thus

$$\alpha = \frac{0.25 \times 10^{-3}}{25 \times 10^{-2}} = 10^{-3} \text{ rad}$$

Making use of the formula (see Problem 6.26)

$$\Delta x = \frac{\lambda_f}{2\alpha} = \frac{\lambda_0}{2n_f\alpha}$$

we get for the fringe separation

$$\Delta x = \frac{694.3 \times 10^{-9}}{2(1)10^{-3}} = 347.15 \times 10^{-6} \text{ m}$$

The number of fringes per centimeter is $(1/\Delta x)10^{-2}$ or 28.8.

6.28. A metal ring is dipped into a soap solution ($n_f = 1.340$) and held in a vertical plane so that a wedge-shaped film forms under the influence of gravity. At near-normal illumination with green light ($\lambda_0 = 514.53$ nm) from an argon laser, one can see 12 fringes per centimeter. Determine the wedge angle of the soap film.

The number of fringes per cm is given by

$$\frac{10^{-2}}{\Delta x} = 12$$

and so $\Delta x = 8.333 \times 10^{-4}$ m. Accordingly,

$$\alpha = \frac{\lambda_0}{2n_f \Delta x} = \frac{514.53 \times 10^{-9}}{2(1.340)(8.333 \times 10^{-4})} = 2.30 \times 10^{-4} \text{ rad}$$

6.29. (a) How does a spherical lens resting on an optical flat produce interference fringes? Describe the resulting pattern. (b) Write an expression for the fringe radii in terms of the radius of curvature of the lens, assuming near-normal illumination. (c) Show that the radii of the dark bands are proportional to the square roots of the integers. (d) How does this pattern differ from the Haidinger system of Problem 6.23?

(a) The setup in question is illustrated in Fig. 6-19. Once again, only two reflected beams will be considered, although multiple reflection does generate many more faint contributing waves. The two beams reflecting off the top and bottom of the circular wedge interfere constructively or destructively depending on the phase angle difference and, hence, on the O.P.D. Because of the circular symmetry of the film's thickness, the fringes in turn are symmetric about the central axis. These fringes of equal thickness are the famous *Newton's rings*.

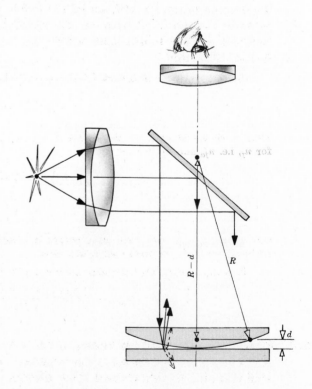

Fig. 6-19

(b) The relationship between the radius x of any one fringe and the radius of curvature of the lens R is simply

$$x^2 = R^2 - (R - d)^2 = 2Rd - d^2$$

For $R \gg d$ this is expressible as $x^2 = 2Rd$.

(c) There is again a phase difference of π radians due to reflection and, as in Problem 6.26, maxima occur when

$$\left(m + \frac{1}{2}\right)\lambda_0 = 2n_f d_m$$

The mth *bright ring* therefore has a radius of

$$x_m = \left[\left(m + \frac{1}{2}\right)\frac{\lambda_0}{n_f}R\right]^{1/2}$$

Minima occur when $\delta = (2m + 1)\pi$. Hence, since $\theta_i \approx 0$,

$$\delta = \frac{4\pi d_m}{\lambda_0}n_f \pm \pi = (2m + 1)\pi$$

or

$$2d_m = m\frac{\lambda_0}{n_f}$$

But from (b), $2d_m = x_m^2/R$ and so the radius of the mth dark ring is given by

$$x_m = \left(m\frac{\lambda_0}{n_f}R\right)^{1/2}$$

(d) As the order m increases, the radius of the fringe increases. This is just the opposite of what happens in the Haidinger pattern, where the central fringe has the highest order.

6.30. A positive lens with a radius of curvature of 20 cm rests on an optical flat and is illuminated normally with sodium D light, $\lambda_0 = 589.29$ nm. The gap between the two surfaces is then filled with carbon tetrachloride ($n = 1.461$). What is the ratio of the radius of the 23rd dark band before introducing the liquid to the radius after introducing the liquid?

The radius of the mth dark fringe is, by Problem 6.29,

$$x_m = \left(m\frac{\lambda_0}{n_f}R\right)^{1/2}$$

Thus if we subscript x_m so that it reads x_{ma} and $x_{m\ell}$ for air and liquid, respectively, and similarly for n_f, i.e. n_{fa} and $n_{f\ell}$, we can write

$$\frac{x_{ma}}{x_{m\ell}} = \left(\frac{m\lambda_0 R/n_{fa}}{m\lambda_0 R/n_{f\ell}}\right)^{1/2}$$

$$= \left(\frac{n_{f\ell}}{n_{fa}}\right)^{1/2}$$

and since $n_{fa} \approx 1$, the ratio quite generally equals $\sqrt{n_{f\ell}}$ regardless of the order m. The fringes shrink in towards the center as n_f increases.

Just out of curiosity let's determine x_{23} with the carbon tetrachloride in place:

$$x_{23} = \left[\frac{23(589.29 \times 10^{-9})0.2}{1.461}\right]^{1/2} = 1.36 \text{ mm}$$

6.31. Some dust separates the vertex of the lens from the optical flat in a Newton's rings setup. If that separation, Δ, is unknown, show that the radius of curvature of the lens can still be determined from directly observable quantities.

Following the development of Problem 6.29(c) dealing with the mth dark fringe, we now require that

$$2(d_m + \Delta) = m\lambda_f$$

Hence, with $2d_m = x_m^2/R$,

$$x_m^2 = (m\lambda_f - 2\Delta)R$$

Similarly, for the ℓth dark fringe,

$$x_\ell^2 = (\ell\lambda_f - 2\Delta)R$$

By subtracting these we arrive at

$$x_m^2 - x_\ell^2 = (m - \ell)\lambda_f R$$

which is independent of Δ. One need only count the orders and then measure the radii of any two fringes to determine R via

$$R = \frac{(x_m^2 - x_\ell^2)n_f}{(m - \ell)\lambda_0}$$

6.32. Describe the Newton's rings system seen in transmitted light as compared to the reflected pattern. Assume near-normal illumination.

The transmitted and reflected patterns are complementary in that a dark band in one corresponds to a bright band in the other. Where the surfaces are in contact and the film thickness goes to zero, a minimum exists in reflected light and a maximum in transmitted light. The two interfering waves in the case of reflection have fairly similar electric field amplitudes, yielding a well-defined fringe system of high visibility. Remember that each of these waves is reflected once at nearly the same angle.

In contradistinction, one of the two interfering transmitted beams is about twenty times weaker than the other, since it passes through the system after undergoing two reflections, while the stronger wave is not reflected at all in its sojourn. Because most of the light incident on the film passes through it the transmitted pattern has very low visibility, particularly at normal incidence.

6.33. Light having two constituent vacuum wavelengths of 650 nm and 520 nm shines at near-normal incidence on a positive spherical lens which rests horizontally on an optical flat. The radius of curvature of the lens is 85 cm and air pervades the gap between it and the flat. If the mth dark band at 650 nm is coincident with the $(m + 1)$th dark band at 520 nm, determine the band's diameter.

We are given that x_m for λ_1 equals x_{m+1} for λ_2, or,

$$\left(\frac{m\lambda_1 R}{n_f}\right)^{1/2} = \left[\frac{(m + 1)\lambda_2 R}{n_f}\right]^{1/2}$$

which simplifies down to

$$m\lambda_1 = (m + 1)\lambda_2 \quad \text{or} \quad m650 = (m + 1)520$$

from which $m = 4$. The radius of the fringe in question is then

$$x_4 = \left[\frac{4(650 \times 10^{-9})(0.85)}{1}\right]^{1/2} = 1.49 \text{ mm}$$

and the diameter is 2.97 mm.

6.5   AMPLITUDE-SPLITTING INTERFEROMETERS

In the several thin-film arrangements examined thus far, the two beams were coincident for most of the way from S to P. This present section treats amplitude-splitting interferometric devices which, through the use of additional mirrors, separate the two beams so that they may be manipulated individually prior to being recombined.

The *Michelson interferometer* (1881) of Fig. 6-20 is representative of this fairly large group of devices and is at the same time one of the most important optical instruments,

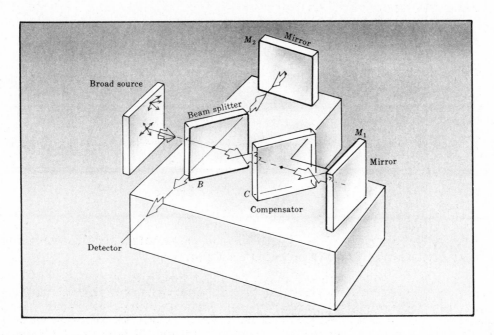

Fig. 6-20

certainly historically.   As illustrated, light from a broad source strikes the beam splitter and is amplitude-sheared into two waves.   One wave reflects off mirror M_2, passes through the beam splitter B and on to the detector.   The other wave reflects first off mirror M_1 and then off the beam splitter and on toward the detector.   The compensator plate is identical in thickness and orientation to the beam splitter.   Its function is to equalize the optical path lengths when M_1 and M_2 are the same distance from B.   In other words, each beam traverses the same total thickness of glass as indicated in Fig. 6-21.

In effect, the region between M_2 and M_1', the image of M_1 in the beam splitter, acts as an air film much like those of Section 6.4. Consequently, when M_2 and M_1 are precisely perpendicular, the enclosed air film is planar and of thickness d.   The pattern consists of a series of concentric circular fringes of equal inclination, as in Fig. 6-17. We can imagine two virtual sources, which are images of S, located on the central axis through M_2 and M_1'.   These generate circular fringes just as in Fig. 6-3.   Furthermore, when M_2 and M_1' are close together and inclined with respect to each other, the contained air film is a thin wedge and the parallel straight-line fringes of Fig. 6-18 result.

Keep in mind that a phase angle difference between the two waves is introduced as a result of internal and external reflections at the beam splitter.   The magnitude of the phase shift depends on the nature of the interfaces.   We shall assume the beam splitter to be a sheet of glass, in which case the phase shift is π rad.

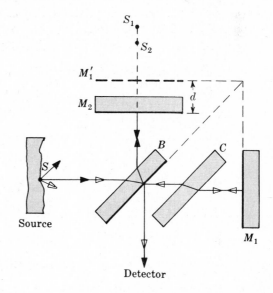

Fig. 6-21

SOLVED PROBLEMS

6.34. The mirrors in a Michelson interferometer are arranged so that by looking into the beam splitter one sees a 3 cm by 3 cm illuminated field corresponding to the area of overlap of the view of the two mirrors, M_1 and M_2. The field displays 24 vertical bright fringes under 600 nm illumination. Compute the angle by which the planes of M_1 and M_2 deviate from perpendicularity.

By Problem 6.26 the fringe spacing for a thin wedge can be written as $\Delta x = \lambda_f/2\alpha$. But from the data we have

$$\Delta x = \frac{3 \times 10^{-2}}{24} = 1.25 \text{ mm}$$

This means that

$$\alpha = \frac{\lambda_f}{2\,\Delta x} = \frac{600 \times 10^{-9}}{2(1.25 \times 10^{-3})} = 2.4 \times 10^{-4} \text{ rad} = 0.0138°$$

6.35. One of the two mirrors in a Michelson interferometer, say M_1, is generally movable. Assume then that the device is set up to display circular fringes and M_1 is slowly shifted so that M_1' approaches M_2, i.e. d is gradually decreased. The fringes sweep in toward the center of the field and, in particular, suppose 850 bright bands pass by when M_1 is displaced through 3.142×10^{-4} m. Assuming quasi-monochromatic illumination, determine the wavelength.

Going back to Problem 6.23 we can show, and logic would attest, that as the planar film increases in thickness by an amount $\lambda_f/2$, the order m increases by one. In other words, an increase in O.P.D. of λ_f results from an increase in d of $\lambda_f/2$, since the beam traverses the film twice. Thus, 850 bright bands correspond to a motion of the mirror through a distance of $\Delta d = 850(\lambda_f/2)$. Hence,

$$\Delta d = \frac{850\lambda_f}{2} = 3.142 \times 10^{-4} \text{ m}$$

and so $\lambda_f = 739.3$ nm.

6.36. Suppose that a Michelson interferometer is illuminated by a source emitting a doublet of vacuum wavelengths, λ_1 and λ_2. As one of the mirrors is moved, the fringes periodically disappear and then reappear. If a displacement Δd of the mirror causes a one-cycle variation in the visibility, write an expression for Δd in terms of $\Delta\lambda \equiv \lambda_1 - \lambda_2$, λ_1 and λ_2.

The fringe visibility will be high when the bright bands of λ_1 nearly overlie those of λ_2 and, of course, it will be poor when the bright bands of λ_1 coincide with the dark fringes of λ_2. The latter situation obtains when the O.P.D. at once equals a whole number of wavelengths of λ_1 and an odd number of half-wavelengths of λ_2. Thus,

$$\text{O.P.D.} = 2d = m_1\lambda_1 = \left(m_2 + \frac{1}{2}\right)\lambda_2$$

is the condition for a minimum in visibility. Hence

$$m_1 = \frac{2d}{\lambda_1} \qquad m_2 + \frac{1}{2} = \frac{2d}{\lambda_2}$$

and, subtracting, we can form

$$m_2 - m_1 + \frac{1}{2} = \frac{2d\,\Delta\lambda}{\lambda_1\lambda_2}$$

where $\Delta\lambda \equiv \lambda_1 - \lambda_2$. The integer $m_2 - m_1$ increases by 1 as d goes to $d + \Delta d$, i.e. at the next occurrence of minimum visibility. We then have

$$m_2 - m_1 + \frac{3}{2} = \frac{2(d + \Delta d)\,\Delta\lambda}{\lambda_1\lambda_2}$$

Subtracting the last two equations gives

$$1 \; = \; \frac{2\,\Delta d\,\Delta\lambda}{\lambda_1\lambda_2} \qquad \text{or} \qquad \Delta d \; = \; \frac{\lambda_1\lambda_2}{2\,\Delta\lambda}$$

6.37. A thin transparent sheet of index n and thickness L is inserted perpendicular to the beam axis in one arm of a Michelson interferometer. If the plate is withdrawn, determine the distance through which the mirror in that arm must be moved in order to duplicate the fringes observed prior to removal.

By inserting the sheet we change the optical path length in that local region from $1L$ for air to nL, and so the change in the O.P.D. introduced for two traversals of the plate is just $2(n-1)L$. If the mirror's motion, Δd, is to compensate for this shift it must introduce an identical O.P.D $= 2\,\Delta d$. Therefore,

$$2\,\Delta d \; = \; 2(n-1)L \qquad \text{or} \qquad \Delta d \; = \; (n-1)L$$

6.38. A thin sheet of fluorite (CaF$_2$) of index 1.434 is inserted normally in one arm of a Michelson interferometer. At $\lambda_0 = 589$ nm, 35 fringes are seen to be displaced. What is the thickness of the sheet?

As we found in Problem 6.37, $\Delta d = (n-1)L$, and, in this case, $2\,\Delta d = 35\lambda_0$. Accordingly,

$$35\,\frac{\lambda_0}{2} \; = \; (n-1)L$$

or
$$L \; = \; \frac{35\lambda_0}{2(n-1)} \; = \; \frac{35(589 \times 10^{-9})}{2(1.434-1)} \; = \; 23.75 \times 10^{-6}\ \text{m}$$

6.39. Looking into a Michelson interferometer one sees a dark central disc surrounded by concentric bright and dark rings. One arm of the device is 2 cm longer than the other and $\lambda_0 = 500$ nm. Determine (a) the order of the central disc, (b) the order of the sixth dark ring.

(a) As in Section 6.4 we find that for fringes of equal inclination with a phase shift from reflections of π radians

$$\delta \; = \; \frac{4\pi d \cos\theta}{\lambda_0} + \pi$$

The minima correspond to those angles θ_m which make $\delta = (2m+1)\pi$; i.e.

$$2d \cos\theta_m \; = \; m\lambda_0$$

In this formula m is the *order* of the dark fringe that subtends the half angle θ_m. This may or may not be the same as the fringe's *counting order*. To distinguish the two we shall write m_j for the order of the dark fringe whose counting order is j. Thus, the central disc has order m_0 and $\theta_{m_0} \approx 0$, whereupon the above formula simplifies to

$$2d \; = \; m_0\lambda_0$$

or
$$m_0 \; = \; \frac{2d}{\lambda_0} \; = \; \frac{2(2 \times 10^{-2})}{500 \times 10^{-9}} \; = \; 80{,}000$$

(b) The pth dark ring (not counting the zeroth) has order $m_p = m_0 - p$, since each consecutive minimum means an O.P.D. decreased by one wavelength. Specifically this tells us that the order of the sixth ring, m_6, is

$$m_6 \; = \; m_0 - 6 \; = \; 79{,}994$$

6.40. The *Jamin interferometer* (1856), as illustrated in Fig. 6-22, consists of two identical thick (~ 2.5 cm) planar glass plates, opaquely silvered on one side. Illumination is

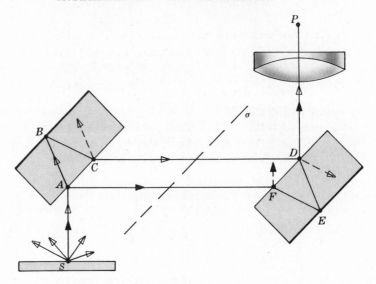

Fig. 6-22

from a broad source and is incident at about 45°. (a) Discuss the pattern resulting when the plates are perfectly parallel to each other. Are the interfering beams nearly equal or widely different in amplitude? (b) Describe the fringes that arise when the plates are tilted, in effect forming a narrow wedge whose apex edge is parallel to the dashed line σ.

(a) With the two plates set parallel, consider the incident ray arriving at A. The wave is amplitude-split with the weaker portion reflected off to point F. But the reflection at D is also weak and the two beams, having followed the routes $SAFED$ and $SABCD$ respectively, emerge with equal amplitudes. They are in phase, having undergone equivalent paths and reflections, and overlap and interfere constructively. This obtains for all rays; consequently, the field of view is uniformly bright.

(b) When the plates are tilted to form a wedge, we can consider Fig. 6-22 as being a cross-sectional view in any horizontal plane. All rays parallel to $\overline{SA}$, in any and all such horizontal planes, will arrive at the same central point, there to form a bright spot on the image plane of the lens. For any other horizontal ray entering at some other angle, constructive interference again results, and all such rays in all horizontal planes arrive at a string of points forming a horizontal bright line — the zeroth-order maximum. For rays not in horizontal planes, the two sheared segments undergo a relative phase shift. The final result is a series of horizontal, equally-spaced, bright and dark fringes.

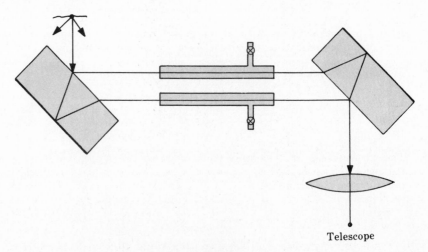

Telescope

Fig. 6-23

6.41. Suppose that two identical transparent chambers are introduced within a Jamin interferometer, as in Fig. 6-23. The chambers, bounded by optical plates, have inner
lengths of 23 cm and both contain argon with a refractive index of 1.000281. Under
sodium illumination ($\lambda_0 = 589$ nm) a fringe system appears and as the gas is pumped
out of one chamber the pattern changes. How many bright fringes will sweep by a
cross hair in the telescope?

One fringe will move past a fixed reference line whenever the O.P.D. changes by one wavelength. For a chamber of length L the change in O.P.D. is

$$\Delta\Lambda = (n-1)L$$

and the number of fringes N which this corresponds to is

$$N = \frac{(n-1)L}{\lambda_0} = \frac{(0.281 \times 10^{-3})(0.23)}{589 \times 10^{-9}} = 109.7$$

6.42. The *Twyman-Green interferometer* (1916) of Fig. 6-24 has become an extremely valu-
able tool in the testing of optical elements. The game is played by replacing mirror
M_2 with the element being examined and an additional mirror, such that the combination, if perfect, would be equivalent to M_2. With quasi-monochromatic light describe
how the interferometer works and how it differs from Michelson's device. Show how
it can be used to test a lens, a prism or an optical flat.

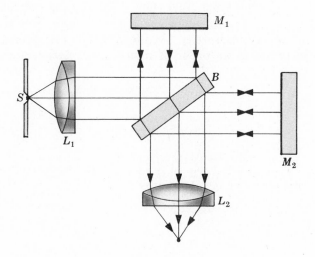

Fig. 6-24

The distinguishing characteristic of the Twyman-Green system is the use of collimated light.
When the Michelson interferometer is set up with a planar air film between M_1' and M_2 in Fig. 6-21,
page 143, a concentric ring pattern appears because light from the broad source contributes at all
angles via the $\cos\theta$ dependence. Contrarily, the collimated light of the Twyman-Green system
generates a uniformly illuminated field under similar conditions ($\theta = 0$).

Suppose now that M_2 is replaced by a test lens L_t and a perfect spherical mirror M_3, with the
latter centered on the focal point of the former as shown in Fig. 6-25(a). Plane waves enter the
lens, reflect off the mirror, and emerge as plane waves, provided the lens under test is perfect.
Any deviation from planar wavefronts is indicative of some imperfection in the element and is quite
discernable in the fringe pattern.

A prism can be studied in the manner indicated in Fig. 6-25(b), and an optical flat need only be
inserted in either arm in order for its imperfections to be observable.

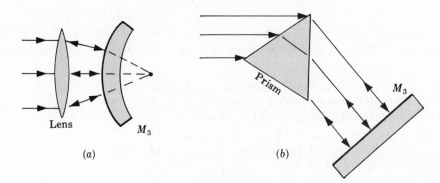

Fig. 6-25

6.6   COHERENCE

The concept of a monochromatic point source is a mathematical idealization. The very best source emits a finite range of wavelengths, albeit a narrow one. Indeed, as will be seen in Chapter 8, the mere fact that an emitter has not been turned on forever means that its signal must be polychromatic. Equivalently, we can picture the emission as composed of finite wave trains rather than an infinitely long, single-frequency sinusoid. Figure 6-26(b) is a representation of the field of a real wave. The time interval over which the phase is fairly constant is the *coherence time* Δt, a concept already introduced. The corresponding spatial interval $\Delta \ell = c\,\Delta t$ is known as the *coherence length*. Notice that for a monochromatic wave Δt is infinite but $\Delta \nu$ is zero. While Δt decreases the frequency bandwidth $\Delta \nu$ increases, and we can write as an order of magnitude relationship:

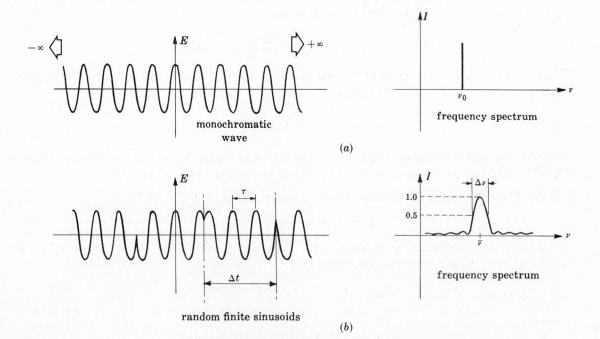

Fig. 6-26

$$\Delta\nu \sim \frac{1}{\Delta t}$$

When the frequency spectrum broadens, the length of the wave train in space decreases, as does the coherence time, and one speaks of this as a decrease in *longitudinal* or *temporal coherence*.

Imagine that we have a point source of quasi-monochromatic light S in Fig. 6-27. The electric fields at P_1 and P_2 will evidently be related when $\Delta\ell \gg \overline{P_1P_2}$, i.e. when a single wave train easily spans the gap between the points. In the same way, it is clear that the two fields must be totally unrelated when $\overline{P_1P_2} > \Delta\ell$, since different independent wave trains would be at P_1 and P_2 at any instant. For a given source the degree to which the time-varying fields are correlated is indicative of the degree of temporal coherence.

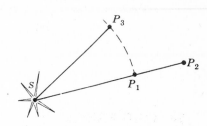

With a perfect point source the fields at P_1 and P_3 would be identical, i.e. totally correlated. But as the source is extended this correlation diminishes and one speaks of a decrease in *lateral spatial coherence*.

Fig. 6-27

Quite generally, the fringe visibility $\mathcal{V}$ in an interferometric system (see Problem 6.6) is a measure of the *degree of coherence*. Accordingly, by moving the mirror in a Michelson interferometer we can examine temporal coherence, and by varying the hole separation in Young's experiment we can measure spatial coherence. For example, consider P_1 and P_3 in Fig. 6-27 to correspond to the two pinholes S_1 and S_2 in Young's setup of Fig. 6-6, page 129. Since any real source would have a finite extent, suppose S is disc-shaped with a diameter D and at a distance R from the aperture plane it subtends an angle ϕ. By varying the separation a between S_1 and S_2, it can be shown that the fringes will disappear (this is the first zero value of $\mathcal{V}$) when $a = a_0$, where

$$a_0 = 1.22\frac{\overline{\lambda_0}}{\phi}$$

$\overline{\lambda_0}$ being the mean wavelength of the quasi-monochromatic source. And "good coherence" with a visibility of 0.88 or better results when

$$a \leq 0.32\frac{\overline{\lambda_0}}{\phi}$$

Actually the visibility drops from its maximum peak value to zero and then continues to oscillate between very small, ever diminishing secondary maxima and zero.

The irradiance distribution for Young's experiment can be shown to be

$$I = I_1 + I_2 + 2\sqrt{I_1 I_2}\,\mathrm{Re}\,\tilde{\gamma}_{12}(\tau)$$

very much like the result of Section 6.2. Here $\tilde{\gamma}_{12}(\tau)$ is the *complex degree of coherence*, whose magnitude specifies the coherence at P as follows:

$$|\tilde{\gamma}_{12}| = 1 \qquad \text{coherent limit}$$
$$|\tilde{\gamma}_{12}| = 0 \qquad \text{incoherent limit}$$
$$0 < |\tilde{\gamma}_{12}| < 1 \qquad \text{partial coherence}$$

The variable τ is the difference in time between $\overline{S_1P}/c$ and $\overline{S_2P}/c$. The irradiance can be recast in a form even more like that of Section 6.2:

$$I = I_1 + I_2 + 2\sqrt{I_1 I_2}\, |\tilde{\gamma}_{12}(\tau)|\cos\delta(\tau)$$

Note that in that earlier discussion we assumed S to be a monochromatic point source, which implies the coherent limit, $|\tilde{\gamma}_{12}(\tau)| = 1$.

SOLVED PROBLEMS

6.43. Determine the frequency bandwidth of white light. Compute the associated coherence length and coherence time.

According to Table 2-1, page 28, white light ranges from 384 THz to 769 THz, so the frequency bandwidth is

$$\Delta\nu = 769 \times 10^{12} - 384 \times 10^{12} = 385 \times 10^{12} \text{ Hz}$$

The relationship between $\Delta\nu$ and Δt,

$$\Delta\nu \sim \frac{1}{\Delta t}$$

then indicates a coherence time of $\Delta t = 2.597 \times 10^{-15}$ s and a coherence length $c\,\Delta t = 779$ nm. Observe that the wave trains of white light are roughly one wavelength long.

6.44. The spectral purity of a source can be appreciated via the quantity $(\Delta\nu)/\bar{\nu}$, the *frequency stability*. For example, a Hg^{198} low-pressure isotope lamp ($\lambda_{\text{air}} = 546.078$ nm) has a bandwidth of $\Delta\nu = 1000$ MHz. Compute the coherence length and coherence time of the light, as well as the frequency stability.

The coherence time is $\Delta t = 1/\Delta\nu = 10^{-9}$ s. The coherence length, $\Delta\ell = c\,\Delta t$, equals 29.9 cm. The mean wavelength was provided; hence, since $\Delta\lambda$ is small,

$$\bar{\nu} \approx \frac{c}{\bar{\lambda}} = \frac{2.9979 \times 10^8}{546.08 \times 10^{-9}} = 5.5 \times 10^{14} \text{ Hz}$$

Finally, the frequency stability is

$$\frac{\Delta\nu}{\bar{\nu}} = \frac{10^9}{5.5 \times 10^{14}} = 1.82 \times 10^{-6}$$

or about 2 parts per million.

6.45. Derive an expression for the coherence length of a wave in terms of the *linewidth* $\Delta\lambda_0$ corresponding to a frequency bandwidth of $\Delta\nu$.

The speed of the wave in vacuum is $c = \nu\lambda_0$. Hence, $\lambda_0 = c/\nu$ and differentiation yields

$$\Delta\lambda_0 = -\frac{c\,\Delta\nu}{\nu^2} = -\frac{\lambda_0^2\,\Delta\nu}{c}$$

The minus sign merely shows that the change in $\Delta\nu$ is opposite to the change in $\Delta\lambda_0$. Dropping the sign and using $\Delta\nu \sim 1/\Delta t$,

$$\Delta\lambda_0 \sim \frac{\lambda_0^2}{c\,\Delta t}$$

or, inasmuch as $\Delta\ell = c\,\Delta t$,

$$\Delta\ell \sim \frac{\lambda_0^2}{\Delta\lambda_0}$$

6.46. A Michelson interferometer is illuminated by red cadmium light with a mean wavelength of 643.847 nm and a linewidth of 0.0013 nm. The initial setting is for zero

O.P.D., i.e. $d = 0$. One mirror is then slowly moved until the fringes disappear — by how much must it be shifted? How many wavelengths does this correspond to?

The needed mirror displacement d is related to the coherence length and thence to the line-width. Consequently, we first compute $\Delta\ell$ from $\Delta\lambda_0 = 0.0013$ nm:

$$\Delta\ell \sim \frac{\lambda_0^2}{\Delta\lambda_0} = \frac{(643.847 \times 10^{-9})^2}{0.0013 \times 10^{-9}} = 31.89 \text{ cm}$$

Once the O.P.D., which equals $2d$, exceeds $\Delta\ell$ the fringes must vanish; hence

$$d \sim 15.94 \text{ cm}$$

This is equivalent to 24.76×10^4 waves.

6.47. Suppose the experiment described in Problem 6.46 were repeated with light ($\lambda_0 = 632.8$ nm) from a He-Ne laser having a frequency stability of 2 parts per 10^{10}. What mirror displacement would now be needed to cause the fringes to vanish?

We are given that

$$\frac{\Delta\nu}{\bar{\nu}} = \frac{2}{10^{10}} \quad \text{or} \quad \Delta\nu = 2 \times 10^{-10} \, \bar{\nu}$$

Hence

$$\Delta\ell = c\,\Delta t \sim \frac{c}{2 \times 10^{-10} \, \bar{\nu}} = \frac{\bar{\lambda}_0}{2 \times 10^{-10}}$$

Substituting the value of $\bar{\lambda}_0$, we obtain $\Delta\ell \sim 3164$ m. Consequently, $2d = \Delta\ell$ and $d \sim 1582$ m. (That's a long interferometer!)

6.48. Suppose that a quasimonochromatic source S ($\bar{\lambda}_0 = 589.3$ nm) consisting of a discharge lamp behind a 0.1-mm diameter circular hole in a screen illuminates Young's experiment. With a distance of 2 m between S and the plane of the two apertures, what is the separation a_0 of the apertures at which the fringes first vanish?

The visibility arising from a circular source becomes zero when

$$a_0 = 1.22 \frac{\bar{\lambda}_0}{\phi}$$

where ϕ is the angle subtended by the extended source. This angle is approximately equal to the ratio of the source diameter to its distance from Σ_a, i.e.

$$\phi = \frac{D}{R} = \frac{0.1 \times 10^{-3}}{2} = 5 \times 10^{-5} \text{ rad}$$

Hence,

$$a_0 = 1.22\left(\frac{589.3 \times 10^{-9}}{5 \times 10^{-5}}\right) = 14.38 \text{ mm}$$

Note that if Σ_a is moved farther from S, i.e. R is increased, a_0 increases. This implies that if you go far enough from even the largest source (e.g. a star) a_0 will be a measurable quantity.

6.49. If the sun's diameter subtends an angle of $\frac{1}{2}°$ at the earth's surface, determine the *area of coherence*, i.e. the circular area into which one could introduce a set of apertures and obtain good clear fringes. Assume a mean wavelength of 550 nm.

The area of "good coherence" A_c is evidently expressible as

$$A_c = \pi\left(\frac{a}{2}\right)^2 = \pi\left(\frac{0.32 \,\bar{\lambda}_0}{2\phi}\right)^2$$

In radians $\phi = 0.0087$ and so

$$A_c = \pi \left[\frac{0.32(550 \times 10^{-9})}{2(0.0087)} \right]^2 = 3.2 \times 10^{-10} \text{ m}^2$$

6.50. Figure 6-28 depicts a quasimonochromatic line source S illuminating Young's apparatus. Write an expression for a_0, the aperture separation which first results in $\mathcal{V} = 0$ (i.e. for which the fringes first vanish).

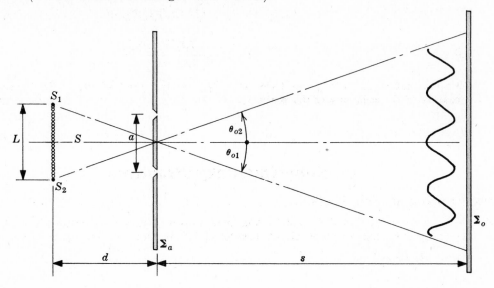

Fig. 6-28

Let us examine the waves emanating from the independent points S_1 and S_2 at either end of S. The cosine fringes arising independently from S_1 and S_2 are centered at θ_{o1} and θ_{o2} on Σ_o, respectively. These angles are approximately given by $\theta_{o2} = L/2d$ and $\theta_{o1} = -L/2d$, provided that $d \gg L$. The fringe system first disappears when the first minimum of one pattern overlies the central maximum of the other. As we saw in Fig. 6-7, the first pair of minima occur at distances from the center of

$$y = \pm \frac{\lambda_0 s}{2a}$$

or since $y/s = \theta$, they are found for S_1 at angles

$$\theta_{o1} \pm \frac{\lambda_0}{2a}$$

The separation between fringes is very large when a is very small and as a increases, the first minimum of S_1 moves toward the zeroth maximum of S_2 until they meet when $a = a_0$, whereupon

$$\theta_{o2} = \theta_{o1} + \frac{\lambda_0}{2a_0}$$

Hence, $\dfrac{L}{2d} = -\dfrac{L}{2d} + \dfrac{\lambda_0}{2a_0}$ or $a_0 = \dfrac{\lambda_0 d}{2L}$

6.51. Compose an expression giving the visibility of Young's experiment in terms of the *degree of coherence* $|\tilde{\gamma}_{12}|$, i.e. the magnitude of the complex degree of coherence $\tilde{\gamma}_{12}$. (The little squiggle on top of the γ_{12} is just to remind us that it's complex.)

The visibility

$$\mathcal{V} = \frac{I_{max} - I_{min}}{I_{max} + I_{min}}$$

can be evaluated by using

$$I = I_1 + I_2 + 2\sqrt{I_1 I_2}\,|\tilde{\gamma}_{12}(\tau)|\,\cos\delta(\tau)$$

Thus

$$I_{max} = I_1 + I_2 + 2\sqrt{I_1 I_2}\,|\tilde{\gamma}_{12}(\tau)|$$

$$I_{min} = I_1 + I_2 - 2\sqrt{I_1 I_2}\,|\tilde{\gamma}_{12}(\tau)|$$

and

$$\mathcal{V} = \frac{4\sqrt{I_1 I_2}\,|\tilde{\gamma}_{12}(\tau)|}{2(I_1 + I_2)}$$

Observe that when $I_1 = I_2$, we have $\mathcal{V} = |\tilde{\gamma}_{12}|$, so that the visibility of the fringes is a direct measure of the coherence of the waves.

Supplementary Problems

INTERFERENCE OF TWO WAVES

6.52. Two point sources, S_1 and S_2, emitting 3-m radiowaves in phase are separated by 3 m. How far must one move out along the perpendicular bisector of $\overline{S_1 S_2}$ to encounter a minimum in the irradiance?

Ans. 2.25 m

6.53. Two equal-amplitude point sources radiating $\mathcal{P}$-states at the same wavelength are locked in phase. Under what circumstances will the irradiance measured on a distant screen equal the sum of the constituent irradiances?

Ans. When $\mathbf{E}_{01} \cdot \mathbf{E}_{02} = 0$, i.e. when the two planes of vibration are normal.

6.54. Two point sources of microwaves, each having a frequency of 10^9 Hz, are separated by 60 cm and transmit in phase. Describe the radiation pattern.

Ans. Lobes occur at $\theta = 0$, $30°$, $90°$, $150°$, $180°$, $210°$, $270°$ and $330°$ measured from the normal to the line of centers.

6.55. Show that even though a fringe pattern exists for two coherent point sources of equal wavelength, energy is conserved. In other words, verify that when the source separation $a \gg \lambda_0$, I_{12} averaged over a large region of space is zero. How do things change when $a < \lambda_0$?

Ans. $\cos\delta$ spatially averages to zero when $a \gg \lambda_0$. When $a < \lambda_0$ the two sources behave much like a single source of double the strength of either one.

6.56. Write an expression for the radiation pattern of two equal-strength point sources, i.e. the irradiance as a function of θ, if the sources are separated by a distance a, have the same frequency, and are initially out of phase ($\varepsilon_1 - \varepsilon_2 \neq 0$).

Ans. $I(\theta) = 4I_0 \cos^2\left(\dfrac{\pi}{\lambda_0} a \sin\theta + \dfrac{\varepsilon_1 - \varepsilon_2}{2}\right)$

6.57. What would happen to the forward lobe ($\theta = 0$) of Problem 6.54 if a relative phase difference of $30°$ were introduced between the two sources?

Ans. It would rotate through $2°23'$. This, incidentally, is a means of steering an antenna array without actually rotating it.

6.58. What phase angle difference should be introduced between S_1 and S_2 of Problem 6.8 in order to rotate the lobe pattern by $20°$?

   Ans. $\varepsilon_1 - \varepsilon_2 = 61.56°$

WAVEFRONT-SPLITTING INTERFEROMETERS

6.59. If β is the angular separation of the sources S_1 and S_2 as seen from a point P on the plane of observation in Young's experiment, show that the fringe separation is $\Delta y = \lambda_0/\beta$.

6.60. Two narrow parallel slits illuminated by yellow helium light ($\lambda_0 = 5875.618$ Å; 1 Å $= 10^{-10}$ m) are found to produce fringes with a separation of 0.50 mm on a screen 2.25 m away. What is the distance between the slits?

   Ans. $a = 2.64$ mm

6.61. Write an expression for the angular separation ($\angle S_1 P_o S_2$) of the two virtual sources as seen from P in Fig. 6-10, page 132, in terms of R, d, and α.

   Ans. $(\angle S_1 P_o S_2) = 2R\alpha/(R + d)$

6.62. A Fresnel double mirror having an angle $\alpha = 0.667°$ has its line of intersection 0.1 m from the source and 1 m from the plane of observation. For light of wavelength 600 nm locate the seventh bright fringe with respect to the central axis.

   Ans. $y_7 = 1.98$ mm

6.63. A point source of wavelength 500 nm is one meter away from a Fresnel biprism which, in turn, is 5 meters from an observing screen. If the prism index is 1.5 and the resulting fringe separation is 0.5 mm, determine the prism angle.

   Ans. $\alpha = 0.343°$

6.64. Obtain an expression for the fringe separation generated by a Fresnel biprism of index n_g immersed in a liquid of index n_ℓ. Begin with the fact that

$$\frac{n_g}{n_\ell} = \frac{\sin \frac{1}{2}(\delta_m + \alpha)}{\sin \frac{1}{2}\alpha}$$

and assume monochromatic point source illumination.

   Ans. $\Delta y = \dfrac{n_\ell(R + d)\lambda_0}{2R(n_g - n_\ell)\alpha}$

6.65. Suppose that a thin transparent plate of index n is inserted in the direct beam of a Lloyd's mirror setup. (*a*) Describe what would happen to the central dark band. (*b*) How might you actually locate this central fringe?

   Ans.   (*a*) An O.P.D. of zero corresponds to the central band and it would move upward by $(n - 1)d/\lambda_0$ fringes, where d is the plate thickness. (*b*) Use white light: the central fringe will show no coloration.

6.66. A line source of quasi-monochromatic light ($\lambda_0 = 500$ nm) generates a fringe system on a screen 2 m away when the source is just above and parallel to a polished glass plate. If the bright fringes have a peak-to-peak separation of 0.667 mm, how high is the source above the plate?

   Ans. $a/2 = 0.75$ mm

6.67. Figure 6-29 shows a point source in front of a convex lens which has been cut in half and separated slightly (*Billet's split lens*). Discuss how it produces interference fringes.

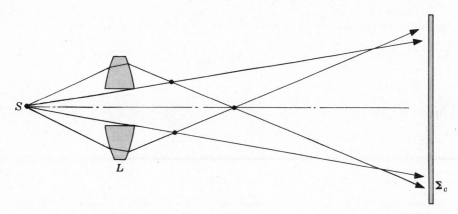

Fig. 6-29

Ans.   The two real images of S serve as coherent in-phase emitters and the whole region where the beams overlap is occupied by fringes.

AMPLITUDE SPLITTING BY THIN FILMS

6.68. A ray of light of wavelength 500 nm is incident at 30° on a film of thickness 0.002 mm and index 1.5 immersed in air. Determine the phase angle difference for the two reflected rays.

Ans. $\delta = 22.63\,\pi \pm \pi$, which is equivalent to $1.63\,\pi$

6.69. Show that for equal-inclination fringes maximum irradiance corresponds to

$$d \cos\theta_t \;=\; (2m+1)\frac{\lambda_f}{4}$$

and minimum irradiance to

$$d \cos\theta_t \;=\; 2m\frac{\lambda_f}{4}$$

where $m = 0, 1, 2, \ldots$.

6.70. A parallel glass plate of index 1.5, 2.5 mm thick, generates a concentric ring system of fringes under normal illumination at a vacuum wavelength of 750 nm. Determine the order of the central fringe. Is it a maximum, minimum or neither?

Ans. $m = 10{,}000$; a minimum $(d = 2m\lambda_f/4)$

6.71. Design a single-layer antireflection coating for a Fabulite (SrTiO$_3$) plate ($n_s = 2.409$) in air at $\lambda_0 = 589$ nm. Assume normal incidence.

Ans. $n_f = 1.552$, $d = 94.88$ nm

6.72. Magnesium fluoride ($n_f = 1.38$) is a common material used for single-layer antireflection coatings on glass. How thick should the thinnest coating be for light of vacuum wavelength 589 nm assuming near-normal incidence?

Ans. $d = 106.7$ nm

6.73. A thin wedge-shaped film of methyl alcohol ($n_f = 1.3290$) is formed between two flat plates of glass. Yellow sodium light of vacuum wavelength 589 nm falls nearly normally on the film, generating fringes separated by 0.2 mm. Determine the wedge angle in degrees.

Ans. $\alpha = 0.0635°$

6.74. How thick is the alcohol film of Problem 6.73 at the location of the fourth maximum counting from the apex side of the wedge? How far from the apex is it?

Ans. $d_3 = 7.76 \times 10^{-7}$ m, $x_3 = 0.7$ mm.   (The apex marks the middle of a dark fringe.)

6.75. Imagine that you have a Newton's rings arrangement and a bottle of unknown clear liquid. Measure the diameter of any ring, for example the fourth bright one. Now pour the liquid into the gap and remeasure the diameter. If it goes from 2.52 cm to 2.21 cm, calculate the liquid's index.

Ans.   1.30

6.76. A Newton's rings apparatus under quasi-monochromatic, normal illumination at 550 nm displays a bright central fringe. If the radius of the fifth dark band is 1.414 cm and that of the eighty-fifth dark band is 1.871 cm, determine the radius of curvature of the convex lens.

Ans. $R = 3.41$ m

6.77. Suppose the optical flat in Newton's apparatus is replaced by a concave spherical surface with a radius of curvature R_2. If a positive lens with a radius of curvature R_1 is placed in contact with it $(R_2 > R_1)$, write an expression for the radius of the mth dark fringe, assuming normally incident, quasi-monochromatic illumination.

Ans. $x_m = [R_1 R_2 m \lambda_f / (R_2 - R_1)]^{1/2}$

AMPLITUDE-SPLITTING INTERFEROMETERS

6.78. A Michelson interferometer displays fringes of equal inclination for which

$$2d \cos \theta_m = m \lambda_0$$

describes a dark band of order m whose radius subtends an angle θ_m. Prove that as the two arm lengths approach each other, a given fringe will collapse into the center.

6.79. A Michelson interferometer is illuminated by the sodium doublet with vacuum wavelengths of 5895.923 Å and 5889.953 Å. One mirror is moved continuously and the fringe pattern fades in and out periodically. Compute the mirror travel corresponding to a shift in visibility from maximum to minimum.

Ans. $\Delta d/2 = 0.1454$ mm

6.80. By how much must one of the mirrors in a Michelson interferometer be moved if a photocell detector is to count off 10,000 bright bands as they sweep by? The light source is the well-investigated orange-red line of the krypton 86 isotope at a vacuum wavelength of 605.7802105 nm.

Ans. $\Delta d = 3.0289$ mm

6.81. A Michelson interferometer is adjusted to display a circular pattern. Assuming the central or zeroth fringe to be a minimum, show that the half angle subtended by the pth dark fringe, i.e. θ_{m_p}, can be computed from the expression

$$2d(1 - \cos \theta_{m_p}) = p \lambda_0$$

Arrive at a small-angle approximation of θ_{m_p}.

Ans. $\theta_{m_p} \approx (p\lambda_0/d)^{1/2}$

6.82. Compute the half angle subtended by the 15th dark ring surrounding a dark central disc as viewed in a Michelson interferometer. The two mirror-supporting arms differ in length by 1 cm and $\lambda_0 = 400$ nm.

Ans. $1°24'$

6.83. A Jamin interferometer arranged to examine gases has a test chamber 35 cm long in each arm. One of these cells is left evacuated while the other is gradually filled with the sample gas. If, with $\lambda_0 = 650$ nm, 75 bright fringes pass a given point on the field of view as the cell is filled, what is the refractive index of the gas?

Ans. $n = 1.000139$

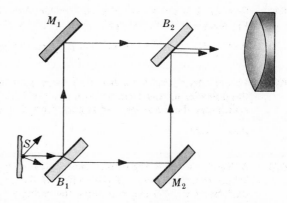

6.84. The *Mach-Zehnder interferometer* (1891) pictured in Fig. 6-30 is yet another amplitude-splitting device. It is formed of two mirrors and two beam splitters. How does it work? Is it similar to any previously examined system? To what purpose might it be applied?

Fig. 6-30

Ans. M_1 is tilted slightly with respect to M_2, B_1 and B_2 to cause a wedge pattern. The setup is very much like the Jamin interferometer. It is quite useful for determining large-volume nonuniformities, as for example in a wind tunnel.

COHERENCE

6.85. In 1963 Jaseija, Javan and Townes attained a short-term frequency stability of roughly 8 parts per 10^{14} with a He-Ne gas laser at $\lambda_0 = 1153$ nm. Compute the coherence time and coherence length.

Ans. $\Delta t = 4.8 \times 10^{-2}$ s, $\Delta \ell = 14.4 \times 10^6$ m

6.86. Roughly what is the linewidth of a hypothetical source if it has an uninterrupted transition time of 10^{-8} s (i.e. assume $\Delta t = 10^{-8}$ s)? Compute the coherence length as well. The vacuum wavelength equals 650 nm.

Ans. $\Delta \lambda_0 = 1.41 \times 10^{-13}$ m, $\Delta \ell = 2.99$ m

6.87. Red light ($\lambda_0 = 650$ nm) emerging from an ordinary filter is comprised of wave trains about $50\lambda_0$ in length. What is the linewidth, $\Delta\lambda_0$, passed by the filter? Determine the maximum range over which the mirror in a Michelson interferometer can be moved before the fringes in this case become unobservable.

Ans. $\Delta\lambda_0 = 13$ nm, 0.01625 mm

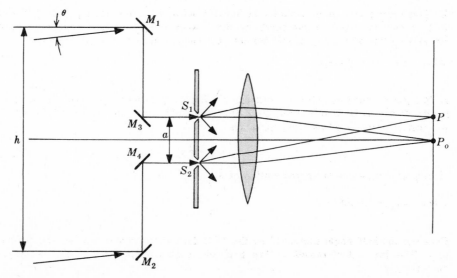

Fig. 6-31

6.88. Show that the inverse of the frequency stability of a source is of the order of magnitude of the number of wavelengths in an emitted wave train.

6.89. A mythical optical filter has a passband of only 1.5 nm centered at 550 nm. With white light incident, compute the coherence length of the emerging beam and the number of wavelengths in the wave train.

    Ans. $\Delta\ell = 2.02 \times 10^{-4}$ m, $\Delta\ell/\lambda_0 = 366.7$

6.90. Figure 6-31, page 157, shows a device known as a *Michelson stellar interferometer*. The mirrors M_1 and M_2 are movable, thereby varying h. The apparatus is directed toward a star and h is adjusted until at some value h_0 the fringes vanish. When Michelson did this for the star Betelgeuse he found h_0 to be 121 inches at $\lambda_0 = 570$ nm. What is the angular diameter of the star?

    Ans. $\phi = 2.26 \times 10^{-7}$ rad $= 1.296 \times 10^{-5}$ deg

6.91. V. P. Chebotayev reported (1973) that the frequency of a methane-stabilized He-Ne laser had been maintained to 6 parts in 10^{16} for a duration of 100 seconds. Assuming the same wavelength as in Problem 6.85, what progress occurred in extending the coherence time during the ten-year period?

    Ans. $\Delta t(1973) = 6.4$ s, $\Delta t(1963) = 4.8 \times 10^{-2}$ s

Chapter 7

Diffraction

7.1 INTRODUCTION

The essential feature of *diffraction* is a deviation from rectilinear propagation arising when a wave is obstructed in some way. Roughly speaking, the wave will bend around an obstacle, thereby forming fringe patterns in what might otherwise be assumed to be the region of uniform geometrical shadow. There is little or no real distinction between this phenomenon and the phenomenon of interference; both are the product of the superposition of several wavelets.

The simplest technique for analyzing diffraction problems, and the one we shall adopt, is based on the *Huygens-Fresnel principle* (see Section 5.8), which we restate as follows: *Every point on a wavefront serves as a source of spherical secondary wavelets of the same frequency as the primary wave. The optical field at any point beyond an obstruction is the superposition of all such wavelets reaching that point.* This principle can be mathematically obtained from the differential wave equation.

Envision an aperture in an opaque screen being illuminated by normally incident plane waves from a He-Ne laser and suppose that the resulting shadow pattern is examined on a piece of white paper. Quite near the screen a bright spot matching the configuration of the aperture will appear on the paper. Moving it farther away will cause a fine fringe pattern surrounding the edges of the spot to become evident. With the paper still farther from the aperture, an extensive system of fringes evolves which is roughly restricted to the geometrical projection of the hole. The pattern continues to vary and spread out as the observing screen moves away from the aperture. This is spoken of as *Fresnel* or *near-field diffraction*. Ultimately the irradiance distribution will metamorphose into a symmetric, vastly extended fringe system bearing practically no resemblance to the aperture. Beyond a certain distance any pattern changes will become imperceptible and, except for a continuing increase in size, the fringe system will appear unaltered. This is *Fraunhofer* or *far-field diffraction* and, being a limiting case, is somewhat simpler to deal with mathematically than is near-field diffraction.

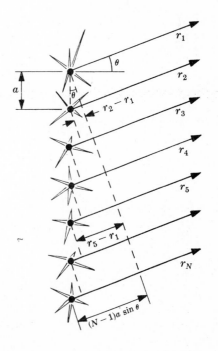

Fig. 7-1

7.2 RADIATION FROM A COHERENT LINE SOURCE

Consider an array of N coherent identical oscillators, as in Fig. 7-1. Assume that the plane of observation is so far away that the rays meeting on it at some point P are essentially parallel. All the waves have nearly the same amplitude at P, so the resultant field is the real part of

$$E_P = E_0(r)e^{i(kr_1-\omega t)} + E_0(r)e^{i(kr_2-\omega t)} + \cdots + E_0(r)e^{i(kr_N-\omega t)}$$

$$= E_0(r)e^{-i\omega t}e^{ikr_1}[1 + e^{ik(r_2-r_1)} + e^{ik(r_3-r_1)} + \cdots + e^{ik(r_N-r_1)}]$$

From the figure we see that the phase differences are given by $\delta = ka\sin\theta = k(r_2-r_1)$, $2\delta = k(r_3-r_1)$, etc. Hence,

$$E_P = E_0(r)e^{-i\omega t}e^{ikr_1}[1 + (e^{i\delta}) + (e^{i\delta})^2 + (e^{i\delta})^3 + \cdots + (e^{i\delta})^{N-1}]$$

But the geometric series in brackets is known to equal

$$e^{i(N-1)\delta/2}\left[\frac{\sin(N\delta/2)}{\sin(\delta/2)}\right]$$

the use of which leads to

$$E_P = E_0(r)e^{-i\omega t}e^{i[kr_1+(N-1)\delta/2]}\left[\frac{\sin(N\delta/2)}{\sin(\delta/2)}\right]$$

One further simplification is possible if we define R as the distance from the center of the array to P:

$$R \equiv \frac{1}{2}(N-1)a\sin\theta + r_1$$

and that yields

$$E_P = E_0(r)e^{i(kR-\omega t)}\left[\frac{\sin(N\delta/2)}{\sin(\delta/2)}\right]$$

This is a wave from the center of the line of emitters whose amplitude is modulated by the bracketed function of θ. The corresponding irradiance, which is proportional to $EE^*/2$, is just

$$I = I_0\frac{\sin^2(N\delta/2)}{\sin^2(\delta/2)}$$

We can think of this situation as relating to a linear array of antennas either transmitting or receiving. Practically, we shall make extensive use of this result when we examine diffraction at a narrow slit.

Suppose now that we let N become exceedingly large and the separation between neighboring point sources vanishingly small. With $\mathcal{E}_0$ as the individual *source strength*, each emitted wave is expressible as

$$E = \frac{\mathcal{E}_0}{r}\sin(\omega t - kr)$$

A segment of the array, Δy_i, as shown in Fig. 7-2, will contain $\Delta y_i(N/D)$ emitters and there are assumed to be M such segments. Then the ith segment contributes a field at P of

$$E_i = \left(\frac{N\,\Delta y_i}{D}\right)\frac{\mathcal{E}_0}{r_i}\sin(\omega t - kr_i)$$

If we let N approach infinity and $\mathcal{E}_0$ approach zero, we can define a *source strength per unit length* as

$$\mathcal{E}_L \equiv \frac{1}{D}\lim_{N\to\infty}(\mathcal{E}_0 N)$$

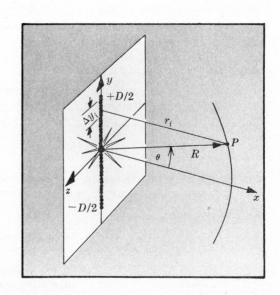

Fig. 7-2

whereupon the total field at P becomes

$$E_P = \sum_{i=1}^{M} \frac{\mathcal{E}_L}{r_i} \sin(\omega t - kr_i)\, \Delta y_i$$

For a continuous line source Δy_i is infinitesimal, M goes to infinity and

$$E_P = \mathcal{E}_L \int_{-D/2}^{+D/2} \frac{\sin(\omega t - kr)}{r}\, dy$$

in which r is a function of y.

We now have descriptions of linear coherent arrays which are either constituted of a finite number of discrete emitters or are continuous in nature.

SOLVED PROBLEMS

7.1.   (a) Write an expression for the irradiance as an explicit function of θ very far from a line source consisting of N identical emitters.   (b) For what angles θ_m will maxima appear?   (c) Compute I_{max} in terms of the irradiance of a single emitter I_0.

   (a) Since $\delta = ka \sin\theta$,

$$I = I_0 \frac{\sin^2(N\delta/2)}{\sin^2(\delta/2)} = \frac{\sin^2[(Nka/2)\sin\theta]}{\sin^2[(ka/2)\sin\theta]}$$

   (b) The pattern consists of strong principal maxima which are separated by series of weak subsidiary maxima. The principal peaks occur in directions for which the waves in Fig. 7-1 are all in phase. In other words, $\delta = 2m\pi$ where $m = 0, \pm1, \pm2$, etc. Knowing that $\delta = ka \sin\theta$, and of course $k = 2\pi/\lambda$, maxima must occur at angular directions θ_m satisfying

$$a \sin\theta_m = m\lambda$$

   (c) To find I_{max} or equivalently $I(\theta_m)$ we must evaluate

$$\frac{\sin(N\delta/2)}{\sin(\delta/2)}$$

   when $\delta = 2m\pi$. From L'Hospital's rule this ratio is equal to N for these values of δ, and consequently

$$I_{max} = N^2 I_0$$

   This is not surprising, for the waves are all in phase at these orientations and the resultant amplitude must then be NE_0.

7.2.   Atoms are separated from each other by roughly 1 Å, i.e. 10^{-10} m, and radiate in the visible at around 5000 Å. What can be expected of a line source of this sort, insofar as $I(\theta)$ is concerned? How is this result related to specular reflection?

   By Problem 7.1, the equation

$$a \sin\theta_m = m\lambda$$

specifies the angular orientations of the principal maxima. For visible atomic radiation λ is five thousand times greater than a. Moreover, $\sin\theta_m \leqq 1$, which means that

$$a \sin\theta_m \ll m\lambda$$

for all m other than $m = 0$. In other words, there is only the possibility of a zeroth-order principal maximum (at $\theta_0 = 0$ and π) whenever $\lambda > a$.

   This explains why the angle of incidence equals the angle of reflection. A wave arriving at an interface at $\theta_i \neq 0$ stimulates one surface atom after the next as it sweeps by, with the effect that each atomic emitter lags the preceding one by a fixed phase. This shifts the reradiated wave, i.e. the central lobe, by an angle θ_r away from the normal to the line source, where, of course, $\theta_i = \theta_r$. (See Problem 7.57.)

7.3. (a) Locate the zeros of irradiance for a coherent linear array of N emitters. (b) For the case of $N = 20$, explain the result of (a) from the standpoint of how the individual waves interact, assuming a distant point of observation.

(a) Fixing the orientations of the minima is easily done from the relation

$$I = I_0 \frac{\sin^2(N\delta/2)}{\sin^2(\delta/2)}$$

Evidently, $\sin(N\delta/2) = 0$ whenever $N\delta/2 = \pm\pi, \pm 2\pi, \pm 3\pi$, etc. Another way to say this is that the minima obtain when

$$(N/2)ka \sin\theta_{m'} = m'\pi$$

where $m' = \pm 1, \pm 2, \pm 3$, and so on, omitting integer multiples of N which would correspond to the principal maxima.

(b) For the first minimum, $m' = \pm 1$ and

$$10ka \sin\theta_1 = \pm\pi$$

But from Fig. 7-1 we can see that $10ka \sin\theta_1$ is precisely the difference in phase between the 1st and 11th waves. Consequently these two disturbances, being out of phase by π rad at P, must cancel each other. Moreover, the 2nd and 12th waves also differ by π, as do the 3rd and 13th, etc., all the way down to the 10th and 20th. Obviously each such pair contributes nothing and $I(\theta_1) = 0$.

Similarly, the second minimum on either side of $\theta = 0$ occurs when $m' = \pm 2$ and

$$5ka \sin\theta_2 = \pm\pi$$

Here the 1st and 6th waves cancel, as do the 11th and 16th. Thus waves 1–5 cancel 6–10, and 11–15 cancel 16–20.

Notice that the process would not have been nearly so obvious if N were an odd integer.

7.4. Assuming that the separation a between each of four identical emitters in a coherent line source is greater than λ, make a rough plot of $I(\theta)$. Compare this with an array of eight equally spaced sources. In both cases we require that the plane of observation be very far away.

The locations of the principal maxima are given by

$$a \sin\theta_m = m\lambda$$

quite independently of N. The zeros of irradiance are determined by way of

$$Na \sin\theta_{m'} = m'\lambda$$

as discussed in Problem 7.3.

Thus, for $N = 4$, principal maxima correspond to $(a/\lambda)\sin\theta$ equal to $0, \pm 1, \pm 2$, etc., while minima occur where $(a/\lambda)\sin\theta$ equals $\pm 1/4, \pm 1/2, \pm 3/4, \pm 5/4$, etc. Between any two consecutive irradiance zeros there must be a maximum. Figure 7-3 shows the pattern (with little concern for the heights of the subsidiary maxima). When $N = 8$ the positions of the main maxima are unchanged. There are again $N-1$ minima between pairs of major peaks, where $(a/\lambda)\sin\theta$ equals $\pm 1/8, \pm 1/4, \pm 3/8, \pm 1/2, \pm 5/8, \pm 3/4, \pm 7/8, \pm 9/8$, etc.

7.5. Make a rough polar sketch of the radiation pattern at a great distance from four coherent emitters spaced $a = \lambda/2$ apart. Compare your results with Fig. 6-5, page 128.

For $N = 4, a = \lambda/2$, we have

$$\delta = ka \sin\theta = \pi \sin\theta$$

and

$$I(\theta) = I_0 \frac{\sin^2(2\pi \sin\theta)}{\sin^2(\tfrac{1}{2}\pi \sin\theta)}$$

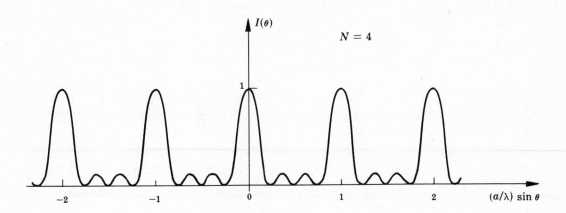

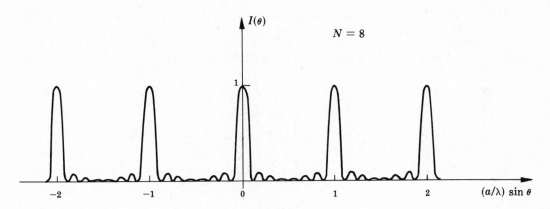

Fig. 7-3

Principal maxima occur only when $m = 0$, i.e. when $\theta = 0$ and $180°$.   Minima arise when

$$\sin\theta_{m'} = \frac{m'}{2}$$

that is, $\theta_1 = \pm30°$ and $\theta_2 = \pm90°$.   Figure 7-4 is a sketch summarizing all of this.   Keep in mind that the pattern is three-dimensional; the lobes shown are just a cross-sectional view in any plane containing the array.

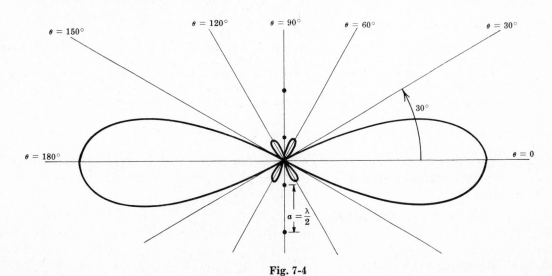

Fig. 7-4

7.3 FRAUNHOFER DIFFRACTION BY ONE AND TWO NARROW SLITS

One of the most practical arrangements for observing Fraunhofer diffraction is illustrated in Fig. 7-5. Light from a monochromatic point source S is collimated by lens L_1, diffracted at Σ_a and brought to a focus on Σ_o by lens L_2. In effect the light entering and leaving the aperture can be thought of as consisting of plane waves. The lenses allow S and Σ_o to both be fairly nearby Σ_a and still generate the same far-field pattern which would prevail were they both very far removed from the aperture.

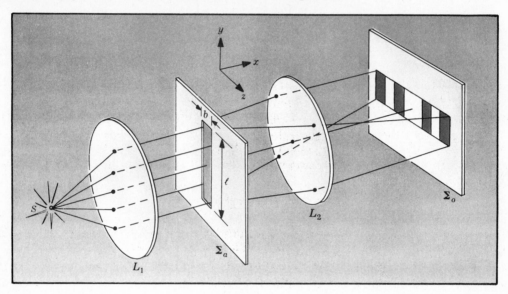

Fig. 7-5

The diagram shows a long narrow slit in Σ_a which appears in somewhat more detail in Fig. 7-6. The Huygens-Fresnel principle suggests that a normally incident plane wave effectively fills the aperture with in-phase point sources. The differential strip (dz by ℓ) can therefore be envisioned as a coherent line source. According to Section 7.2, each segment dy of this line source makes a contribution to the far field of

$$dE_P = \frac{\mathcal{E}_L}{R} \sin(\omega t - kr)\, dy$$

where r in the amplitude has been approximated by R, the latter being assumed constant. However, the sine function is far more sensitive than the amplitude to variations in distance; therefore, since

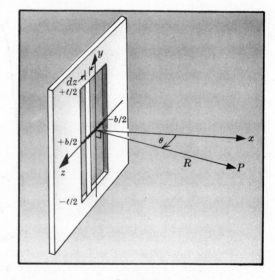

Fig. 7-6

$$r = R - y \sin\theta + \frac{y^2}{2R}\cos^2\theta + \cdots$$

we retain the first *two* terms in the expansion when dealing with the phase. This linear dependence of r on y constitutes the Fraunhofer approximation. The total field at P due to the differential strip is then

$$E_P = \frac{\mathcal{E}_L}{R} \int_{-\ell/2}^{+\ell/2} \sin\left[\omega t - k(R - y\sin\theta)\right] dy$$

or
$$E_P = \frac{\mathcal{E}_L \ell}{R}\left(\frac{\sin\beta''}{\beta''}\right) \sin(\omega t - kR) \qquad \text{where} \qquad \beta'' \equiv \frac{k\ell}{2}\sin\theta$$

The function $(\sin u)/u$ is known as sinc u; values of it are tabulated in the Appendix.

Because $I(\theta) = c\epsilon_0 \langle E^2\rangle$ the *far-field irradiance distribution from a coherent line source* is

$$I(\theta) = I(0) \operatorname{sinc}^2 \beta''$$

where
$$I(0) = \frac{c\epsilon_0}{2}\left(\frac{\mathcal{E}_L \ell}{R}\right)^2$$

When $\ell \gg \lambda$, as it is here, $I(\theta)$ drops off very rapidly as θ moves away from zero, and the line source emits predominantly in the xz-plane. Moreover, E_P has the form of the field of a point source a distance R from P. In other words, a differential strip (dz by ℓ) behaves as if it were a point emitter on the central z-axis. All such strips spanning the width of the aperture (from $-b/2$ to $+b/2$) correspond to a linear array of point sources lying along the z-axis. This line source, in turn, generates a pattern equivalent to that of the entire aperture, which is given by

$$I(\theta) = I(0) \operatorname{sinc}^2 \beta$$

provided that now $\beta = (kb/2)\sin\theta$. Thus the Fraunhofer pattern for a single narrow slit has the form of the sinc function squared, as shown in Fig. 7-7. The pattern consists of a broad central bright band accompanied by a series of narrow fringes all parallel to the slit itself. The corresponding result for two slits is obtained in Problem 7.11.

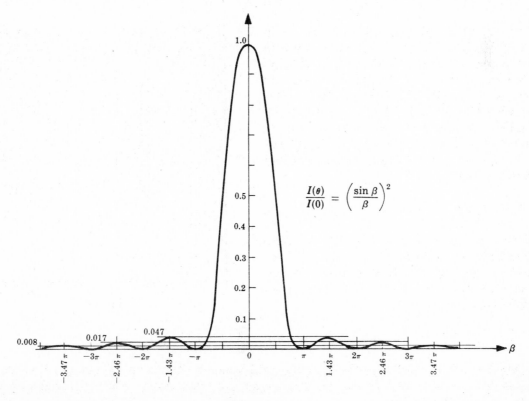

Fig. 7-7

SOLVED PROBLEMS

7.6. Locate via β both the minima and subsidiary maxima in the far-field diffraction pattern for a narrow slit.

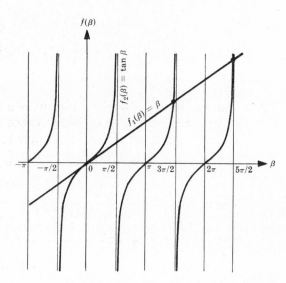

Extrema of $I = I(0) \operatorname{sinc}^2 \beta$ correspond to values of β for which

$$\frac{dI}{d\beta} = I(0) \frac{2 \sin \beta (\beta \cos \beta - \sin \beta)}{\beta^3} = 0$$

Minima occur when $\sin \beta = 0$ and $\beta \neq 0$; that is, when

$$\beta = \pm m\pi \quad \text{with} \quad m = 1, 2, 3, \ldots$$

Contrastingly, the subsidiary maxima exist for nonzero β satisfying

$$\beta \cos \beta - \sin \beta = 0$$

or $\qquad\qquad \tan \beta = \beta$

Fig. 7-8

This last equation is most simply solved graphically by superimposing the straight line $f_1(\beta) = \beta$ on the curves for $f_2(\beta) = \tan \beta$. The points of intersection other than the origin (Fig. 7-8) locate the subsidiary maxima at $\beta = \pm 1.4303\,\pi$, $\pm 2.4590\,\pi$, $\pm 3.4707\,\pi$, etc. Notice that these peaks are not quite midway between irradiance zeros.

7.7. Determine the irradiances of the first three subsidiary maxima in terms of $I(0)$, the principal peak value, for the far-field pattern of a single slit.

From Problem 7.6, we know that the secondary peaks occur at $\beta_1 = 1.43\,\pi$, $\beta_2 = 2.46\,\pi$ and $\beta_3 = 3.47\,\pi$. Furthermore,

$$I(\theta) = I(0) \operatorname{sinc}^2 \beta$$

and so we need only look up the sinc function in the Appendix to find

$$\operatorname{sinc}^2 \beta_1 = (-0.217)^2 = 0.047$$

$$\operatorname{sinc}^2 \beta_2 = (0.128)^2 = 0.016$$

$$\operatorname{sinc}^2 \beta_3 = (-0.091)^2 = 0.008$$

The desired irradiances are then $0.047\,I(0)$, $0.016\,I(0)$ and $0.008\,I(0)$.

7.8. Show that the irradiance, I_m, of the mth subsidiary peak, as discussed in Problem 7.6, can be well approximated by

$$I_m = I(0) \left[\frac{1}{(m + \frac{1}{2})\pi} \right]^2$$

If we assume for simplicity that the peaks occur dead center between minima, then

$$\beta_1 = 1.5\,\pi, \quad \beta_2 = 2.5\,\pi, \quad \ldots, \quad \beta_m = (m + \tfrac{1}{2})\pi, \quad \ldots$$

It follows that $\sin \beta_m = \pm 1$ for all values of m, and so

$$I_m = I(0) \left(\frac{\sin \beta_m}{\beta_m} \right)^2 = I(0) \frac{1}{\beta_m^2} = I(0) \left[\frac{1}{(m + \frac{1}{2})\pi} \right]^2$$

As a check, let us compute I_2:

$$I_2 = I(0)\left[\frac{1}{(2+\frac{1}{2})\pi}\right]^2 = 0.0162\,I(0)$$

which isn't too bad — the actual value is $0.0169\,I(0)$.

7.9. The slit in Fig. 7-5 is 0.5 mm wide and 3 cm long. Both lenses have focal lengths of 50 cm and $\lambda = 650$ nm. Compute the locations of the first minimum and the first subsidiary maximum in terms of their linear displacements from the central axis on Σ_o.

We saw in Problem 7.6 that minima occur when

$$\beta = \pm m\pi \qquad m = 1, 2, 3, \ldots$$

that is, when

$$b\sin\theta = \pm m\lambda$$

The first dark fringe ($m = 1$) is located by

$$\sin\theta = \pm\frac{\lambda}{b} = \pm\frac{650 \times 10^{-9}}{0.5 \times 10^{-3}} = \pm 0.0013$$

or $\theta = \pm 4.5' = \pm 0.0013$ rad (as you might expect, $\sin\theta \approx \theta$). For so small an angle the focal length times θ closely approximates the displacement Z. Thus

$$Z \approx f\theta = \pm 0.5(0.0013) = \pm 0.65 \text{ mm}$$

The first subsidiary bright band was found in Problem 7.6 to be at $\beta = \pm 1.43\,\pi$. In other words

$$\frac{kb}{2}\sin\theta = \pm 1.43\,\pi$$

$$\sin\theta = \pm\frac{1.43\,\lambda}{b} = \pm 1.859 \times 10^{-3}$$

Hence $\theta = \pm 1.86 \times 10^{-3}$ rad and so the first maxima on either side of center are at

$$Z \approx f\theta = 0.5(\pm 1.86 \times 10^{-3}) = \pm 0.93 \text{ mm}$$

7.10. Derive an approximate expression for the *angular width at half-maximum irradiance,* $\Delta\theta_{1/2}$, of the central peak in the diffraction pattern of a single slit.

We must determine the value of θ for which

$$\frac{I(\theta)}{I(0)} = \text{sinc}^2\beta = \frac{1}{2}$$

i.e.

$$\text{sinc}\,\beta = 0.7071$$

From the sinc table in the Appendix we see that the central peak (which runs from $u = 0$ to π) has its half-maximum value at $\beta_{1/2} = 1.39$ rad. Since $\beta = (\pi b/\lambda)\sin\theta \approx (\pi b/\lambda)\theta$, for small angles, we have

$$1.39 \approx \frac{\pi b}{\lambda}\theta_{1/2} \qquad \text{or} \qquad \theta_{1/2} \approx 0.442\frac{\lambda}{b}$$

The total angular width, $\Delta\theta_{1/2}$, is $2\theta_{1/2}$; hence

$$\Delta\theta_{1/2} \approx 0.885\frac{\lambda}{b}$$

This is generally just rounded off to

$$\Delta\theta_{1/2} \approx \frac{\lambda}{b}$$

Observe that the width of the diffraction peak varies inversely with the width of the slit. As discussed in Section 6.6, the same kind of relationship exists between the temporal signal width and the frequency bandwidth.

7.11. Use the geometry of Fig. 7-9 to arrive at an expression for the irradiance distribution in the far-field pattern of a double slit. How does this result relate to Young's experiment?

Following our earlier procedure for the single slit, we write the field as

$$E_P = \frac{\mathcal{E}_L}{R}\int_{-b/2}^{+b/2} \sin\left[\omega t - k(R - z\sin\theta)\right]dz + \frac{\mathcal{E}_L}{R}\int_{a-(b/2)}^{a+(b/2)} \sin\left[\omega t - k(R - z\sin\theta)\right]dz$$

Integration of this leads to

$$E_P = b\frac{\mathcal{E}_L}{R}(\operatorname{sinc}\beta)[\sin(\omega t - kR) + \sin(\omega t - kR + 2\alpha)]$$

where β as before equals $(kb/2)\sin\theta$ and now

$$\alpha \equiv \frac{ka}{2}\sin\theta$$

A simpler form of the field equation is

$$E_P = 2b\frac{\mathcal{E}_L}{R}\operatorname{sinc}\beta\cos\alpha\sin(\omega t - kR + \alpha)$$

and so

$$I(\theta) = 4I_0(\operatorname{sinc}^2\beta)(\cos^2\alpha)$$

The irradiance I_0 is that of either slit at $\theta = 0$ and $I(0) = 4I_0$.

What we now have is evidently Young's cosine-squared fringes modulated by the single-slit sinc-squared diffraction envelope (Fig. 7-10). When the slits are very narrow, b is very small and the central diffraction peak is quite broad. As b is made to go to zero, $\operatorname{sinc}^2\beta$ approaches one and $I(\theta) \to 4I_0\cos^2\alpha$, which is identical to our findings in Chapter 6 for Young's arrangement.

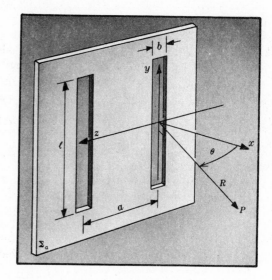

7.12. The bright band of order m in the interference pattern for two slits is said to be *suppressed* or *missing* when it is coincident with a null in the diffraction envelope. (a) Write an expression for the above condition and apply it to Fig. 7-10. (b) How many fringes will exist under the central diffraction maximum in such cases?

Fig. 7-9

(a) Maxima in the interference pattern occur when

$$a\sin\theta = m\lambda$$

where $m = 0, \pm 1, \pm 2, \dots$. As we saw in Problem 7.9, minima in the single-slit diffraction pattern result when

$$b\sin\theta = m'\lambda$$

where this time $m' = \pm 1, \pm 2, \dots$. When both of these equations are satisfied,

$$\frac{a}{b} = \frac{m}{m'} = M$$

In Fig. 7-10 the first missing order corresponds to $m = 3$, where $m' = 1$ and, accordingly, $a = 3b$. The second missing order results when $m = 6$ and $m' = 2$; again $M = 3$.

(b) If we include the two "half-fringes" at the missing orders, there are $2M$ bright bands within the central diffraction peak.

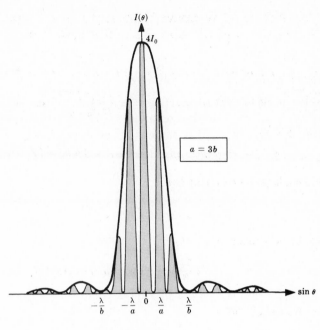

Fig. 7-10

7.13. The Fraunhofer pattern of a double slit under $\lambda = 650$ nm illumination appears in the back focal plane of a lens having an 80-cm focal length. The center-to-center separation between bright fringes is observed to be 1.04 mm and the fifth maximum is missing. Determine the width of each slit and the distance between them.

Problem 6.10 gives the fringe separation in Young's experiment as $\Delta y = s\lambda/a$; and here $s = f = 80$ cm. Thus

$$a = \frac{s\lambda}{\Delta y} = \frac{(80 \times 10^{-2})(650 \times 10^{-9})}{1.04 \times 10^{-3}} = 0.5 \text{ mm}$$

The missing fifth maximum means that $m = 5$, $m' = 1$ (see Problem 7.12). Therefore $M = m/m' = 5$, whence $a = 5b$ and $b = 0.1$ mm.

7.14. Sketch the far-field irradiance distribution for a double slit where each aperture is 0.1 mm wide and the separation between them is 0.6 mm.

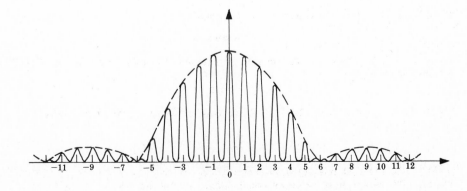

Fig. 7-11

Here $a = 6b$, so that $M = 6$ and the 6th, 12th, 18th, etc., orders are missing. Figure 7-11 illustrates the pattern of sinc²-modulated, equally spaced fringes.

7.4 MULTIPLE NARROW SLITS — THE DIFFRACTION GRATING

The same analysis as was used in the preceding sections for far-field diffraction can be applied to an array of N slits. For the geometrical layout indicated in Fig. 7-12 and normally incident plane waves, the field at some distant point P is

$$E_P = \frac{\mathcal{E}_L}{R}\int_{-b/2}^{+b/2} \sin\left[\omega t - k(R - z\sin\theta)\right] dz$$

$$+ \frac{\mathcal{E}_L}{R}\int_{a-(b/2)}^{a+(b/2)} \sin\left[\omega t - k(R - z\sin\theta)\right] dz + \cdots$$

$$+ \frac{\mathcal{E}_L}{R}\int_{(N-1)a-(b/2)}^{(N-1)a+(b/2)} \sin\left[\omega t - k(R - z\sin\theta)\right] dz$$

$$= b\frac{\mathcal{E}_L}{R}\operatorname{sinc}\beta\left(\frac{\sin N\alpha}{\sin\alpha}\right)\sin\left[\omega t - kR + (N-1)\alpha\right]$$

where again

$$\beta \equiv \frac{kb}{2}\sin\theta$$

$$\alpha \equiv \frac{ka}{2}\sin\theta$$

Consequently, the far-field irradiance distribution for N slits is

$$I(\theta) = I_0\operatorname{sinc}^2\beta\left(\frac{\sin N\alpha}{\sin\alpha}\right)^2$$

The flux density for any one slit at P_0, i.e. $\theta = 0 = \beta = \alpha$, is I_0 and, as we saw in Problem 7.1 for the linear array of N emitters, $I(0)$ equals $N^2 I_0$. The present irradiance distribution differs from the earlier one only in the (sinc² β)-modulation due to the now-finite slit widths.

Any periodic array of diffracting elements, be they apertures or obstacles, which alters the amplitude, phase, or both,

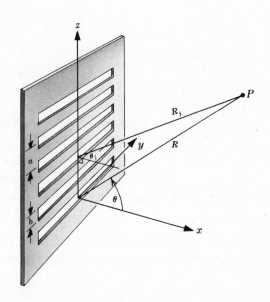

Fig. 7-12

of an incident wave is known as a *diffraction grating*. Clearly, the multiple-slit configuration of Fig. 7-12 is a diffraction grating, although a more common arrangement consists of several thousand parallel grooves cut into the surface of a glass plate.

SOLVED PROBLEMS

7.15. Locate the principal and subsidiary maxima, as well as the minima, in the far-field pattern for an array of N slits under normal monochromatic illumination.

Principal maxima occur whenever

$$\frac{\sin N\alpha}{\sin\alpha} = N$$

that is, whenever $\alpha = 0, \pm\pi, \pm2\pi, \ldots$, or equivalently,

$$a \sin \theta_m = m\lambda \qquad m = 0, \pm1, \pm2, \ldots$$

Flux density minima exist where

$$\frac{\sin N\alpha}{\sin \alpha} = 0$$

or when

$$\alpha = \pm\frac{\pi}{N}, \pm\frac{2\pi}{N}, \ldots, \pm\frac{(N-1)\pi}{N}, \pm\frac{(N+1)\pi}{N}, \ldots$$

The omitted α-values of 0, $N\pi/N$, $2N\pi/N$, etc., yield peak values rather than nulls. Between consecutive principal maxima there are $N-1$ minima and therefore $N-2$ subsidiary maxima. The latter are located approximately midway between nulls at

$$\alpha = \pm\frac{3\pi}{2N}, \pm\frac{5\pi}{2N}, \ldots$$

Actually they are very slightly closer to the nearest principal maximum.

7.16. Approximate the relative irradiances of the first three subsidiary maxima in the monochromatic Fraunhofer pattern of a large array of N slits.

By Problem 7.15 the secondary maxima are positioned where, very nearly,

$$\alpha = \pm\frac{3\pi}{2N}, \pm\frac{5\pi}{2N}, \pm\frac{7\pi}{2N}, \ldots$$

Since $I(0) = N^2 I_0$,

$$I(\theta) = \frac{I(0)}{N^2} \operatorname{sinc}^2 \beta \left(\frac{\sin N\alpha}{\sin \alpha}\right)^2$$

At the points of interest $|\sin N\alpha| = 1$. Moreover, since N is taken to be large, the values of α for the first several subsidiary maxima will be small, whence $\sin^2 \alpha \approx \alpha^2$. Again, β will be small and sinc β will vary little from 1. Thus, at the first secondary peak,

$$I_1 \approx I(0)\left(\frac{2}{3\pi}\right)^2 \approx 0.045 \, I(0)$$

or $I_1/I(0) \approx 1/22$. The irradiance of the second peak is

$$I_2 \approx I(0)\left(\frac{2}{5\pi}\right)^2 \approx 0.016 \, I(0)$$

or $I_2/I(0) \approx 1/62$. And for the third secondary maximum,

$$I_3 \approx I(0)\left(\frac{2}{7\pi}\right)^2 \approx 0.008 \, I(0)$$

or $I_3/I(0) \approx 1/121$.

7.17. Make a rough sketch of the far-field diffraction pattern for an array of 6 parallel slits separated by a distance equal to 4 times the individual slit width. Compute the irradiance of the second secondary peak adjacent to the first principal maximum.

Inasmuch as $a = 4b$, we can expect the 4th, 8th, 12th, etc., principal peaks to be suppressed by a null in the diffraction envelope. Furthermore, there will be $N-2 = 4$ subsidiary maxima between each pair of principal maxima and these will be quite small; how small can be found from

$$I(\theta) = \frac{I(0)}{N^2} \operatorname{sinc}^2 \beta \left(\frac{\sin N\alpha}{\sin \alpha}\right)^2$$

The second subsidiary peak corresponds approximately to

$$\alpha = \frac{5\pi}{2N} = \frac{5\pi}{12}$$

But $a = 4b$, which means that $\alpha = 4\beta$ and so

$$\beta = \frac{5\pi}{48}$$

Consequently, the irradiance in question becomes

$$I(\theta_2) = \frac{I(0)}{36}\left(\mathrm{sinc}^2\frac{5\pi}{48}\right)\left(\frac{\sin 5\pi/2}{\sin 5\pi/12}\right)^2 = \frac{I(0)}{36}(0.965)(1.072) = 0.029\,I(0)$$

Figure 7-13 is the appropriate plot of the diffraction pattern.

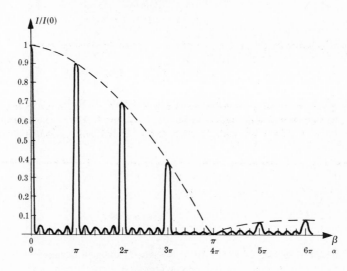

Fig. 7-13

7.18. Usually the source illuminating a grating in a spectroscopic device is configured as a narrow slit and so the principal maxima appear as thin bright bands (hence the name *spectral lines*). Show that the angular width of such a line, $\Delta\theta$, with normally incident monochromatic light is inversely proportional to the width of the grating.

Assume a principal peak to extend from one adjacent minimum to the other, i.e. from $\alpha = -\pi/N$ to $\alpha = +\pi/N$, as in Problem 7.15. Accordingly, the peak width corresponds to

$$\Delta\alpha = \frac{2\pi}{N}$$

But $\alpha = (ka/2)\sin\theta$, so that

$$\Delta\alpha = \frac{ka}{2}(\cos\theta)(\Delta\theta) = \frac{2\pi}{N}$$

Therefore, the angular width of the mth-order spectral line is

$$\Delta\theta = \frac{2\lambda}{Na\cos\theta_m}$$

which varies inversely with Na, the width of the grating. This dependence is known as *instrumental broadening*. The greater the number of grooves and the larger their separation, the sharper will be the spectral lines. Since N is large, the subsidiary maxima have vanishingly small irradiance and one "sees" only the principal peaks.

7.19. (a) Find an expression describing the angular spread for a *small range of wavelengths*, $\Delta\theta/\Delta\lambda$, i.e. the *angular dispersion* or *dispersive power* $\mathscr{D}$. (b) Compute the dispersive power in the first and second orders for a grating with 1500 grooves per inch operating in the visible.

(a) We know (Problem 7.15) that principal maxima occur when

$$a \sin \theta_m = m\lambda \qquad m = 0, \pm1, \pm2, \ldots$$

and this has come to be known as the *grating equation* for normal incidence. Under white light illumination each wavelength component will evidently have a maximum of a given order at a slightly different value of θ. This gives rise to a broad band of colors, or spectrum, for each value of m (see Fig. 7-14).

On differentiating the grating equation, we get

$$\mathcal{D} \equiv \frac{\Delta\theta}{\Delta\lambda} = \frac{m}{a \cos \theta_m}$$

(b) The groove separation or grating constant is given in centimeters as

$$a = \frac{2.54}{1500} = 1.69 \times 10^{-3}$$

which is quite large as compared to $\lambda \approx 5 \times 10^{-5}$ cm. It is clear from the grating equation that in the first and second orders, since $a \gg \lambda$, $\cos \theta_m \approx 1$. Therefore, dispersive powers are $m(5.91 \times 10^2)$, or 5.91×10^2 rad/cm and 11.8×10^2 rad/cm, respectively.

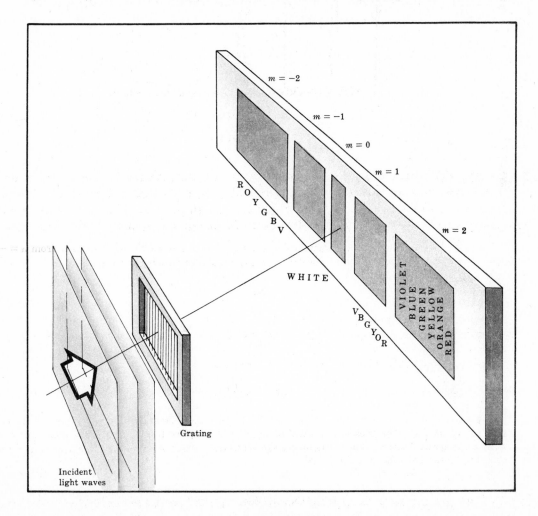

Fig. 7-14

7.20. What is the angular separation between the sodium D lines (589.592 nm and 588.995 nm) in the first-order spectrum generated by a plane transmission grating having 10,000 lines per inch at normal incidence?

The wavelength spread $\Delta\lambda = 0.597$ nm is small enough to allow use of the equation

$$\mathcal{D} = \frac{m}{a \cos\theta_m} = \frac{\Delta\theta}{\Delta\lambda}$$

derived in Problem 7.19. This time

$$a = \frac{2.54 \times 10^{-2}}{10,000} = 2.54 \times 10^{-6} \text{ m}$$

and so from the grating equation with $m = 1$,

$$\sin\theta_1 = \frac{\lambda}{a} = \frac{589.294 \times 10^{-9}}{2.54 \times 10^{-6}} = 0.232$$

Hence, $\theta_1 = 13°25'$ and $\cos\theta_1 = 0.9727$. Finally,

$$\Delta\theta = \frac{m\,\Delta\lambda}{a \cos\theta_m} = \frac{0.597 \times 10^{-9}}{2.54 \times 10^{-6}(0.9727)} = 2.42 \times 10^{-4} \text{ rad}$$

7.21. Find the angular extent of the first-order spectrum for white light (390 nm to 780 nm) normally incident on a transmission grating having 17,000 grooves per inch.

Obviously the wavelength spread is too large to allow use of the equation for $\mathcal{D}$. Instead we calculate the angles directly using the grating equation with $m = 1$, i.e.

$$\sin\theta_1 = \frac{390 \times 10^{-9}}{a} \qquad \sin\theta_1' = \frac{780 \times 10^{-9}}{a}$$

Computing a we get

$$a = \frac{2.54 \times 10^{-2}}{17,000} = 1.49 \times 10^{-6} \text{ m}$$

Accordingly,

$$\sin\theta_1 = 0.261 \qquad \sin\theta_1' = 0.523$$

from which $\theta_1 = 15°8'$, $\theta_1' = 31°32'$ and $\Delta\theta = 16°24'$.

7.22. The chromatic *resolving power* $\mathcal{R}$ of a spectroscopic device is

$$\mathcal{R} \equiv \frac{\lambda}{(\Delta\lambda)_{\min}}$$

wherein λ is the mean wavelength and $(\Delta\lambda)_{\min}$ corresponds to *the least resolvable wavelength difference* between two adjacent lines. For this condition we will use *Rayleigh's criterion* which states that *two fringes are just resolvable when the principal maximum of one coincides with the first minimum of the other.* Derive an expression for the resolving power of a grating.

From Rayleigh's criterion we can say that the minimum angular separation between just resolvable peaks corresponds to half the width of a principal maximum. From Problem 7.18, $(\Delta\theta)_{\min} = \Delta\theta/2$ and so

$$(\Delta\theta)_{\min} = \frac{\lambda}{Na \cos\theta_m}$$

But from Problem 7.19,

$$(\Delta\theta)_{\min} = \mathcal{D}(\Delta\lambda)_{\min} = \frac{(\Delta\lambda)_{\min}\,m}{a \cos\theta_m}$$

By combining these two expressions we obtain

$$\mathcal{R}_m \;=\; \frac{\lambda}{(\Delta\lambda)_{\min}} \;=\; mN$$

for the resolving power in the mth order.

7.23. How many lines must be ruled on a transmission grating so that it will just resolve the sodium doublet (589.592 nm and 588.995 nm) in the first-order spectrum?

According to Problem 7.22, the resolving power of a grating in the first order is

$$\frac{\lambda}{(\Delta\lambda)_{\min}} \;=\; N$$

In this case,

$$\lambda \;=\; \frac{589.592 + 588.995}{2} \;=\; 589.294$$

$$(\Delta\lambda)_{\min} \;=\; 589.592 - 588.995 \;=\; 0.597$$

Hence

$$N \;=\; \frac{589.294}{0.597} \;=\; 987.09$$

and so 988 grooves are needed.

7.24. To get most of the energy out of the zeroth order, where it's wasted from a spectroscopic viewpoint since the constituent wavelengths overlap, modern gratings have shaped or *blazed* grooves. Figure 7-15 shows a blazed reflection grating which shifts the strong specularly reflected peak from the zeroth to some higher order. (*a*) For plane waves incident normal to the grating plane, derive an expression for that reinforced order in terms of the *blaze angle* γ. (*b*) Compute the necessary angle γ in order that normally incident radiant energy at a wavelength of 200 nm be strongly channeled into the second order by a grating having 2000 lines/mm.

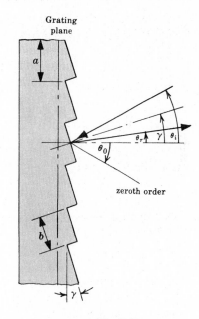

Fig. 7-15

(*a*) Had the facets in Fig. 7-15 not been tilted, specular reflection would have carried most of the energy off into the zeroth order at θ_0; instead it now goes out at θ_r. Notice that here

$$\theta_i \;=\; \gamma + (\gamma + \theta_r) \;=\; 2\gamma + \theta_r$$

where θ_r is a negative number since it is on the same side of the grating normal as is θ_i. At normal incidence, $\theta_i = 0$, the zeroth order ($m = 0$) is at $\theta_0 = 0$, i.e. straight out the normal to the grating plane. Most of the diffracted radiation is now concentrated at $\theta_r = -2\gamma$ and this will correspond to the mth order when

$$a \sin 2\gamma \;=\; m\lambda$$

In this case the mth-order interference peak will reside at the central maximum of the single-slit diffraction pattern.

(*b*) Since

$$a \;=\; \frac{10^{-3}}{2000} \;=\; 0.5 \times 10^{-6} \text{ m}$$

we have from (*a*):

$$2\gamma = \sin^{-1}\left[\frac{2(200 \times 10^{-9})}{0.5 \times 10^{-6}}\right] = 53°8' \quad \text{or} \quad \gamma = 26.6°$$

7.5 RECTANGULAR AND CIRCULAR APERTURES — FRAUNHOFER DIFFRACTION

The far-field diffraction pattern associated with a single rectangular aperture can be calculated with the assistance of Fig. 7-16. Each area element $dS = dy\, dz$ serves as a monochromatic point source of Huygens wavelets whose complex representation is

$$dE_P = \frac{\mathcal{E}_A}{r} e^{i(\omega t - kr)}\, dS$$

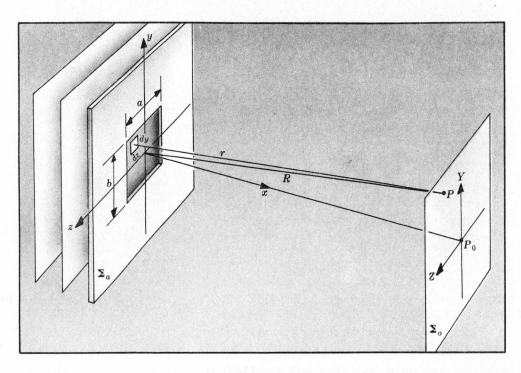

Fig. 7-16

with $\mathcal{E}_A$ denoting the source strength per unit area. The Fraunhofer condition carries r from its precise value of

$$r = [x^2 + (Y - y)^2 + (Z - z)^2]^{1/2}$$

to the approximation

$$r \approx R\left(1 - \frac{Yy + Zz}{R^2}\right)$$

Accordingly, the total field at point P becomes

$$E_P = \frac{\mathcal{E}_A e^{i(\omega t - kR)}}{R} \iint\limits_{\text{aperture}} e^{ik(Yy + Zz)/R}\, dS$$

$$= \frac{A\mathcal{E}_A e^{i(\omega t - kR)}}{R} \operatorname{sinc} \alpha' \operatorname{sinc} \beta'$$

where A is the area of the hole, $\alpha' \equiv kaZ/2R$ and $\beta' \equiv kbY/2R$. Since $I \propto \langle(\text{Re } E)^2\rangle$,

$$I(Y, Z) = I(0) \text{ sinc}^2 \alpha' \text{ sinc}^2 \beta'$$

a result which could have been anticipated from the discussion of the single slit. Figure 7-17 illustrates the irradiance distribution as a function of α' and β', or equivalently of Z and Y.

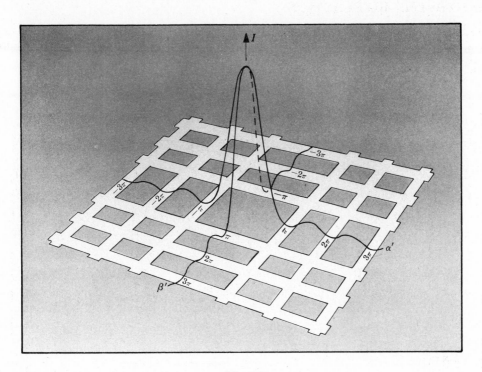

Fig. 7-17

The source strength per unit area of the point emitters, $\mathcal{E}_A$, is related to the electric field of the incident primary wave via

$$\mathcal{E}_A = \frac{E_0}{\lambda}$$

E_0 being the primary field amplitude over the aperture.

The above expression for E_P as a double integral is quite general and can be applied as well to the case of a circular aperture like the one illustrated in Fig. 7-18. After a complicated calculation in polar coordinates the irradiance distribution is found to be

$$I(q) = I(0)\left[\frac{2J_1(kaq/R)}{kaq/R}\right]^2$$

Here $J_1(u)$ is a *first-order Bessel function* defined by the series

$$J_1(u) = \frac{u}{2}\left[1 - \frac{1}{1!\,2!}\left(\frac{u}{2}\right)^2 + \frac{1}{2!\,3!}\left(\frac{u}{2}\right)^4 - \frac{1}{3!\,4!}\left(\frac{u}{2}\right)^6 + \cdots\right]$$

It roughly resembles a decaying sine wave. Because $\sin \theta = q/R$, an alternative expression for the flux density is

$$I(\theta) = I(0)\left[\frac{2J_1(ka \sin \theta)}{ka \sin \theta}\right]^2$$

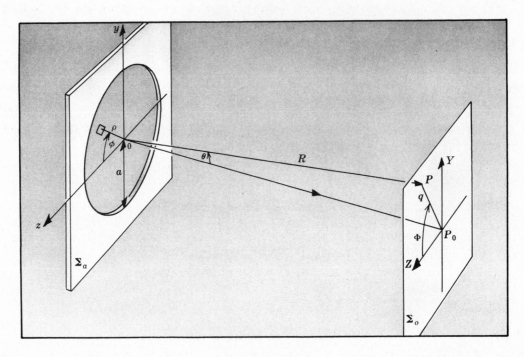

Fig. 7-18

This is the well-known *Airy pattern*, named after the famous British astronomer who first derived the formula. It consists of a central bright disk surrounded by a system of concentric alternately dark and bright rings. As shown in Fig. 7-19, the first zero occurs at $ka \sin \theta = 3.83$, and if we take q_1 as the distance from P_0 to that null, we can think of it as the radius of the Airy disk, viz.

$$q_1 = 1.22 \frac{R\lambda}{2a}$$

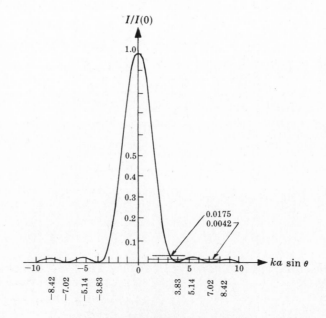

Fig. 7-19

This discussion is of great practical significance in that the image of a point source formed by an ideal optical system consisting of circular lenses or mirrors is not a point but rather an Airy pattern.

SOLVED PROBLEMS

7.25. Prove that a horizontal rectangular opening will generate a Fraunhofer diffraction pattern having a vertical rectangular bright area at its center. How does increasing the wavelength affect the size of the fringe system?

The central maximum is bounded by four nodal lines along which $I = 0$; these occur at $\alpha' = \pm\pi$ and $\beta' = \pm\pi$, as in Fig. 7-17. Consequently, the half-dimensions of the central maximum, Z_0 and Y_0, satisfy

$$\alpha' = \frac{2\pi a Z_0}{\lambda 2R} = +\pi \qquad \beta' = \frac{2\pi b Y_0}{\lambda 2R} = +\pi$$

or

$$Z_0 = \frac{\lambda R}{a} \qquad Y_0 = \frac{\lambda R}{b}$$

For a horizontal aperture $a > b$ and so $Z_0 < Y_0$, i.e. the rectangular bright region is vertical.

When λ increases, both Z_0 and Y_0 increase and the entire pattern enlarges. A like effect is produced by an increase in R.

7.26. Determine, at least approximately, the relative irradiances of the four diagonal off-axis subsidiary maxima nearest the central peak of a rectangular-aperture Fraunhofer pattern. How do these compare with the sixth axial secondary peaks?

Again assuming that the secondary maxima occur midway between consecutive minima, the peaks of present concern correspond to

$$(\alpha', \beta') = \left(\frac{3\pi}{2}, \frac{3\pi}{2}\right); \ \left(-\frac{3\pi}{2}, \frac{3\pi}{2}\right); \ \left(-\frac{3\pi}{2}, -\frac{3\pi}{2}\right); \ \left(\frac{3\pi}{2}, -\frac{3\pi}{2}\right)$$

We wish, therefore, to evaluate

$$I(Y, Z) = I(0) \ \text{sinc}^2 \alpha' \ \text{sinc}^2 \beta'$$

at these points and for that we return to the approximation developed in Problem 7.8. Accordingly, with $m = 1$,

$$\frac{I}{I(0)} \approx \left[\frac{1}{(1 + \frac{1}{2})\pi}\right]^2 \left[\frac{1}{(1 + \frac{1}{2})\pi}\right]^2 \approx 0.0020$$

for each of the first diagonal peaks. The sixth axial peaks at

$$(\alpha', \beta') = \left(0, \frac{13\pi}{2}\right); \ \left(0, -\frac{13\pi}{2}\right); \ \left(\frac{13\pi}{2}, 0\right); \ \left(-\frac{13\pi}{2}, 0\right)$$

all have irradiance ratios of

$$\frac{I}{I(0)} \approx 1\left[\frac{1}{(6 + \frac{1}{2})\pi}\right]^2 \approx 0.0024$$

Evidently, the bright spots along the coordinate axes are considerably more pronounced than the off-axis peaks. In effect, the straight edges of the aperture produce long perpendicular flares in the fringe pattern.

7.27. A rectangular horizontal hole 0.25 mm $\times$ 0.75 mm in an opaque screen is illuminated normally by plane waves of blue light from an argon ion laser at $\lambda = 488$ nm. The diffraction pattern is cast on a screen in the focal plane of a nearby positive lens ($f = 2.5$ m). Describe the resulting central maximum.

As in Problem 7.25, the central rectangular region of the pattern is bounded by the first null lines, such that

$$Z_0 = \frac{\lambda R}{a} \qquad Y_0 = \frac{\lambda R}{b}$$

In this instance $a = 0.75$ mm, $b = 0.25$ mm, $R = f$ and so

$$Z_0 = \frac{(488 \times 10^{-9})(2.5)}{0.75 \times 10^{-3}} = 1.63 \text{ mm}$$

$$Y_0 = \frac{(488 \times 10^{-9})(2.5)}{0.25 \times 10^{-3}} = 4.88 \text{ mm}$$

The central region is a vertical rectangle 9.76 mm $\times$ 3.26 mm.

7.28. A monochromatic plane wave at $\lambda = 500$ nm is normally incident on a **rectangular horizontal hole** 1 mm $\times$ 5 mm in an opaque screen. Centered in the aperture is a 0.1 mm $\times$ 0.5 mm horizontal opaque rectangle. Express the irradiance distribution appearing on the focal plane of a nearby converging lens having a 1-m focal length.

We can imagine that the 1 mm $\times$ 5 mm hole would contribute a field at P of E_{P1} were it unobstructed.   Furthermore, the 0.1 mm $\times$ 0.5 mm rectangle can be thought of as obscuring an array of Huygens-Fresnel emitters which might otherwise contribute a field E_{P2}. The total field at P is then

$$E_P = E_{P1} - E_{P2}$$

where

$$E_{P1} = \frac{A_1 \mathcal{E}_A e^{i(\omega t - kR)}}{R} \text{ sinc } \alpha_1' \text{ sinc } \beta_1'$$

$$E_{P2} = \frac{A_2 \mathcal{E}_A e^{i(\omega t - kR)}}{R} \text{ sinc } \alpha_2' \text{ sinc } \beta_2'$$

The terms A_1 and A_2 are the corresponding aperture areas. Hence,

$$E_P = \frac{\mathcal{E}_A e^{i(\omega t - kR)}}{R} [A_1 \text{ sinc } \alpha_1' \text{ sinc } \beta_1' - A_2 \text{ sinc } \alpha_2' \text{ sinc } \beta_2']$$

and the flux density is

$$I_P = \frac{I(0)}{(A_1 - A_2)^2} [A_1 \text{ sinc } \alpha_1' \text{ sinc } \beta_1' - A_2 \text{ sinc } \alpha_2' \text{ sinc } \beta_2']^2$$

where $I(0)$ corresponds to $\alpha_1' = \alpha_2' = \beta_1' = \beta_2' = 0$. In the particular case at hand $A_1 = 5 \times 10^{-6}$, $A_2 = 5 \times 10^{-8}$, and

$$\alpha_1' = \frac{2\pi(5 \times 10^{-3})Z}{500 \times 10^{-9}(2)1} = Z\pi 10^4$$

$$\beta_1' = \frac{2\pi(10^{-3})Y}{500 \times 10^{-9}(2)1} = 2Y\pi 10^3$$

$$\alpha_2' = \frac{2\pi(0.5 \times 10^{-3})Z}{500 \times 10^{-9}(2)1} = Z\pi 10^3$$

$$\beta_2' = \frac{2\pi(0.1 \times 10^{-3})Y}{500 \times 10^{-9}(2)1} = 2Y\pi 10^2$$

7.29. Formulate a rough estimate of the extent of the Airy disk in the visible **spectrum** for a lens, in terms of its *f-number* (ratio of focal length to diameter).

Beginning with the radius of the Airy disk,

$$q_1 = 1.22 \frac{R\lambda}{2a}$$

we first make use of the fact that $R \approx f$, i.e.

$$q_1 = 1.22 \frac{f\lambda}{2a}$$

Then, since $2a = D$, the diameter of the lens, the disk diameter is

$$2q_1 = 2\left(1.22\,\lambda\frac{f}{D}\right) = 2.44\,\lambda(f/\#)$$

where f/# is the f-number. In the visible we can roughly take $2.44\,\lambda$ to equal 1000 nm. Hence $2q_1 \approx$ f/# in millionths of a meter or microns. My camera lens, with an f/# of 1.4, forms an image of a distant point 1.4×10^{-6} m (or $1.4\,\mu$) in diameter on the film plane.

7.30. A collimated monochromatic beam ($\lambda = 600$ nm) is incident normally on a 1.2-cm diameter converging lens of focal length 50 cm. Compute both the angular and linear extent of the central disk of the diffraction pattern appearing on the focal plane.

The Airy disk has a radius of

$$q_1 = 1.22 \frac{R\lambda}{2a}$$

where now $R = f = 0.5$, and so its diameter is just

$$2q_1 = 1.22 \frac{(0.5)600 \times 10^{-9}}{0.6 \times 10^{-2}} = 6.1 \times 10^{-5} \text{ m}$$

The "angular radius" is generally denoted as θ, where $\theta = q_1/R = q_1/f$. Consequently, the angular diameter of the disk is

$$2\theta = \frac{2q_1}{f} = 1.22 \times 10^{-4} \text{ rad}$$

or 6.99×10^{-3} degrees.

7.31. The second-largest refracting telescope is the 36-inch, 56-foot focal length instrument at the Lick Observatory. Compute the radius of the second bright ring in the Airy pattern of a star formed on the focal plane of the objective.

From Fig. 7-19 we see that the second subsidiary maximum occurs at $ka \sin\theta = 8.42$. Hence

$$\frac{kaq}{R} = 8.42$$

and with $R = f$ and a mean wavelength of 550 nm,

$$q = \frac{8.42\,f}{ak} = \frac{8.42(56 \times 12 \text{ inch})}{(18 \text{ inch})2\pi/(550 \times 10^{-9} \text{ m})} = 0.0275 \times 10^{-3} \text{ m}$$

7.32. Apply Rayleigh's criterion as stated in Problem 7.22 to the case of a circular aperture and thereby arrive at an expression for $(\Delta\phi)_{\min}$, the *minimum resolvable angular separation* between two distant object points.

According to Rayleigh's criterion, two Airy systems will just be resolvable when the central peak of one coincides with the first minimum of the other. But that just corresponds to a separation equivalent to the angular radius of the Airy disk, and so

$$(\Delta\phi)_{\min} = \frac{q_1}{R} = 1.22 \frac{\lambda}{D}$$

D being the diameter of the aperture.

7.33. Determine the smallest angular separation between two equally bright stars that can be resolved (in the sense of the Rayleigh criterion) by the 200-inch Hale telescope on Mount Palomar. What linear separation results if the prime focal length is 666 inches?

From Problem 7.32

$$(\Delta\phi)_{min} = 1.22 \frac{\lambda}{D}$$

and assuming $\lambda = 550$ nm

$$(\Delta\phi)_{min} = 1.22 \frac{550 \times 10^{-9}}{200 \times 2.54 \times 10^{-2}} = 1.32 \times 10^{-7} \text{ rad}$$

or 0.027 second of arc (as compared, say, with the roughly 18-second maximum angular diameter of Mars as seen from Earth). The corresponding linear distance on the image plane, or *limit of resolution*, is

$$(\Delta\ell)_{min} = f(\Delta\phi)_{min} = 1.22 \frac{f\lambda}{D}$$

here equal to 2.2×10^{-6} m.

7.6 FRESNEL DIFFRACTION — CIRCULAR SYSTEMS

At this point the simple Huygens-Fresnel theory needs to be sharpened up a bit — something that wasn't necessary when working at small angles in the far field. Consider the fact that if the secondary wavelets were really spherically symmetric, a primary wavefront would give rise to two disturbances, one propagating forward and one backward, and this, of course, is not the case. The solution to the dilemma, as Fresnel recognized, is to presume that the wavelets drop in amplitude as one moves away from the propagation direction of the primary wave. Each wavelet in the Fraunhofer limit of the previous sections contributed mainly in the forward direction, so this difficulty was of no concern. Now, however, we must multiply the amplitude of any spherical secondary wavelet by an *obliquity* or *inclination factor*, which turns out to have the form

$$K(\theta) = \frac{1}{2}(1 + \cos\theta)$$

as pictured in Fig. 7-20.

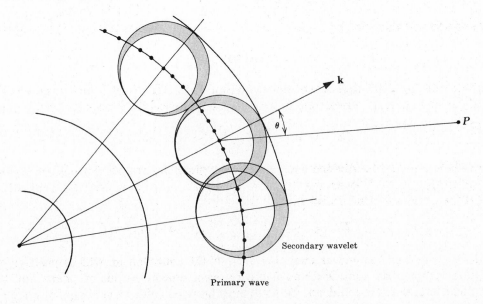

Fig. 7-20

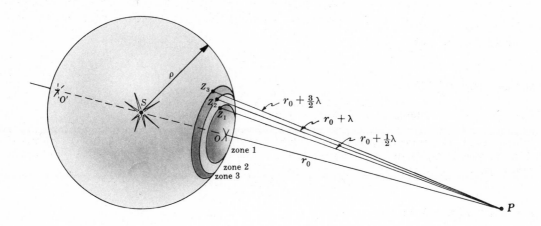

Fig. 7-21

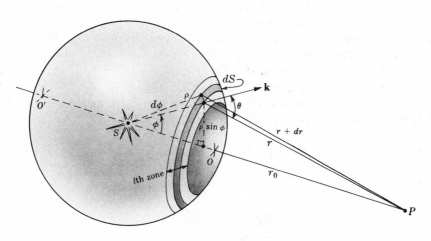

Fig. 7-22

Consider the unobstructed wave emitted from S in Fig. 7-21. Each area element dS on the primary spherical wavefront of radius ρ emits secondary wavelets of the form

$$dE = K\frac{\mathcal{E}_A}{r}\cos\left[\omega\left(t - \frac{\rho}{c}\right) - kr\right]dS$$

The primary wavefront is divided into annular regions known as *Fresnel* or *half-period zones*. By integrating dE over the lth *zone* ($l = 1, 2, 3, \ldots$) (Fig. 7-22) its contribution to the field at point P is found to be

$$E_l = (-1)^{l+1}\frac{2K_l\mathcal{E}_A\rho\lambda}{\rho + r_0}\sin\left[\omega t - k(\rho + r_0)\right]$$

Depending on whether l is odd or even, the sign of the contribution will be positive or negative; which means that contributions from adjacent zones are out of phase and tend to cancel. Such cancellation could not be complete, however, since the obliquity factor weakens successive zones. Adding the field amplitudes for all m zones, i.e.

$$E_0 = E_{01} - E_{02} + E_{03} - \cdots \pm E_{0m}$$

and assuming adjacent zones to have almost equal amplitudes, since K changes slowly with m, it can be shown that

$$E_0 \approx \frac{E_{01}}{2} + \frac{E_{0m}}{2}$$

when m is odd, and

$$E_0 \approx \frac{E_{01}}{2} - \frac{E_{0m}}{2}$$

when m is even. In either event, the contribution from the mth zone, which surrounds O', goes to zero since $K_m = K(\pi) = 0$. Therefore

$$E_0 \approx \frac{E_{01}}{2}$$

i.e., *the optical disturbance at P generated by the entire unobstructed wave approximates half the contribution from just the first zone.*

Imagine that we divide that first zone into N regions bounded by distances to P of

$$r_0 + \frac{1}{N}\frac{\lambda}{2}, \quad r_0 + \frac{2}{N}\frac{\lambda}{2}, \quad \ldots, \quad r_0 + \frac{N}{N}\frac{\lambda}{2}$$

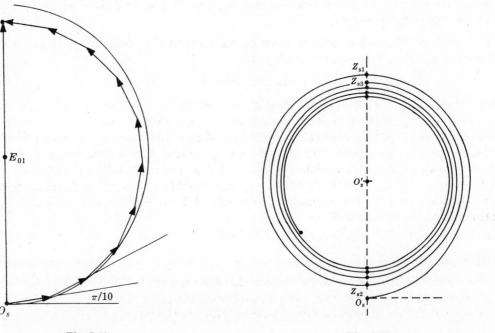

Fig. 7-23   Fig. 7-24

The field contributions from each of these subzones can be added vectorally as in the phasor diagram of Fig. 7-23, where $N = 10$. The obliquity factor causes a slight gradual diminution in the constituent amplitudes, which therefore combine to form something of a spiral. Letting $N \to \infty$ generates a smooth, tight spiral known as the *vibration curve* (Fig. 7-24), which swings through half a turn with the inclusion of each successive zone. Note that the points $O, Z_1, Z_2, \ldots, O'$ (see Fig. 7-21) on the wavefront correspond to points $O_s, Z_{s1}, Z_{s2}, \ldots, O'_s$ on the curve, which spirals around and finally ends on O'_s. The field amplitude of the unobstructed wave, E_0, is equal to the length of the vector from O_s to O'_s. Similarly, the vector from O_s to Z_{s1} gives the contribution from the first zone, E_{01}, and clearly $E_0 \approx E_{01}/2$.

Now suppose we insert an opaque screen with a circular hole of radius R perpendicular to $\overline{SP}$ at O in Fig. 7-21. The area of each zone is shown in Problem 7.34 to be very nearly equal to

$$A = \frac{\pi\lambda\rho r_0}{\rho + r_0}$$

and so the number of zones m within the aperture as seen from P is approximately

$$m = \frac{\pi R^2}{A} = \frac{(\rho + r_0)R^2}{\rho r_0 \lambda}$$

If m is an even integer,

$$E_0 = (E_{01} - E_{02}) + (E_{03} - E_{04}) + \cdots + (E_{0,m-1} - E_{0m}) \approx 0$$

and P resides at a dark spot. If m is an odd integer,

$$E_0 = E_{01} - (E_{02} - E_{03}) - \cdots - (E_{0,m-1} - E_{0m}) \approx E_{01}$$

which corresponds to a bright spot. In either case, m is relatively small compared to the number of zones in the unobstructed wave and $K_m \neq 0$. The vibration curve shows the effect rather nicely since the field for the first zone corresponds to $\overline{O_s Z_{s1}}$, while for the first two zones the field is down to only $\overline{O_s Z_{s2}}$. Obviously the aperture could partially uncover a zone, as well, thereby yielding a gray spot at P. Similar arguments apply off-axis, and because of the symmetry the diffraction pattern on a plane at P is a series of concentric rings of varying flux density.

If a small opaque disk or sphere were placed at O in Fig. 7-21 rather than an aperture, it would obstruct the first l zones such that

$$E_0 = E_{0,l+1} - E_{0,l+2} + \cdots \pm E_{0m}$$

As with the freely propagating wave m is very large, $K_m \to 0$ and $E_0 \approx E_{0,l+1}/2$. The disk might cover only a portion of the lth zone but, in any event, there will be a spot of light everywhere along the axis except immediately behind the obstacle. In other words, there will be an illuminated region right at the center of the shadow, known as *Poisson's spot* after the famous French scientist who insisted that such a result was ludicrous. As for the vibration curve, the periphery of the obstacle would locate a point B_s somewhere on the spiral. The length $\overline{B_s O_s'}$ corresponds to the field at P on the axis and evidently it will be nonzero regardless of where B_s is.

SOLVED PROBLEMS

7.34. Use Fig. 7-22 to derive an expression for the area of the lth Fresnel zone on a spherical wavefront. Show that the ratio of this area to its mean distance from P is independent of l, i.e. the same for all zones. Discuss the physical significance of this fact.

The surface element can be formulated as

$$dS = (2\pi\rho \sin \phi)\rho \, d\phi$$

yielding an area A for the end cap of

$$A = 2\pi\rho^2 \int_0^\phi \sin \phi \, d\phi = 2\pi\rho^2(1 - \cos \phi)$$

From the law of cosines applied to the lth zone

$$r_l^2 = \rho^2 + (\rho + r_0)^2 - 2\rho(\rho + r_0) \cos \phi$$

moreover,

$$r_l = r_0 + \frac{l\lambda}{2}$$

Hence,
$$\cos \phi = \frac{\rho^2 + (\rho + r_0)^2 - (r_0 + l\lambda/2)^2}{2\rho(\rho + r_0)}$$

The area of the lth zone is the area of the cap consisting of the first l zones,

$$\sum_{i=1}^{l} A_i = 2\pi\rho^2 - \frac{2\pi\rho^2[\rho^2 + (\rho + r_0)^2 - (r_0 + l\lambda/2)^2]}{2\rho(\rho + r_0)}$$

minus the area of the cap consisting of the first $l-1$ zones,

$$\sum_{i=1}^{l-1} A_i = 2\pi\rho^2 - \frac{2\pi\rho^2\{\rho^2 + (\rho + r_0)^2 - [r_0 + (l-1)\lambda/2]^2\}}{2\rho(\rho + r_0)}$$

Thus,
$$A_l = \frac{\lambda\pi\rho}{\rho + r_0}\left[r_0 + \frac{(2l-1)\lambda}{4}\right]$$

The second term is generally neglected, yielding an expression independent of l, which means that the zones all have approximately equal areas.

The mean distance from the lth zone to P is denoted as $\overline{r}_l$. The distances to the edges of the zone are $r_l = r_0 + l\lambda/2$ and $r_{l-1} = r_0 + (l-1)\lambda/2$, hence

$$\overline{r}_l = \frac{r_l + r_{l-1}}{2} = r_0 + \frac{(2l-1)\lambda}{4}$$

Consequently,

$$\frac{A_l}{\overline{r}_l} = \frac{\lambda\pi\rho}{\rho + r_0}$$

which is certainly independent of l. We can expect that the field amplitude contributed by the lth zone would depend on $K_l A_l/\overline{r}_l$ and so $|E_l|/K_l$ should be independent of l, as indeed it is (see page 183).

7.35. Derive an expression for the area of the lth Fresnel zone as seen from some point P, where now the incoming waves are planar. Compute the area of the first zone when $\lambda = 600$ nm and the point of observation is 0.5 m from the wavefront. What error results in the area if the λ^2-term is omitted?

From Problem 7.34

$$A_l = \frac{\lambda\pi\rho}{\rho + r_0}\left[r_0 + \frac{2l-1}{4}\lambda\right]$$

for a spherical wavefront of radius ρ. In the situation at hand the waves are planar, that is, $\rho \to \infty$. Hence, $\rho + r_0 \approx \rho$ and

$$A_l = \lambda\pi\left[r_0 + \frac{2l-1}{4}\lambda\right]$$

In particular, when $\lambda = 600$ nm, $r_0 = 0.5$ m and $l = 1$,

$$A_1 = 600 \times 10^{-9}\pi\left[0.5 + \frac{600 \times 10^{-9}}{4}\right] = 9.42 \times 10^{-3} \text{ cm}^2$$

The approximate area, dropping the second term, is

$$A_1 \approx \lambda\pi r_0 \approx 9.42 \times 10^{-3} \text{ cm}^2$$

The difference between the two is 2.83×10^{-9} cm^2, or about 0.00003 percent.

7.36. Derive an expression for the outer radius of the lth Fresnel zone on a planar wavefront as viewed from a distance of r_0.

From Fig. 7-25 the distance to the periphery of the lth zone is evidently

$$r_l = r_0 + l\frac{\lambda}{2}$$

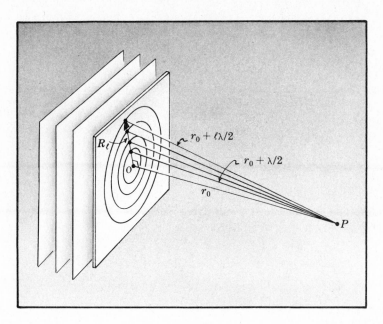

Fig. 7-25

Consequently,

$$R_l^2 \;=\; r_l^2 - r_0^2 \;=\; l r_0 \lambda \;+\; \frac{l^2 \lambda^2}{4}$$

Generally the last term is negligible and R_l is taken to be simply

$$R_l \;=\; \sqrt{l r_0 \lambda}$$

wherein $l = 1, 2, 3, \ldots$.

7.37. A He-Ne laser beam ($\lambda = 632.8$ nm) is expanded and collimated by sending it into the back end of a telescope focused at infinity. Compute the radius of the first half-period zone both exactly and approximately, when the wave is viewed axially from a distance of 1.58 m.

The approximate radius R_1 is given in Problem 7.36 as

$$R_1 \;=\; \sqrt{(1)(1.58)(632.8 \times 10^{-9})} \;=\; \sqrt{9.998 \times 10^{-7}} \;=\; 9.999 \times 10^{-4} \text{ m}$$

or, if you like, $R_1 = 1$ mm. More accurately,

$$R_1^2 \;=\; 9.998 \times 10^{-7} + \frac{\lambda^2}{4} \;=\; 9.998 \times 10^{-7} + 1 \times 10^{-13}$$

7.38. A 3-mm diameter hole in an opaque screen is illuminated normally by plane waves of wavelength 550 nm. A small probe is moved along the central axis recording flux density. Compute the locations of the first three maxima and minima.

Maxima occur when the aperture uncovers odd numbers of zones. Accordingly, since $R_l = \sqrt{l r_0 \lambda}$,

$$R_1 \;=\; 1.5 \times 10^{-3} \;=\; \sqrt{(1) r_0 (550 \times 10^{-9})}$$

which leads to

$$r_0 \;=\; \frac{(1.5 \times 10^{-3})^2}{550 \times 10^{-9}}$$

and only the first zone is uncovered at a distance of $r_0 = 4.09$ m, thereby producing a maximum. The next maximum occurs when three zones fill the aperture, i.e.

$$r_0 = \frac{(1.5 \times 10^{-3})^2}{3(550 \times 10^{-9})} = \frac{4.09}{3}$$

and $r_0 = 1.36\ m$. It should be clear that the next several maxima occur at distances of 0.82 m, 0.58 m and 0.45 m.

Similarly, minima correspond to even numbers of uncovered zones and we need only divide 4.09 m by 2, 4, 6, ..., to locate them. Thus the first minimum resides at 2.05 m.

7.39. Plane waves ($\lambda = 624$ nm) impinge normally on a circular aperture 2.09 mm in radius. A screen 1 m from the hole intercepts the diffraction pattern. Describe the appearance of the pattern at the central point P_0.

Let us first find the number of Fresnel zones uncovered by the aperture of radius R. The area of each zone is, from Problem 7.35,

$$A_l = \lambda \pi r_0$$

while the area of the hole is πR^2. Hence the number of zones N is just

$$N = \frac{\pi R^2}{\lambda \pi r_0} = \frac{R^2}{\lambda r_0}$$

In this specific case

$$N = \frac{(2.09 \times 10^{-3})^2}{(624 \times 10^{-9})1} = 7$$

and the central point is a bright spot.

7.40. Imagine that we again have plane waves perpendicularly incident on a circular hole. If at some axial point P the aperture reveals $\frac{1}{4}$ of the first Fresnel zone, what will the irradiance at P be in terms of I_0, the value with the screen removed?

The vibration curve is drawn in Fig. 7-26. Bear in mind that it is actually very tightly wound, so that we can approximate it over a small region as being circular. The chord $\overline{O_s A_s}$ corresponds to the amplitude in question, where A_s is located at one-quarter of the arc length from O_s to Z_{s1}, i.e. $\frac{1}{4}$ of the first zone. Evidently

$$\overline{O_s O'_s} \approx \overline{O'_s A_s} \approx \frac{\overline{O_s Z_{s1}}}{2}$$

and so

$$\sin 22.5° = \frac{\overline{O_s A_s}/2}{\overline{O_s O'_s}} = \frac{\overline{O_s A_s}}{\overline{O_s Z_{s1}}}$$

where $\sin 22.5° = 0.383$. But this is the ratio of the field amplitudes and its square would then be the desired irradiance ratio.

Recalling that $\overline{O_s Z_{s1}}$ is twice the unobstructed amplitude, we have

$$\frac{I}{4I_0} = 0.147$$

and so $I = 0.587\ I_0$. Keep in mind that I_0 is also the incident irradiance, since the waves are planar.

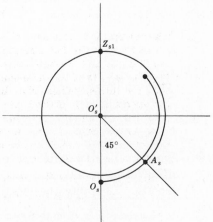

Fig. 7-26

7.41. A plane monochromatic wave ($\lambda = 450$ nm) is incident normally on an opaque screen containing an aperture in the form of an annulus of radii 1.000 mm and 1.414 mm.

Calculate the electric field amplitude at an axial point 2.222 m away, in terms of the incident field amplitude E_0.

A hole of radius 1.414 mm contains N zones, where

$$N \;=\; \frac{R^2}{\lambda r_0} \;=\; \frac{(1.414 \times 10^{-3})^2}{(450 \times 10^{-9})(2.222)} \;=\; 2$$

Similarly, the disk of radius 1 mm excludes only the first zone — hence only the second contributes. The resulting field amplitude is (see Fig. 7-27)

$$\overline{Z_{s1}Z_{s2}} \;\approx\; 2\,\overline{O_sO_s'} \;=\; 2E_0$$

Notice that the two vectors $\overrightarrow{Z_{s1}Z_{s2}}$ and $\overrightarrow{O_sO_s'}$ are oppositely directed, i.e. π rad out of phase.

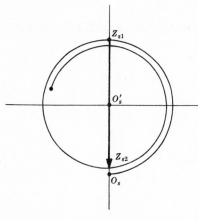

Fig. 7-27

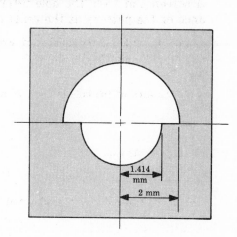

Fig. 7-28

7.42. Plane waves ($\lambda = 500$ nm) of irradiance I_0 arrive perpendicularly on an opaque screen having an aperture as indicated in Fig. 7-28. Compute the irradiance at an axial point 4 m from the screen.

The initial approach in all aperture problems is to find out which zones are contributing. Accordingly, for a hole of radius 2 mm

$$N \;=\; \frac{R^2}{\lambda r_0} \;=\; \frac{(2 \times 10^{-3})^2}{(500 \times 10^{-9})(4)} \;=\; 2$$

whereas for $R = 1.414$ mm, $N = 1$. Evidently, we must determine the field for the combination of the 1st and half of the 2nd zones. This is quite different from the case of a single circular hole uncovering the first one and one-half zones. In the latter circumstance, if point B were on the periphery of the aperture, then B_s, three-quarters of a turn around the spiral, would be the associated point and $\overrightarrow{O_sB_s}$ the corresponding field. By contrast, imagine the second zone, a segment of which is seen in the aperture, to be divided into, say, ten subzones in the manner relating to Fig. 7-22. The fields from the ten annular subzones would add to yield the equivalent of one-half turn of the spiral. Now for the case at hand, we add the ten segments of these subzones uncovered by the outer portion of the hole. Each one of these field vectors is one-half as long as in the previous case, and we get a spiral just as before but now reduced to half size. Rather than a field contribution of $\overrightarrow{Z_{s1}Z_{s2}}$ for a complete 2nd zone, we get $\overrightarrow{Z_{s1}Z_{s2}}/2$ for half that zone.

The entire field is then given by the vector sum

$$\overrightarrow{O_sZ_{s1}} \;+\; \frac{1}{2}\overrightarrow{Z_{s1}Z_{s2}}$$

which has an amplitude, in terms of the incident field amplitude E_0, of

$$2E_0 - \frac{1}{2}(2E_0) \;=\; E_0$$

The irradiance at P is then equal to the incident irradiance I_0.

7.7 FRESNEL DIFFRACTION — STRAIGHT EDGES

We now turn our attention to Fresnel diffraction arising from systems bounded by straight edges, such as rectangular holes, slits, wires, etc. Figure 7-29 depicts a typical arrangement. To find the field at P we once again integrate over all the differential contributions within the aperture, assuming each area element to emit a secondary wavelet.

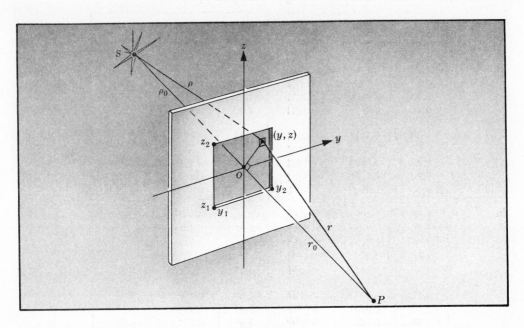

Fig. 7-29

The obliquity factor is taken as one, since the aperture is assumed small compared with ρ_0 and r_0. The term $\rho + r$ in the phase is approximated as

$$\rho + r \approx \rho_0 + r_0 + (y^2 + z^2)\frac{\rho_0 + r_0}{2\rho_0 r_0}$$

which is quadratic in the aperture variables. And the $(1/\rho r)$-dependence of the amplitudes of the wavelets is taken as $1/\rho_0 r_0$. The dimensionless quantities

$$u \equiv y\left[\frac{2(\rho_0 + r_0)}{\lambda \rho_0 r_0}\right]^{1/2} \qquad v \equiv z\left[\frac{2(\rho_0 + r_0)}{\lambda \rho_0 r_0}\right]^{1/2}$$

are introduced and the field at P turns out to be

$$E_P = \frac{\mathcal{E}_0}{2(\rho_0 + r_0)} e^{i[k(\rho_0 + r_0) - \omega t]} \int_{u_1}^{u_2} e^{i\pi u^2/2}\,du \int_{v_1}^{v_2} e^{i\pi v^2/2}\,dv$$

The term multiplying the integrals is one-half the unobstructed disturbance at P; call it $E_U/2$. The integrals themselves can be evaluated in terms of the *Fresnel integrals*

$$C(w) \equiv \int_0^w \cos\frac{\pi w'^2}{2}\,dw' \qquad \mathcal{S}(w) \equiv \int_0^w \sin\frac{\pi w'^2}{2}\,dw'$$

inasmuch as

$$\int_0^w e^{i\pi w'^2/2}\,dw' = C(w) + i\mathcal{S}(w)$$

where w is either u or v. Finally, then,

$$E_P = \frac{E_U}{2}\Big[C(u) + i\mathcal{S}(u)\Big]_{u_1}^{u_2}\Big[C(v) + i\mathcal{S}(v)\Big]_{v_1}^{v_2}$$

Table 7-1. Values of the Fresnel Integrals

w	$C(w)$	$\mathcal{S}(w)$	w	$C(w)$	$\mathcal{S}(w)$
0.00	0.0000	0.0000	4.50	0.5261	0.4342
0.10	0.1000	0.0005	4.60	0.5673	0.5162
0.20	0.1999	0.0042	4.70	0.4914	0.5672
0.30	0.2994	0.0141	4.80	0.4338	0.4968
0.40	0.3975	0.0334	4.90	0.5002	0.4350
0.50	0.4923	0.0647	5.00	0.5637	0.4992
0.60	0.5811	0.1105	5.05	0.5450	0.5442
0.70	0.6597	0.1721	5.10	0.4998	0.5624
0.80	0.7230	0.2493	5.15	0.4553	0.5427
0.90	0.7648	0.3398	5.20	0.4389	0.4969
1.00	0.7799	0.4383	5.25	0.4610	0.4536
1.10	0.7638	0.5365	5.30	0.5078	0.4405
1.20	0.7154	0.6234	5.35	0.5490	0.4662
1.30	0.6386	0.6863	5.40	0.5573	0.5140
1.40	0.5431	0.7135	5.45	0.5269	0.5519
1.50	0.4453	0.6975	5.50	0.4784	0.5537
1.60	0.3655	0.6389	5.55	0.4456	0.5181
1.70	0.3238	0.5492	5.60	0.4517	0.4700
1.80	0.3336	0.4508	5.65	0.4926	0.4441
1.90	0.3944	0.3734	5.70	0.5385	0.4595
2.00	0.4882	0.3434	5.75	0.5551	0.5049
2.10	0.5815	0.3743	5.80	0.5298	0.5461
2.20	0.6363	0.4557	5.85	0.4819	0.5513
2.30	0.6266	0.5531	5.90	0.4486	0.5163
2.40	0.5550	0.6197	5.95	0.4566	0.4688
2.50	0.4574	0.6192	6.00	0.4995	0.4470
2.60	0.3890	0.5500	6.05	0.5424	0.4689
2.70	0.3925	0.4529	6.10	0.5495	0.5165
2.80	0.4675	0.3915	6.15	0.5146	0.5496
2.90	0.5624	0.4101	6.20	0.4676	0.5398
3.00	0.6058	0.4963	6.25	0.4493	0.4954
3.10	0.5616	0.5818	6.30	0.4760	0.4555
3.20	0.4664	0.5933	6.35	0.5240	0.4560
3.30	0.4058	0.5192	6.40	0.5496	0.4965
3.40	0.4385	0.4296	6.45	0.5292	0.5398
3.50	0.5326	0.4152	6.50	0.4816	0.5454
3.60	0.5880	0.4923	6.55	0.4520	0.5078
3.70	0.5420	0.5750	6.60	0.4690	0.4631
3.80	0.4481	0.5656	6.65	0.5161	0.4549
3.90	0.4223	0.4752	6.70	0.5467	0.4915
4.00	0.4984	0.4204	6.75	0.5302	0.5362
4.10	0.5738	0.4758	6.80	0.4831	0.5436
4.20	0.5418	0.5633	6.85	0.4539	0.5060
4.30	0.4494	0.5540	6.90	0.4732	0.4624
4.40	0.4383	0.4622	6.95	0.5207	0.4591

We could divide the wavefront into strip zones and then add the field components (as done for Fig. 7-22) to again form a spiral. That same curve, known as the *Cornu spiral* and depicted in Fig. 7-30, can be generated by plotting the points

$$B(w) = C(w) + i S(w)$$

in the complex plane for all w from $-\infty$ to $+\infty$. For specific values of u, say, u_1 and u_2, there will be corresponding points on the spiral, $B(u_1)$ and $B(u_2)$. The vector, or more accurately the *phasor*, $\mathbf{B}_{12}$ drawn from $B(u_1)$ to $B(u_2)$ is the complex number $B(u_2) - B(u_1)$, i.e.

$$\mathbf{B}_{12} = \left[C(u) + i S(u) \right]_{u_1}^{u_2}$$

The electric field in complex form for the specific case of a rectangular aperture is, therefore,

$$E_P = \frac{E_U}{2}[B(u_2) - B(u_1)][B(v_2) - B(v_1)]$$

Because the element of arc length on the spiral is given by $d\ell^2 = dC^2 + dS^2$, it follows from the definitions of the Fresnel integrals that

$$d\ell^2 = \left(\cos^2 \frac{\pi w^2}{2} + \sin^2 \frac{\pi w^2}{2} \right) dw^2$$

Therefore $d\ell = dw$ and w corresponds to arc length measured along the curve.

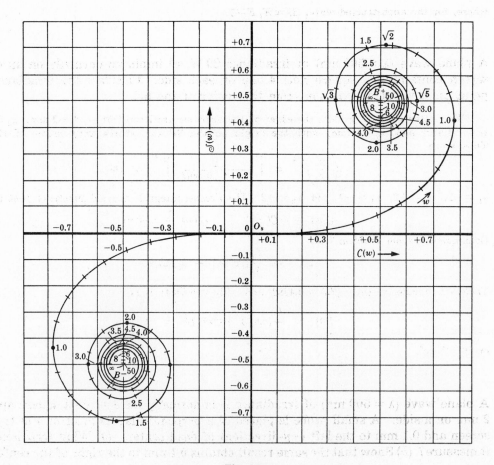

Fig. 7-30

Recall that for simplicity the obliquity factor was taken as one and the amplitudes of the secondary wavelets were made independent of ρ and r. Had this not been done, the resulting vibration curve would have been somewhat more tightly wound than the Cornu spiral. Provided that $\rho_0 \gg \lambda$ and $r_0 \gg \lambda$, the effect is quite negligible.

SOLVED PROBLEMS

7.43. Derive an expression for the irradiance in the near-field pattern of the rectangular aperture of Fig. 7-29, where, this time, the incident waves are planar.

With incoming plane waves $\rho_0 = \infty$, $\rho_0 + r_0 \approx \rho_0$ and so

$$u = y\left(\frac{2}{\lambda r_0}\right)^{1/2} \qquad v = z\left(\frac{2}{\lambda r_0}\right)^{1/2}$$

in the expression for E_P. The latter can be reformulated more explicitly as

$$E_P = \frac{E_U}{2}\{[C(u_2) - C(u_1)] + i[\mathcal{S}(u_2) - \mathcal{S}(u_1)]\}$$

$$\times \{[C(v_2) - C(v_1)] + i[\mathcal{S}(v_2) - \mathcal{S}(v_1)]\}$$

Since $I \propto E_P E_P^*/2$,

$$I = \frac{I_0}{4}\{[C(u_2) - C(u_1)]^2 + [\mathcal{S}(u_2) - \mathcal{S}(u_1)]^2\}$$

$$\times \{[C(v_2) - C(v_1)]^2 + [\mathcal{S}(v_2) - \mathcal{S}(v_1)]^2\}$$

where, for the unobstructed wave, $I_0 \propto E_U E_U^*/2$.

7.44. A plane wave ($\lambda = 500$ nm) of irradiance 20 W/m² impinges normally on an opaque screen containing a square hole 4 mm on each side. Calculate the irradiance at a point on the central axis 4 m from the center of the hole.

With reference to Fig. 7-29, the edges of the aperture are located at $y_1 = -2$ mm, $y_2 = 2$ mm, $z_1 = -2$ mm and $z_2 = 2$ mm, with the origin on the $\overline{SP}$ line at the very center of the hole. Thus, since

$$\left(\frac{2}{\lambda r_0}\right)^{1/2} = \left[\frac{2}{(500 \times 10^{-9})(4)}\right]^{1/2} = 10^3$$

$u_1 = -2$, $u_2 = +2$, $v_1 = -2$ and $v_2 = +2$. The Fresnel integrals are odd functions; that is,

$$C(w) = -C(-w) \qquad \mathcal{S}(w) = -\mathcal{S}(-w)$$

Consequently, from Problem 7.43,

$$I(0) = \frac{I_0}{4}\{[2C(2)]^2 + [2\mathcal{S}(2)]^2\}^2$$

Table 7-1 provides us with $C(2) = 0.4882$ and $\mathcal{S}(2) = 0.3434$; thus

$$I(0) = \frac{I_0}{4}(0.9534 + 0.4717)^2 = 0.508\, I_0$$

or 10.2 W/m².

7.45. A plane wave ($\lambda = 500$ nm) of irradiance I_0 is normally incident on a square aperture 2 mm on a side. A small probe is placed at a perpendicular distance of 4 m from the screen and 0.1 mm to the left ($-y$-direction) of dead center. (a) What irradiance will it measure? (b) Show that the same result obtains 0.1 mm to the right of the center line.

(a) The line from S to P is now 0.1 mm left of center on the y-axis. Measuring from the intersection point O of $\overline{SP}$ and the aperature plane, we have $y_1 = -0.9$ mm, $z_1 = -1$ mm, $y_2 = 1.1$ mm and $z_2 = 1$ mm. Inasmuch as

$$\left(\frac{2}{\lambda r_0}\right)^{1/2} = \left[\frac{2}{(500 \times 10^{-9})(4)}\right]^{1/2} = 10^3$$

$u_1 = -0.9$, $v_1 = -1.0$, $u_2 = 1.1$ and $v_2 = 1.0$. From Problem 7.43,

$$I = \frac{I_0}{4}\{[C(1.1) + C(0.9)]^2 + [\mathcal{S}(1.1) + \mathcal{S}(0.9)]^2\}$$

$$\times \{[C(1.0) + C(1.0)]^2 + [\mathcal{S}(1.0) + \mathcal{S}(1.0)]^2\}$$

or, using Table 7-1,

$$I = \frac{I_0}{4}\{[0.7638 + 0.7648]^2 + [0.5365 + 0.3398]^2\}$$

$$\times \{[2(0.7799)]^2 + [2(0.4383)]^2\} = 2.48\, I_0$$

(b) To the right of center, $y_1 = -1.1$ mm, $z_1 = -1.0$ mm, $y_2 = 0.9$ mm and $z_2 = 1.0$ mm. Therefore $u_1 = -1.1$, $v_1 = -1.0$, $u_2 = 0.9$ and $v_2 = 1.0$, and so

$$I = \frac{I_0}{4}\{[C(0.9) + C(1.1)]^2 + [\mathcal{S}(0.9) + \mathcal{S}(1.1)]^2\}$$

$$\times \{[C(1.0) + C(1.0)]^2 + [\mathcal{S}(1.0) + \mathcal{S}(1.0)]^2\}$$

which is identical to the result in (a).

7.46. Apply the Cornu spiral representation (a) to Problem 7.44, (b) to Problem 7.45.

(a) In Problem 7.44 there were the usual two sets of variables, $u_1 = -2$, $u_2 = +2$ and $v_1 = -2$, $v_2 = +2$. Because the two sets are equal in this instance, $\mathbf{B}_{12}(u) = \mathbf{B}_{12}(v)$. Quite generally, though,

$$I_P = \frac{I_0}{4}|\mathbf{B}_{12}(u)|^2\,|\mathbf{B}_{12}(v)|^2$$

for a rectangular aperture. To evaluate $\mathbf{B}_{12}(u)$ one need only locate $B(u_1)$ and $B(u_2)$ and connect them with a line of length $|\mathbf{B}_{12}(u)|$, as in Fig. 7-31. Measure the extent of $|\mathbf{B}_{12}|$ with any convenient scale and then read off the corresponding value on either the C- or $\mathcal{S}$-axis; in this case it's just about 1.2 units. A more precise calculation of

$$(\mathbf{B}_{12}\mathbf{B}_{12}^*)^{1/2} = |\mathbf{B}_{12}|$$

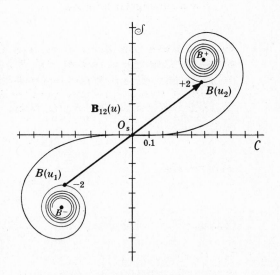

Fig. 7-31

proceeds from

$$\mathbf{B}_{12} = [C(u_2) - C(u_1)] + i[\mathcal{S}(u_2) - \mathcal{S}(u_1)]$$

whereupon

$$|\mathbf{B}_{12}| = 2[(0.4882)^2 + (0.3434)^2]^{1/2} = 1.19$$

Because of the symmetry,

$$I(0) = \frac{I_0}{4}(1.2)^2\,(1.2)^2 = 0.5\, I_0$$

(b) We locate $u_1 = -0.9$ and $u_2 = 1.1$ and then again $v_1 = -1.0$ and $v_2 = 1.0$ in the manner shown in Fig. 7-32. The length of $B_{12}(u)$ is about 1.75, while that of $B_{12}(v)$ is roughly 1.78. Hence

$$I = \frac{I_0}{4}(1.75)^2\,(1.78)^2 = 2.4\, I_0$$

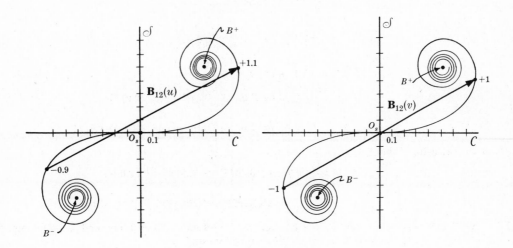

Fig. 7-32

7.47. A monochromatic point source S of vacuum wavelength $\lambda_0 = 400$ nm resides at a perpendicular distance of 1 m from an opaque planar screen in which there is a long narrow horizontal slit of width $\Delta z = 0.2$ mm. Calculate the irradiance at a point P on the center line $\overline{SP}$ of the slit and 4 m from it, using the Cornu spiral.

For a rectangular aperture

$$I_P = \frac{I_0}{4}|\mathbf{B}_{12}(u)|^2\,|\mathbf{B}_{12}(v)|^2$$

In the case of a long slit, $y_1 \rightarrow -\infty$, $y_2 \rightarrow +\infty$, and so $u_1 \rightarrow -\infty$ and $u_2 \rightarrow +\infty$. The limiting values of $B(\infty)$ and $B(-\infty)$ are B^+ and B^-, respectively (see Fig. 7-30). Thus in the y-direction we have $|\mathbf{B}_{12}(u)| = \overline{B^-B^+}$, which is equal, from the Cornu spiral, to $\sqrt{2}$. In other words, the line from B^- through O_s to B^+, i.e. from point $(-0.5, -0.5)$ to $(0.5, 0.5)$, is $\sqrt{2}$ units long. Hence the above expression applied to a narrow slit becomes

$$I_P = \frac{I_0}{4}|\mathbf{B}_{12}(v)|^2$$

For the data,

$$v = z\left[\frac{2(\rho_0 + r_0)}{\lambda \rho_0 r_0}\right]^{1/2}$$

becomes

$$v = z\left[\frac{2(1 + 4)}{400 \times 10^{-9}(1)4}\right]^{1/2} = z(2.5 \times 10^3)$$

For P opposite the middle of the slit, $z_1 = -0.1$ mm, $z_2 = +0.1$ mm, and therefore $v_1 = -0.25$, $v_2 = +0.25$. The arc length from $B(v_1)$ to $B(v_2)$ is $\Delta v = v_2 - v_1 = 0.5$; therefore, the chord $|\mathbf{B}_{12}|$ is very slightly less than 0.5, and

$$I(0) \approx \frac{I_0}{2}(0.5)^2 \approx 0.125\,I_0$$

7.48. In Problem 7.47, the slit was quite small and the fringe system resembled the Fraunhofer case. Supposing the slit now has a width of 1.6 mm, make a plot of I, leaving everything else unaltered.

Once again $v = z(2.5 \times 10^3)$, but now, opposite dead center, $z_1 = -0.8$ mm, $z_2 = +0.8$ mm, hence $v_2 = +2$, $v_1 = -2$ and $\Delta v = 4$. The quantity Δv is independent of the location of P, provided P stays a constant distance behind the screen. It can be thought of as a length of string lying atop the spiral. Moving P vertically merely causes the string to slide up or down the spiral, thereby changing the straight-line distance $|\mathbf{B}_{12}|$ between its endpoints. Rather than plotting I against z, let us graph $|\mathbf{B}_{12}(v)|^2$ versus $(v_1 + v_2)/2$, which is the midpoint location of the string.

Table 7-2

$(v_1 + v_2)/2$	$|\mathbf{B}_{12}|$	$|\mathbf{B}_{12}|^2$
0	1.19	1.42
0.1	1.23	1.51
0.2	1.33	1.77
0.3	1.44	2.07
0.4	1.56	2.43
0.5	1.59	2.53
0.6	1.57	2.46
0.7	1.54	2.37
0.8	1.55	2.40
0.9	1.63	2.66
1.0	1.67	2.79
1.1	1.65	2.72
1.2	1.45	2.10
1.3	1.27	1.61
1.4	1.15	1.32
1.5	1.13	1.28
1.6	1.11	1.23
1.7	1.02	1.04
1.8	0.87	0.76
1.9	0.70	0.49
2.0	0.65	0.42

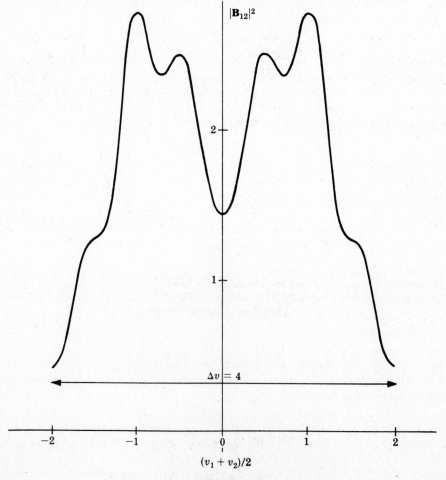

Fig. 7-33

The results, which are applicable to any slit for which $\Delta v = 4$, are indicated in Table 7-2 and Fig. 7-33. This is the pattern as it would be measured by a small probe moving vertically across the slit. What would actually be seen is a gray central horizontal band followed symmetrically on either side by bright and dark regions.

7.49. Write an expression for the irradiance at a point beyond a semi-infinite planar opaque screen and discuss the near-field pattern.

First, envision a narrow horizontal slit in a vertical screen, for which (see Problem 7.47)

$$I_P = \frac{I_0}{2}|\mathbf{B}_{12}(v)|^2$$

$$= \frac{I_0}{2}\{[C(v_2) - C(v_1)]^2 + [\mathcal{S}(v_2) - \mathcal{S}(v_1)]^2\}$$

Now suppose the upper opaque portion of the screen is removed leaving one straight edge. Clearly $z_2 \to \infty$, as does v_2, and so $C(v_2) \to 1/2$, as does $\mathcal{S}(v_2)$. Hence

$$I_P = \frac{I_0}{2}\left\{\left[\frac{1}{2} - C(v_1)\right]^2 + \left[\frac{1}{2} - \mathcal{S}(v_1)\right]^2\right\}$$

The spiral of Fig. 7-34 displays $\mathbf{B}_{12}$ for five different vertical locations of P, ranging from (1) below the edge to (5) well above the edge. Figure 7-35 is the corresponding irradiance as might be measured by a probe moving vertically. Of course, the fringes are horizontal bands.

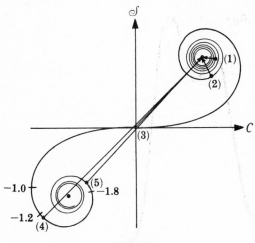

Fig. 7-34

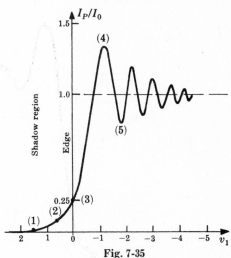

Fig. 7-35

7.50. A large opaque sheet of cardboard is held vertically so that its top edge is horizontal. Plane waves ($\lambda_0 = 640$ nm) impinge on it normally. Calculate the irradiance at a point 2 m behind the screen and 0.8 mm below its edge.

From Problem 7.49

$$I_P = \frac{I_0}{2}\left\{\left[\frac{1}{2} - C(v_1)\right]^2 + \left[\frac{1}{2} - \mathcal{S}(v_1)\right]^2\right\}$$

where

$$v_1 = z_1\left[\frac{2}{\lambda r_0}\right]^{1/2}$$

Here

$$v_1 = 0.8 \times 10^{-3}\left[\frac{2}{(640 \times 10^{-9})2}\right]^{1/2} = 1.0$$

and so

$$I_P = \frac{I_0}{2}[(0.5 - 0.7799)^2 + (0.5 - 0.4383)^2]$$

This yields

$$I_P = \frac{I_0}{2}[0.0783 + 0.0038] = 0.041\,I_0$$

7.51. A He-Ne laser beam ($\lambda_0 = 632.8$ nm) passes through a positive lens of focal length 25 cm. A 0.4-mm diameter straight wire is held in the beam vertically 225 cm from the lens and the near-field diffraction pattern is examined on a screen 3 m from the wire. Use the Cornu spiral to calculate the irradiance at a central point, $I(0)$.

As viewed from a central point, $y_1 = -0.2$ mm and $y_2 = 0.2$ mm. Hence, since

$$u = y\left[\frac{2(\rho_0 + r_0)}{\lambda \rho_0 r_0}\right]^{1/2}$$

and $\lambda = 632.8 \times 10^{-9}$ m, $\rho_0 = 2$ m, and $r_0 = 3$ m, we get

$$u_1 = -0.325 \qquad u_2 = +0.325$$

Thus, $\Delta u = 0.650$ and the spiral of Fig. 7-36 shows the two contributions to the field,

$$\overrightarrow{B^- B(u_1)} + \overrightarrow{B(u_2)B^+}$$

both to equal about 0.51 in length. Hence,

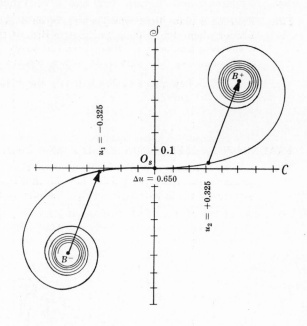

Fig. 7-36

$$I(0) = \frac{I_0}{2}(1.02)^2 = 0.52\,I_0$$

Supplementary Problems

RADIATION FROM A COHERENT LINE SOURCE

7.52. Verify that $I(\theta)$ for Young's experiment is a special case of the linear array where $N = 2$.

7.53. A coherent linear array contains N elements. Prove that there will be $N - 1$ minima between each pair of principal maxima. How many secondary maxima exist in that region?

   Ans. $N - 2$

7.54. A radio-frequency interferometer at the University of Sydney consists of thirty-two, 2-m diameter parabolic antennas spaced 7 m apart. It operates at a wavelength of 21 cm. Compute the angular separation between principal maxima, as well as the width of the zeroth-order central peak.

   Ans. $1°43'$; $6'$

7.55. Suppose that an intrinsic phase difference of ε exists between adjacent emitters in a coherent line source. Write an expression for the angular orientation of the zeroth-order principal maximum.

   Ans. $\theta_0 = \sin^{-1}\dfrac{\varepsilon\lambda}{2\pi a}$

7.56.   Use the result of Problem 7.55 to determine the angular displacement which would occur in the orientation of the zeroth-order lobe with the introduction of a 30° phase shift between successive emitters in a linear coherent array. Assume there to be 20 point sources with a frequency of 10^9 Hz, the spacing being 60 cm.

   Ans.   See Problem 6.57.

7.57.   Imagine a plane light wave impinging on a planar smooth interface at an angle θ_i. Consider the row of atoms lying along the intersection of the surface and the plane of incidence to constitute a coherent line source. Show that the single principal maximum reradiated by the atoms will be at an angle $\theta_0 = \theta_i$; this is, of course, the law of reflection.

   Ans.   The key consideration is that ε, the induced phase shift between successive atoms, is equal to $(2\pi a \sin \theta_i)/\lambda$.

FRAUNHOFER DIFFRACTION BY ONE AND TWO NARROW SLITS

7.58.   Use the geometry of Fig. 7-37 to show that the location Z_m of the mth irradiance minimum is independent of the position of lens L_2 in Fig. 7-5. In other words, the irradiance pattern is un-altered by displacements of the lens (assuming the latter to be very large in diameter).

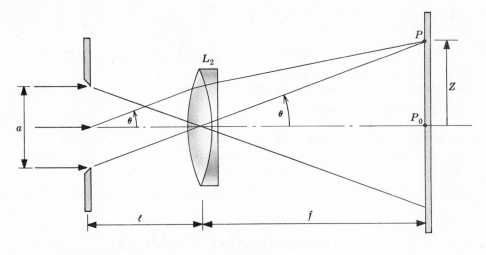

Fig. 7-37

   Ans. $Z_m = m\lambda f/(a^2 - m^2\lambda^2)^{1/2}$ and therefore independent of ℓ.

7.59.   How is the far-field single-slit pattern altered if the incident plane waves arrive at an angle of 30°?

   Ans.   Central fringe shifted to $\theta = 30°$, fringes widened by a factor of 1.15.

7.60.   Consider the far-field diffraction pattern of a single slit under polychromatic illumination. If the first minimum with λ_1 is found to be coincident with the third minimum at λ_2, determine the rela-tionship between these two wavelengths.

   Ans. $\lambda_1 = 3\lambda_2$

7.61.   Plane waves from a He-Ne laser at 632.8 nm impinge normally on a narrow slit 1 mm wide. What is the width W of the central peak at half-maximum irradiance on a screen 1 kilometer away?

   Ans. $W = 632.8$ mm

7.62. Plane waves at 550 nm are incident normally on a narrow slit having a 0.25 mm width. The Fraunhofer diffraction pattern resides in the focal plane of a large, 60-cm focal length lens. Compute the distance between the first minima on either side of the central axis.

Ans.   2.64 mm

7.63. Plane waves ($\lambda = 550$ nm) impinge normally on a 0.25-mm wide slit, thereby generating a Fraunhofer pattern in the focal plane of a collecting lens. If the separation between the two fourth-order minima is measured to be 1.25 mm, compute the focal length of the lens.

Ans. $f = 7.1$ cm

7.64. Show that there will be $2M$ bright fringes within the central diffraction peak in the far-field pattern of two slits each of width b and separated by a, provided that $a = Mb$.

7.65. Consider the far-field pattern of a double-slit arrangement. Fifteen bright fringes appear within the central diffraction peak. If each slit is 0.25 mm wide, by how much are they separated?

Ans. $a = 3.75$ mm. The seventh interference maximum coincides with the first diffraction minimum.

7.66. Two narrow slits separated by 0.4 mm are illuminated normally by plane waves of wavelength 550 nm. Nine bright fringes are observed on a screen 3 m away. On either side of these bright bands, several very weak fringes are visible. Determine the distance separating consecutive maxima. Calculate the width of each slit.

Ans. $\Delta Z = 4.125$ mm, $b = 0.089$ mm

MULTIPLE NARROW SLITS — THE DIFFRACTION GRATING

7.67. Show that the formula for N slits,

$$I(\theta) \;=\; I_0(\text{sinc}^2\,\beta)\left(\frac{\sin N\alpha}{\sin \alpha}\right)^2$$

reduces to the previously studied equations for one and two slits when $N = 1$ and 2, respectively.

7.68. Is there any limit to the number of principal maxima in the far-field pattern of an N-slit array?

Ans.   Yes, $m \leqq a/\lambda$.

7.69. Imagine that you have an array of N slits, where N is an odd number. Show that the irradiance of the subsidiary maximum residing midway between a pair of consecutive principal peaks equals $I(0)/N^2$, provided θ is small enough so that $\text{sinc}^2\,\beta \approx 1$.

Ans. $(\sin N\alpha/\sin \alpha)^2 = 1$, since α is odd multiple of $\pi/2$, as is $N\alpha$.

7.70. A beam of polychromatic light ranging in wavelength from 450 nm to 650 nm impinges normally on a 12,000-line-per-inch transmission grating. The pattern appears in the focal plane of a collecting lens which follows the grating. What focal length must the lens have if the second-order spectrum is to be 1.25 cm in extent?

Ans. $f = 5.63$ cm

7.71. Prove that the resolving power of a grating cannot be greater than aN/λ.

Ans.   See Problem 7.68.

7.72. A transmission grating having 16,000 lines per inch is 2.5 inches wide. Operating in the green at about 550 nm, what is the resolving power in the third order? Calculate the minimum resolvable wavelength difference in the second order.

Ans. $\mathcal{R}_3 = 120,000$, $(\Delta\lambda)_{\min} = 6.88 \times 10^{-3}$ nm

7.73. A one-meter long He-Ne laser operating at a mean wavelength of 632.8 nm has longitudinal cavity modes separated by 150 MHz. How large must a 200 line/mm grating be if it is blazed to operate in the second order, where it must resolve the laser's mode structure?

   Ans. $aN = 79$ cm

7.74. Can the first and second orders of a diffraction grating under visible illumination ever substantially overlap? Can the second and third orders overlap?

   Ans.   The 1st and 2nd just about miss (depending on your definition of visible), but the 2nd and 3rd certainly can overlap.

7.75. A diffraction grating has its third- and fourth-order spectra overlapping. What wavelength in the third order coincides with the 490 nm line in the fourth order?

   Ans. $\lambda = 653.3$ nm

RECTANGULAR AND CIRCULAR APERTURES — FRAUNHOFER DIFFRACTION

7.76. Suppose that a point source S of wavelength λ lies along the central axis a perpendicular distance L from an aperture whose maximum extent is d. Show that Fraunhofer diffraction will obtain on a distant screen when

$$L \gg \frac{d^2}{8\lambda}$$

(As a rule of thumb $L > d^2/\lambda$ is generally used to define the far field both in this instance and in the reverse, where the incident waves are planar and the observation screen is a distance L from the aperture.)

7.77. A circular hole of radius 1.25 mm in an opaque screen is illuminated perpendicularly by plane waves from a He-Ne laser ($\lambda = 632.8$ nm). Roughly how far away must a viewing screen be in order to see a Fraunhofer diffraction pattern without using any lenses?

   Ans. L greater than about 10 m (see Problem 7.76)

7.78. In the far-field pattern of a square aperture, determine the peak irradiance of the third bright spot lying along a diagonal ($\alpha' = \beta'$) as compared to $I(0)$.

   Ans. $I/I(0) \approx 0.000068$

7.79. How does the central peak in the far-field irradiance pattern of a rectangular aperture depend on the wavelength and hole area?

   Ans. $I(0) \propto A^2 \mathcal{E}_A^2$ and therefore $I(0) \propto A^2/\lambda^2$.

7.80. A telescope objective is 12 cm in diameter and has a focal length of 150 cm. Light of mean wavelength 550 nm from a distant star enters the scope as a nearly collimated beam. Compute the radius of the central disk of light forming the image of the star on the focal plane of the lens.

   Ans. $q_1 = 8.39 \times 10^{-3}$ mm

7.81. A laser beam can be so well collimated that it spreads out only as a result of diffraction. Suppose that we have just such a diffraction-limited He-Ne laser emitting a 2-mm diameter beam at 632.8 nm. Determine the beam diameter at a distance of 1 kilometer from the laser.

   Ans. ≈ 77 cm

7.82. Light from a very distant point is filtered so that only 450 nm enters a perfect collecting lens of focal length 225 mm. How large must the lens be if the image is to consist of a central spot $1\ \mu$ (i.e., 1 micron $= 10^{-6}$ m) in diameter?

Ans. $D\ =\ 20.25$ cm

7.83. The 140-ft diameter parabolic disk of the National Radio Astronomy Observatory in Green Bank is the largest fully steerable radio telescope. Compute its angular resolution for the 1420 MHz line emitted by interstellar hydrogen

Ans. 6×10^{-3} rad or 0.344 degrees

7.84. The expression for $(\Delta\phi)_{\min}$, the minimum angular source separation derived in Problem 7.32, applies to the eye as well, although because of the vitreous humor the angular separation on the retina is $(\Delta\phi)_{\min}/n$. In any event, assuming the eye to be diffraction-limited and having a pupil diameter of 2.5 mm, determine $(\Delta\phi)_{\min}$ for a mean wavelength of 550 nm. How far apart must two small balls be if they are to be just resolvable at 1 kilometer?

Ans. $(\Delta\phi)_{\min}\ =\ 2.68 \times 10^{-4}$ rad, 26.8 cm

7.85. The two headlights on my '69 Toyota are about 43 inches apart. How far away must I stand if the lights are to be just resolvable as two separate sources? Assume $\lambda = 550$ nm and a nighttime pupil diameter of 4 mm.

Ans. 6.5 km

FRESNEL DIFFRACTION — CIRCULAR SYSTEMS

7.86. A plane wave ($\lambda = 500$ nm) impinges normally on an opaque screen containing a 1-cm diameter hole. How many Fresnel zones will be uncovered by the aperture when viewed on-axis from 0.5 m away?

Ans. $\dfrac{R^2}{r_0\lambda}\ =\ 100$

7.87. A small irradiance probe sits on the central axis 2.25 m from an opaque screen containing a circular hole. Under normally incident plane wave illumination at $\lambda = 500$ nm, what hole radii will generate readings which are maxima and minima?

Ans. maxima: $R\ =\ 1.06$ mm, 1.84 mm, 2.37 mm

   minima: $R\ =\ 1.50$ mm, 2.12 mm, 2.60 mm

7.88. Collimated He-Ne laser light ($\lambda = 632.8$ nm) impinges normally on a circular aperture of radius 0.7955 mm. What is the irradiance at an axial point 2 m from the hole in terms of the incident irradiance?

Ans. $I\ =\ 2I_0$

7.89. A collimated beam of wavelength 500 nm and an irradiance of 40 W/m² is incident normally on a screen having an aperture as indicated in Fig. 7-38. Determine the irradiance at an axial point 4 m away.

Ans. $I\ =\ 90$ W/m²

7.90. As viewed from an axial point P, a circular aperture uncovers the first one and one-half Fresnel zones. What is the irradiance at P as compared to the unobstructed value of I_0? Assume incident monochromatic plane waves.

Ans. From the vibration curve, $I = 2I_0$.

7.91. Light from a very distant point source ($\lambda = 500$ nm) arrives normally on the aperture shown in Fig. 7-39. Compute the irradiance at an axial point 4 m from the hole if the incident irradiance is 25 W/m².

Ans. $I_0\ =\ 100$ W/m²

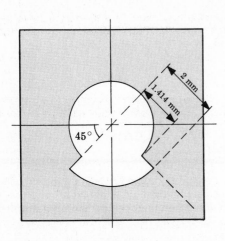

Fig. 7-38

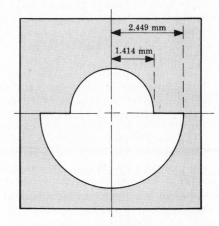

Fig. 7-39

7.92. The obstruction shown in Fig. 7-40 is illuminated by normally incident plane waves ($\lambda = 500$ nm). Discuss how the vibration curve can be used to compute the flux density at point P, 4 m away on the central axis.

> *Ans.* Field for obstruction by central disk only, $\overrightarrow{Z_{s1}O'_s}$; for two-zone opaque disk, $\overrightarrow{Z_{s2}O'_s}$. Therefore second-zone obstruction contributes $2\,\overrightarrow{Z_{s2}O'_s}$. Hence field here is $\overrightarrow{Z_{s1}O'_s} - 2\,\overrightarrow{Z_{s2}O'_s}/2$ and $I \approx 0$.

7.93. One form of *zone plate* is a sheet of clear material with alternately opaque and transparent concentric rings painted on it. The arrangement is such that when placed a distance ρ_0 from a point source S and viewed at P a distance r_0 away, it will obscure alternate Fresnel zones on the wavefront. Show that

$$ f = \frac{R_m^2}{m\lambda} $$

is the equivalent focal length of the device inasmuch as

$$ \frac{1}{\rho_0} + \frac{1}{r_0} = \frac{1}{f} $$

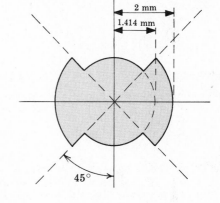

Fig. 7-40

Compute the radius of the first zone of the plate when $\rho_0 = r_0 = 5\ m$ and $\lambda = 500$ nm.

> *Ans.* $R_1 = 0.11$ cm

7.94. Compute the irradiance at the focus of a zone plate in terms of the unobstructed value I_0, when all the zones with the exception of the first are obscured.

> *Ans.* $I \approx 4I_0$

7.95. What is the focal length of a zone plate for light of wavelength 650 nm if the radius of the 8th ring is 4.5 mm? Compute the image distance for a point source 7.788 m from the plate.

> *Ans.* $f = 3.89$ m, $r_0 = 7.788$ m

FRESNEL DIFFRACTION — STRAIGHT EDGES

7.96. Show that the near-field irradiance $I(0)$ at a point opposite the center of a wide slit approaches I_0, the unobstructed value.

7.97. The angle β formed by a tangent to the Cornu spiral and the C-axis through O_s is the relative phase. Show that the slope of the spiral is $\tan(\pi w^2/2)$ and from that, $\beta = \pi w^2/2$. Locate the points on the spiral corresponding to the boundaries of the Fresnel strip zones, i.e. find the horizontal tangent points. Locate the vertical tangents as well.

Ans.   horizontal: $w = \sqrt{2}, \sqrt{4}, \sqrt{6}, \ldots$

vertical: $w = \sqrt{1}, \sqrt{3}, \sqrt{5}, \ldots$

7.98. A horizontal line source of vacuum wavelength 600 nm is positioned 2 m from and parallel to a long narrow horizontal slit 0.25 mm wide. What is the irradiance at a point on the central axis 3 m from the aperture? Compute the arc length on the Cornu spiral corresponding to Δz.

Ans. $I(0) = 0.09\,I_0$, $\Delta v = 0.417$

7.99. A plane wave ($\lambda = 640$ nm) impinges normally on a horizontal, narrow slit of width 0.4 mm. Calculate the irradiance on a screen 2 m away at a point 1 mm below the center line, in terms of the unobstructed value of I_0.

Ans. $I = 0.0896\,I_0$

7.100. Determine the greatest possible value of irradiance which will be measured by a small probe directly opposite the center of a long narrow slit whose width is variable.

Ans. $I = 1.8\,I_0$

7.101. Using the Cornu spiral, explain how a very narrow slit generates a pattern approaching the Fraunhofer situation.

Ans.   Uncovering a fraction of a zone corresponds to Fraunhofer diffraction and a small Δv. Only after many turns around either B^- or B^+ will the endpoints of the arc Δv approach each other, yielding a near-zero minimum. Look at $\mathbf{B}_{12} \approx 0$ for $\Delta v = 0.5$ extending from $v_1 = 3.8$ to $v_2 = 4.3$.

7.102. A horizontal line source of $\lambda = 500$ m, 1 m from a parallel narrow slit, produces a near-field pattern on a screen 4 m from the aperture. Roughly what must the slit width be if the irradiance at an axial point is to be a maximum?

Ans. $\Delta z \approx 1.13$ mm, $\Delta v \approx 2.53$

7.103. Prove that the irradiance for a semi-infinite screen (Fig. 7-35) has a value of $I_0/4$ opposite the edge of the obstruction.

7.104. A plane wave ($\lambda_0 = 400$ nm) is incident normally on an opaque half-plane with a horizontal upper edge. Locate the positions of the first maximum and minimum on a screen 10 m away.

Ans.   1st min., $z = 2.66$ mm; 1st max., $z = 1.78$ mm

7.105. A narrow opaque strip 1.766 mm wide is under plane wave illumination from a ruby laser at 693.4 nm. The diffraction pattern appears on a screen 1 m from the strip. Calculate the irradiance at a central point and sketch I/I_0.

Ans. $I(0) = 0.08\,I_0$
(See Fig. 7-41.)

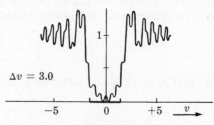

Fig. 7-41

Chapter 8

Introduction to Fourier Optics

8.1 PERIODIC WAVES AND FOURIER SERIES

An interesting thing happens when you add two sinusoidal waves of different frequencies, such as those of Fig. 8-1. The resultant is itself not sinusoidal; which rather sug-

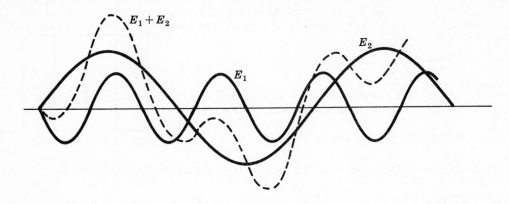

Fig. 8-1

gests that complicated waveforms can be generated by the judicious selection of harmonic contributions of different frequencies, amplitudes and phases. Indeed *Fourier's theorem* states that *a function f(x), of spatial period λ, can be synthesized as a sum of harmonic functions whose wavelengths are integral submultiples of λ (i.e. λ, λ/2, λ/3, etc.).* In other words, if $f(x)$ is a periodic function of wavelength λ, it can be represented by a *Fourier series* of the form

$$f(x) = C_0 + C_1 \cos\left(\frac{2\pi}{\lambda}x + \varepsilon_1\right) + C_2 \cos\left(\frac{2\pi}{\lambda/2}x + \varepsilon_2\right) + \cdots$$

wherein the C-terms are constants specifying the amplitudes of the various contributions. Notice that changing x to $x - vt$ creates a traveling wave out of the profile and we can think of anharmonic disturbances as sums of sinusoidal waves.

An equivalent restatement of the above series is the more common representation fabricated of both sines and cosines, namely

$$f(x) = \frac{A_0}{2} + \sum_{m=1}^{\infty} A_m \cos mkx + \sum_{m=1}^{\infty} B_m \sin mkx$$

The amplitude cofficients are computed from the following integral expressions:

$$A_m = \frac{2}{\lambda} \int_0^\lambda f(x) \cos mkx \, dx \qquad (m = 0, 1, 2, \ldots)$$

$$B_m = \frac{2}{\lambda} \int_0^\lambda f(x) \sin mkx \, dx \qquad (m = 1, 2, 3, \ldots)$$

The integrals could also be evaluated over any other one-period interval.

If the function being synthesized is *even*, it is mirror-symmetrical about the $x = 0$ axis, i.e. $f(x) = f(-x)$. In that instance the Fourier series representation contains only even functions, that is, only cosine terms, and so $B_m = 0$ for all m. On the other hand, should the function be odd, i.e. $f(x) = -f(-x)$, the Fourier series contains only the odd sine functions; $A_m = 0$ for all m. To be sure, $f(x)$ does not have to be either odd or even, in which case the series consists of both sines and cosines.

Keep in mind that for now we are dealing only with periodic waves of infinite extent.

SOLVED PROBLEMS

8.1. Compute the Fourier series representation of the periodic function

$$f(x) = \begin{cases} +1 \text{ when } 0 < x < \lambda/2, \text{ etc.} \\ -1 \text{ when } \lambda/2 < x < \lambda, \text{ etc.} \end{cases}$$

as depicted in Fig. 8-2.

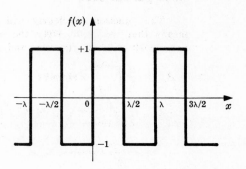

Fig. 8-2

Evidently $f(x)$ is odd and therefore all cosine terms are absent; $A_m = 0$. The coefficients B_m are computed from

$$B_m = \frac{2}{\lambda} \int_0^\lambda f(x) \sin mkx \, dx$$

which upon substitution of the actual value of $f(x)$ becomes

$$B_m = \frac{2}{\lambda} \int_0^{\lambda/2} (+1) \sin mkx \, dx + \frac{2}{\lambda} \int_{\lambda/2}^\lambda (-1) \sin mkx \, dx$$

$$= \frac{1}{m\pi} \left[-\cos mkx \right]_0^{\lambda/2} + \frac{1}{m\pi} \left[\cos mkx \right]_{\lambda/2}^\lambda = \frac{2}{m\pi} (1 - \cos m\pi)$$

In other words,

$$B_1 = \frac{4}{\pi}, \quad B_2 = 0, \quad B_3 = \frac{4}{3\pi}, \quad B_4 = 0, \quad \text{etc.}$$

and the desired series is

$$f(x) = \frac{4}{\pi} \left(\sin kx + \frac{1}{3} \sin 3kx + \frac{1}{5} \sin 5kx + \cdots \right)$$

Incidentally, when a function looks the same above and below the axis, its series representation will contain only *odd harmonics*, i.e. only odd multiples of k, the *fundamental angular spatial frequency*. If the curve of Fig. 8-2 is rotated 180° about the x-axis and advanced $\lambda/2$, it is unchanged, and this sort of behavior is sometimes spoken of as *screw symmetry*.

8.2. Graphically add the first three contributions to the Fourier series for the square wave of Problem 8.1.

We call the sum of the first N terms of a Fourier series its Nth *partial sum*. For the square wave the first partial sum, Σ_1, is just

$$\frac{4}{\pi} \sin kx = 1.3 \sin kx$$

as plotted in Fig. 8-3(a). The second partial sum,

$$\Sigma_2 = \Sigma_1 + \frac{4}{3\pi} \sin 3kx$$

is plotted in Fig. 8-3(b). In it the third harmonic (with frequency $3k$) has an amplitude of only 0.4. The third partial sum,

$$\Sigma_3 \;=\; \Sigma_2 + \frac{4}{5\pi}\,\sin 5kx$$

is pictured in Fig. 8-3(c). Notice that each term in the series is a positive sine, all of which are therefore in phase at $x = 0$.

8.3. Derive the Fourier series representation of the periodic function depicted in Fig. 8-4. Plot each of the first six harmonics, as well as the sixth partial sum.

The function is clearly odd, which means that $A_m = 0$. Over the interval from $-\lambda/2$ to $+\lambda/2$, $f(x) = x$ and so

$$B_m \;=\; \frac{2}{\lambda}\int_{-\lambda/2}^{+\lambda/2} x\,\sin mkx\; dx$$

$$= \frac{2}{\lambda}\left[\frac{\sin mkx}{(mk)^2} - \frac{x\cos mkx}{mk}\right]_{-\lambda/2}^{\lambda/2}$$

$$= -\frac{2}{mk}\cos m\pi$$

and so

$$f(x) \;=\; \frac{2}{k}\Bigg(\sin kx - \frac{1}{2}\,\sin 2kx$$

$$+ \frac{1}{3}\,\sin 3kx - \cdots\Bigg)$$

Figure 8-5 shows the first six harmonics, as well as the sixth partial sum. Observe that the minus signs in the series are equivalent to a phase shift.

8.4. Figure 8-6, page 208, shows a periodic function, the repeating element of which corresponds to $f(x) = x^2$. Compute the appropriate Fourier series.

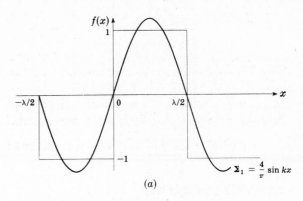

(a)

$$\Sigma_1 = \frac{4}{\pi}\,\sin kx$$

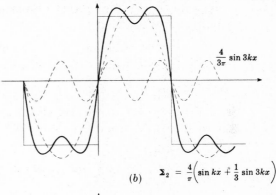

$$\frac{4}{3\pi}\,\sin 3kx$$

(b) $$\Sigma_2 = \frac{4}{\pi}\left(\sin kx + \frac{1}{3}\,\sin 3kx\right)$$

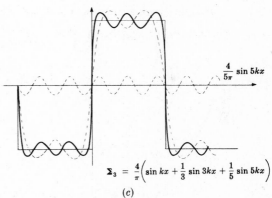

$$\frac{4}{5\pi}\,\sin 5kx$$

$$\Sigma_3 = \frac{4}{\pi}\left(\sin kx + \frac{1}{3}\,\sin 3kx + \frac{1}{5}\,\sin 5kx\right)$$

(c)

Fig. 8-3

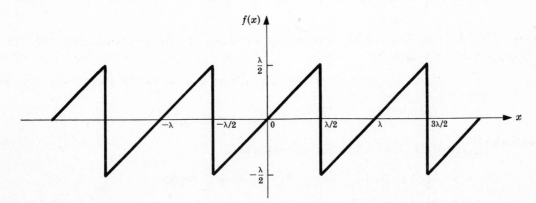

Fig. 8-4

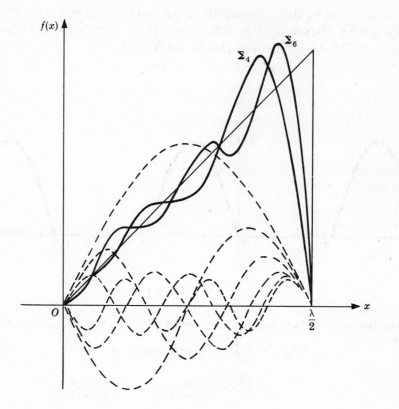

Fig. 8-5

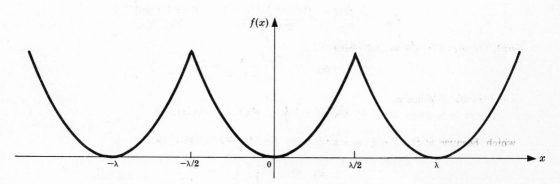

Fig. 8-6

The function is evidently even, hence $B_m = 0$ and

$$A_m = \frac{2}{\lambda} \int_{-\lambda/2}^{+\lambda/2} x^2 \cos mkx \, dx$$

We need first to evaluate A_0:

$$A_0 = \frac{2}{\lambda} \int_{-\lambda/2}^{+\lambda/2} x^2 \, dx = \frac{2}{3}\left(\frac{\lambda}{2}\right)^2$$

Integrating the expression for A_m $(m > 0)$ yields

$$A_m = \frac{2}{\lambda}\left[\frac{2x}{(mk)^2} \cos mkx - \frac{2}{(mk)^3} \sin mkx + \frac{x^2}{mk} \sin mkx \right]_{-\lambda/2}^{+\lambda/2} = \frac{4(-1)^m}{(mk)^2}$$

whence

$$f(x) = \frac{1}{3}\left(\frac{\lambda}{2}\right)^2 + \sum_{m=1}^{\infty} \frac{4(-1)^m}{(mk)^2} \cos mkx$$

8.5. Imagine the electric field component of an electromagnetic wave to have the rather unlikely profile shown in Fig. 8-7. This is the familiar half-wave rectified sine function. Compute its Fourier series representation.

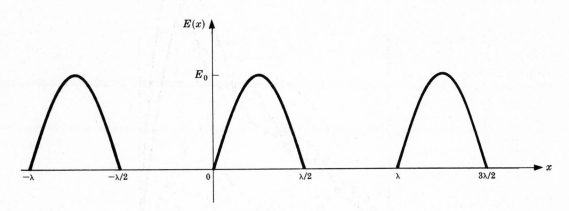

Fig. 8-7

This time the function, with the vertical axis at the given location, is neither odd nor even. We have:

$$A_m = \frac{2}{\lambda} \int_0^{\lambda/2} E_0 \sin kx \cos mkx \, dx$$

$$= \frac{2E_0}{\lambda} \left[-\frac{\cos k(1-m)x}{2k(1-m)} - \frac{\cos k(1+m)x}{2k(1+m)} \right]_0^{\lambda/2}$$

except when $m = 1$, in which case the integral is zero (i.e., $A_1 = 0$). Carrying on,

$$A_m = \frac{2E_0}{\lambda} \left\{ -\frac{\cos[\pi(1-m)] - 1}{2k(1-m)} - \frac{\cos[\pi(1+m)] - 1}{2k(1+m)} \right\}$$

which is zero for odd m and yields

$$A_m = \frac{E_0}{\pi} \left[\frac{1}{1-m} + \frac{1}{1+m} \right]$$

for even m. Similarly,

$$B_m = \frac{2}{\lambda} \int_0^{\lambda/2} E_0 \sin kx \sin mkx \, dx$$

which, because of the limits, is zero for all $m \neq 1$. When, however, $m = 1$,

$$B_1 = \frac{2E_0}{\lambda} \left[\frac{x}{2} \right]_0^{\lambda/2} = \frac{E_0}{2}$$

Hence

$$E(x) = \frac{E_0}{\pi} + \frac{E_0}{2} \sin kx - \frac{2E_0}{\pi} \left(\frac{1}{3} \cos 2kx + \frac{1}{15} \cos 4kx + \cdots \right)$$

8.6. Envision a plane wave of amplitude E_0 impinging normally on a large horizontal *Ronchi ruling*. The latter is a grating formed of alternately transparent and opaque strips, each of width b. The emerging electric field over the screen or aperture is a step function — compute its Fourier series representation, assuming it to have effectively infinite extent.

Figure 8-8 is the field as seen along a vertical line across the ruling. By locating the origin mid-peak, we create an even function for which $B_m = 0$. Thus, with $b = \lambda/2$,

$$A_0 = \frac{2}{\lambda} \int_{-\lambda/4}^{+\lambda/4} E_0 \, dy = E_0$$

Bear in mind that here λ is the spatial period of the aperture function and not the wavelength of the incident wave. Continuing,

$$A_m = \frac{2}{\lambda} \int_{-\lambda/4}^{+\lambda/4} E_0 \cos mky \, dy$$

$$= \frac{2E_0}{mk\lambda} \left[\sin mky \right]_{-\lambda/4}^{+\lambda/4}$$

$$= \frac{2E_0}{m\pi} \sin \frac{m\pi}{2} = E_0 \operatorname{sinc} \frac{m\pi}{2}$$

Consequently, the field across the aperture can be thought of as having a specific harmonic content, namely

$$E(y) = \frac{E_0}{2} + \sum_{m=1}^{\infty} E_0 \operatorname{sinc} \frac{m\pi}{2} \cos mky$$

As in Problem 8.5, the function resides above the axis and, therefore, has a nonzero *average value*, i.e. $A_0 \neq 0$. In effect, this means that to go from a situation such as that of Fig. 8-2 to the present case, we must raise up the function by the inclusion of a constant or "dc" term $(E_0/2)$.

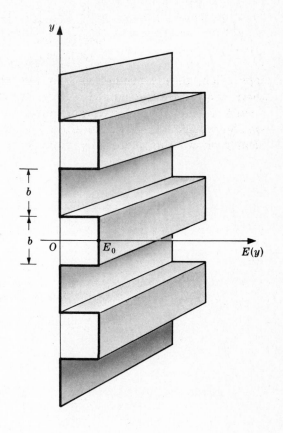

8.7. Derive a complex exponential form of the Fourier series.

Using the identities

$$e^{iu} + e^{-iu} = 2 \cos u$$

$$e^{iu} - e^{-iu} = 2i \sin u$$

Fig. 8-8

the trigonometric statement

$$f(x) = \frac{A_0}{2} + \sum_{m=1}^{\infty} A_m \cos mkx + \sum_{m=1}^{\infty} B_m \sin mkx$$

can be reformulated as

$$f(x) = \frac{A_0}{2} + \sum_{m=1}^{\infty} A_m \frac{e^{imkx} + e^{-imkx}}{2} - i \sum_{m=1}^{\infty} B_m \frac{e^{imkx} - e^{-imkx}}{2}$$

Shifting terms around a bit leads to

$$f(x) = \frac{A_0}{2} + \sum_{m=1}^{\infty} \frac{A_m - iB_m}{2} e^{imkx} + \sum_{m=1}^{\infty} \frac{A_m + iB_m}{2} e^{-imkx}$$

or

$$f(x) = C_0 + \sum_{m=1}^{\infty} C_m e^{imkx} + \sum_{m=1}^{\infty} C_{-m} e^{-imkx}$$

where $C_0 \equiv A_0/2$, and, for $m = 1, 2, 3, \dots$,

$$C_m \equiv \frac{A_m - iB_m}{2} \qquad C_{-m} \equiv \frac{A_m + iB_m}{2}$$

A still more concise form comes from allowing negative spatial frequencies, i.e. negative values of mk, whereupon

$$f(x) = \sum_{m=-\infty}^{m=+\infty} C_m e^{imkx}$$

The complex coefficients C_m are now given for all m by

$$C_m = \frac{1}{\lambda} \int_0^{\lambda} f(x) e^{-imkx} \, dx$$

8.2 FOURIER TRANSFORMS

Come back for a moment to the periodic square wave of Fig. 8-8. The coefficients of the appropriate Fourier series, i.e. the A_m, vary as a sinc function. Since these terms are weighting factors which specify how much of each harmonic component at any one spatial frequency is present in the synthesis, a plot of their values, as in Fig. 8-9, is known as the *frequency spectrum*. Notice in the figure that as the peaks of the square wave increase in separation, they represent smaller and smaller fractions of the wavelength, even though their widths are unchanged. Now, as the details of the function being reproduced get smaller in comparison to λ, the Fourier components themselves must have correspondingly smaller wavelengths and, therefore, higher spatial frequencies. The spectrum of Fig. 8-9(c) clearly shows this increasing number of frequency components.

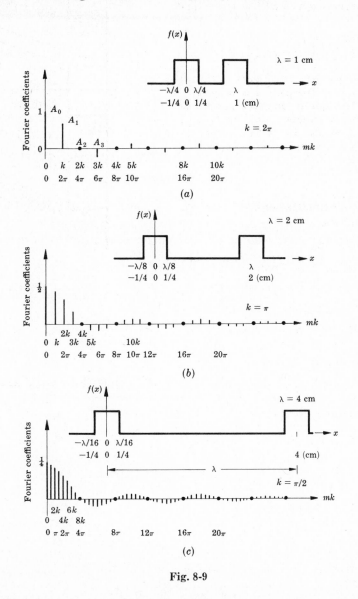

Fig. 8-9

When λ is envisioned as extending to infinity, leaving a *single pulse* rather than a periodic function, the discrete set of spatial frequencies, i.e. all the mk-values, smooth out into a continuous distribution. The Fourier series representation of the periodic function changes to the *Fourier integral* for a *nonperiodic function*:

$$f(x) = \frac{1}{\pi}\left[\int_0^\infty A(k)\cos kx\,dk + \int_0^\infty B(k)\sin kx\,dk\right]$$

where
$$A(k) = \int_{-\infty}^{+\infty} f(x)\cos kx\,dx \qquad B(k) = \int_{-\infty}^{+\infty} f(x)\sin kx\,dx$$

The latter two quantities are known as the *Fourier cosine and sine transforms* of the function $f(x)$.

Once again a convenient complex representation can also be generated and one finds it to be
$$f(x) = \frac{1}{2\pi}\int_{-\infty}^{+\infty} F(k)\,e^{-ikx}\,dk$$

where
$$F(k) = \int_{-\infty}^{+\infty} f(x)\,e^{ikx}\,dx$$

The function $F(k)$ is spoken of as the *Fourier transform* of $f(x)$, and we write
$$F(k) = A(k) + iB(k) = \mathcal{F}\{f(x)\}$$

Adopting a similar notation for the cosine and sine transforms, we have
$$\mathcal{F}\{f(x)\} = \mathcal{F}_C\{f(x)\} + i\mathcal{F}_S\{f(x)\}$$

The quantity $f(x)$ is said to be the *inverse Fourier transform* of $F(k)$, or
$$f(x) = \mathcal{F}^{-1}\{F(k)\}$$

Our particular present interest in the *Fourier pair*, $f(x)$ and $F(k)$, lies far more with the transform itself than with the process of actually synthesizing the function. Moreover, as the functions of concern are usually two-dimensional in the spatial domain, we generalize the Fourier transform as follows:

$$f(x, y) = \frac{1}{(2\pi)^2}\iint_{-\infty}^{+\infty} F(k_x, k_y)\,e^{-i(k_x x + k_y y)}\,dk_x\,dk_y$$

and
$$F(k_x, k_y) = \iint_{-\infty}^{+\infty} f(x, y)\,e^{i(k_x x + k_y y)}\,dx\,dy$$

In Section 7.5 we saw that the diffracted field in the Fraunhofer case was given by
$$E(Y, Z) = \frac{e^{i(\omega t - kR)}}{R}\iint_{\text{aperture}} \mathcal{E}_A(y, z)\,e^{ik(Yy + Zz)/R}\,dy\,dz$$

where now we allow the possibility that the source strength is a variable over the aperture. The exponential in front of the integral contributes only to the phase of the wave at (Y, Z), while the $1/R$ term, which generates the drop-off in amplitude in going from the aperture to the plane of observation, is constant over that plane in the Fraunhofer approximation. Therefore, as far as the amplitude distribution in the diffracted field is concerned, the factor multiplying the integral is inessential. Defining the spatial frequency parameters k_Y and k_Z by

$$k_Y \equiv \frac{kY}{R} \qquad k_Z \equiv \frac{kZ}{R}$$

and introducing the *aperture function*, $\mathcal{A}(y, z)$, we then have for the diffracted field:

$$E(k_Y, k_Z) = \iint_{-\infty}^{+\infty} \mathcal{A}(y, z)\,e^{i(k_Y y + k_Z z)}\,dy\,dz$$

This, of course, just means that

$$E(k_y, k_z) = \mathcal{F}\{\mathcal{A}(y, z)\}$$

The diffracted field in the Fraunhofer case is the Fourier transform of the field distribution over the aperture as expressed by $\mathcal{A}(y, z)$. In other words, each spot of light in the diffraction pattern signifies the presence of a particular spatial frequency component in the synthesis of the aperture function. This is just one of many examples of the applicability of Fourier methods in optics.

SOLVED PROBLEMS

8.8. Calculate the Fourier transform of the square pulse in Fig. 8-10.

The origin is located so as to make the function even and thereby nullify the sine transform. Hence, the cosine transform is

$$\mathcal{F}_C\{f(x)\} = A(k) = \int_{-\infty}^{+\infty} f(x) \cos kx \, dx = \int_{-L/2}^{+L/2} E_0 \cos kx \, dx$$

or

$$A(k) = \left[\frac{E_0}{k} \sin kx\right]_{-L/2}^{+L/2} = \frac{2E_0}{k} \sin \frac{kL}{2} = E_0 L \operatorname{sinc} \frac{kL}{2}$$

Just as in Problem 8.6 we have a sinc function, and it is the envelope of the frequency spectrum of Fig. 8-9 as well. If you would like to know how strongly a given frequency contributes to making up $f(x)$, you need only plug that value of k into the sinc function and get the specific $A(k)$.

Notice that $f(x)$ can be related to the aperture function of a long narrow slit, whereupon its transform resembles the diffracted field magnitude as studied in Section 7.3.

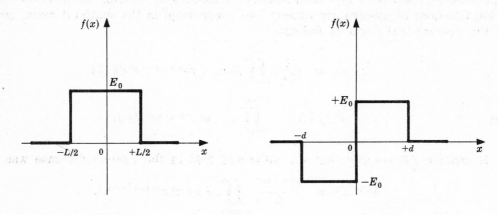

Fig. 8-10   Fig. 8-11

8.9. Compute and plot the complex Fourier transform of the function depicted in Fig. 8-11.

Inasmuch as

$$F(k) = \int_{-\infty}^{+\infty} f(x) \, e^{ikx} \, dx$$

we must evaluate

$$\mathcal{F}\{f(x)\} = \int_{-d}^{0} -E_0 e^{ikx} \, dx + \int_{0}^{+d} E_0 e^{ikx} \, dx$$

This is simply

$$\mathcal{F}\{f(x)\} = \left[-\frac{E_0}{ik} e^{ikx}\right]_{-d}^{0} + \left[\frac{E_0}{ik} e^{ikx}\right]_{0}^{+d}$$

$$= -\frac{2E_0}{ik} + \frac{2E_0}{ik}\left(\frac{e^{ikd} + e^{-ikd}}{2}\right)$$

$$= \frac{2iE_0}{k}(1 - \cos kd)$$

$$= 2iE_0 d\,\frac{\sin^2(kd/2)}{kd/2}$$

Figure 8-12 is a plot of the transform.

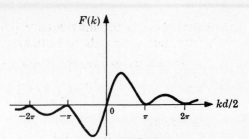

Fig. 8-12

8.10. The pulse illustrated in Fig. 8-13 can be expressed as

$$E(x) = \begin{cases} E_0 \cos k_p x & \text{when } -L \le x \le L \\ 0 & \text{when } |x| > L \end{cases}$$

Or, if $P(x)$ is a unit-amplitude square pulse, as in Fig. 8-14, $E(x) = P(x)\,E_0 \cos k_p x$. In either event, k_p is the spatial frequency of the oscillatory region of the pulse. Compute $\mathcal{F}\{E(x)\}$.

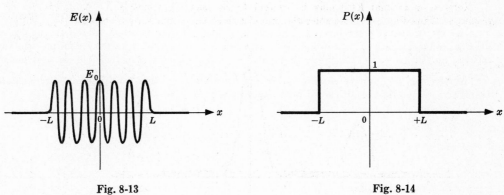

Fig. 8-13   Fig. 8-14

Because $E(x)$ is even,

$$\mathcal{F}\{E(x)\} = A(k) = \int_{-L}^{+L} E_0 \cos k_p x \cos kx\,dx$$

Using the identity

$$\cos(\alpha \pm \beta) = \cos\alpha\cos\beta \mp \sin\alpha\sin\beta$$

the integral becomes

$$\mathcal{F}\{E(x)\} = \int_{-L}^{+L} E_0 \frac{1}{2}\left[\cos(k_p + k)x + \cos(k_p - k)x\right]dx$$

and hence

$$\mathcal{F}\{E(x)\} = E_0 L\left[\frac{\sin(k_p + k)L}{(k_p + k)L} + \frac{\sin(k_p - k)L}{(k_p - k)L}\right]$$

$$= E_0 L[\operatorname{sinc}(k_p + k)L + \operatorname{sinc}(k_p - k)L]$$

If $L \gg \lambda_p$, $(k_p + k)L \gg 2\pi$ and

$$\mathcal{F}\{E(x)\} \approx E_0 L \operatorname{sinc}(k_p - k)L$$

8.11. Determine the Fourier transform of the wave train given by

$$E(x) = P(x)\cos^2 k_p x$$

where $P(x)$ is the unit square pulse of Fig. 8-14.   Sketch the transform in the limit as the pulse extends to infinity.

We will use the complex transform and so represent $E(x)$ in exponential form as follows:

$$\cos^2 k_p x \;=\; \frac{1}{2} + \frac{1}{2}\cos 2k_p x \;=\; \frac{1}{2} + \frac{1}{4}\left(e^{i2k_p x} + e^{-i2k_p x}\right)$$

Consequently,

$$\mathcal{F}\{E(x)\} \;=\; \frac{1}{2}\int_{-L}^{+L} e^{ikx}\,dx \;+\; \frac{1}{4}\int_{-L}^{+L} e^{i(k+2k_p)x}\,dx \;+\; \frac{1}{4}\int_{-L}^{+L} e^{i(k-2k_p)x}\,dx$$

$$=\; k\sin kL \;+\; \frac{1}{2(k+2k_p)}\sin(k+2k_p)L$$

$$+\; \frac{1}{2(k-2k_p)}\sin(k-2k_p)L$$

$$=\; L\operatorname{sinc} kL \;+\; \frac{L}{2}\operatorname{sinc}(k+2k_p)L \;+\; \frac{L}{2}\operatorname{sinc}(k-2k_p)L$$

Figure 8-15 is a plot of the transform and its limiting case as $L \to \infty$.   Notice that although we began this computation with the complex transform, the ultimate solution is real, just as it was in Problem 8.10.   The function here is even and $\mathcal{F}\{E(x)\} = \mathcal{F}_C\{E(x)\}$.

Keep in mind that $E(x)$ may be related to an aperture function (in this instance, of a long grating).   Thus Fig. 8-15 resembles the magnitude of the Fraunhofer diffracted field.

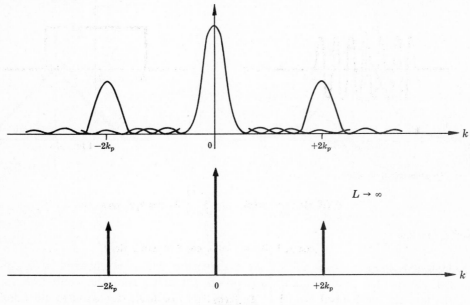

Fig. 8-15

8.12.   Calculate and plot the complex Fourier transform of the function $E(x)$ depicted in Fig. 8-16.   Observe that it can be expressed as

$$E(x) \;=\; U(x)e^{-ax}$$

wherein a is a positive constant and $U(x)$ is the *unit step function* equal to zero for $x < 0$ and one for $x > 0$.

From the definition of the transform

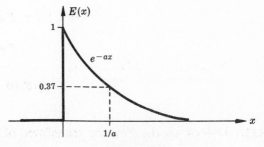

Fig. 8-16

$$\mathcal{F}\{E(x)\} \;=\; \int_0^\infty e^{-ax}\,e^{ikx}\,dx \;=\; \int_0^\infty e^{-(a-ik)x}\,dx$$

$$=\; \left[-\frac{1}{a-ik}\,e^{-(a-ik)x}\right]_0^\infty \;=\; \frac{1}{a-ik}$$

In order to plot this complex frequency spectrum, we first express it in terms of its magnitude and phase, i.e.

$$\mathcal{F}\{E(x)\} \;=\; F(k) \;=\; |F(k)|\,e^{i\varphi(k)}$$

To that end multiply top and bottom by $(a-ik)^*$, yielding

$$F(k) \;=\; \frac{a+ik}{(a-ik)(a+ik)} \;=\; \frac{a}{a^2+k^2} + i\,\frac{k}{a^2+k^2}$$

As in Fig. 1-6 it follows that

$$|F(k)|^2 \;=\; \left(\frac{a}{a^2+k^2}\right)^2 + \left(\frac{k}{a^2+k^2}\right)^2$$

$$\tan\varphi(k) \;=\; \left(\frac{k}{a^2+k^2}\right)\bigg/\left(\frac{a}{a^2+k^2}\right)$$

Consequently,

$$\mathcal{F}\{E(x)\} \;=\; \frac{1}{\sqrt{a^2+k^2}}\,e^{\,i\,\tan^{-1}(k/a)}$$

and this is plotted in Fig. 8-17.

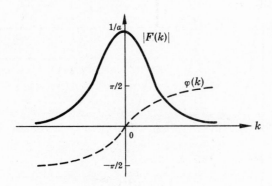

Fig. 8-17

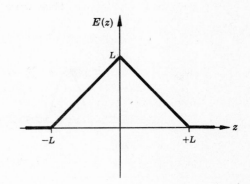

Fig. 8-18

8.13. Consider a long narrow slit in the y-direction which is covered with an amplitude mask so that the field is made to drop off linearly from the center as in Fig. 8-18. Calculate the diffracted Fraunhofer field for normally incident monochromatic light.

This is an even function and so

$$\mathcal{F}\{E(z)\} \;=\; \int_{-L}^0 (L+z)\cos k_z z\,dz + \int_0^L (L-z)\cos k_z z\,dz$$

$$=\; \frac{2}{k_z^2}(1-\cos k_z L)$$

By trigonometric transformation we get

$$\mathcal{F}\{E(z)\} \;=\; \frac{4}{k_z^2}\sin^2\frac{k_z L}{2} \;=\; L^2\,\mathrm{sinc}^2\frac{k_z L}{2}$$

and this squared is proportional to the irradiance distribution. The process of masking an aperture, generally to reduce the diffracted secondary peaks, is known as *apodization*.

8.14. The *Dirac delta function* defined by

$$\delta(x) = \begin{cases} 0 & x \neq 0 \\ \infty & x = 0 \end{cases} \qquad \text{and} \qquad \int_{-\infty}^{+\infty} \delta(x)\,dx = 1$$

is also known as the *unit impulse function*. One of its most interesting character-istics is the *sifting property*

$$\int_{-\infty}^{+\infty} \delta(x - x_0)\,f(x)\,dx = f(x_0)$$

Show that $\mathcal{F}\{\delta(x - x_0)\} = e^{ikx_0}$. In addition, determine the transform of the two delta functions comprising $f(x)$ in Fig. 8-19.

From the definition of the transform

$$\mathcal{F}\{\delta(x - x_0)\} = \int_{-\infty}^{+\infty} \delta(x - x_0)e^{ikx}\,dx$$

If we then think of e^{ikx} as $f(x)$, the shifting property dictates that the integral equal $f(x_0)$, or in this case, e^{ikx_0}. Hence

$$\mathcal{F}\{\delta(x - x_0)\} = e^{ikx_0}$$

It should be evident from the form of the trans-form that if a function $f(x)$ consists of a sum of indi-vidual functions, $\mathcal{F}\{f(x)\}$ is, in turn, the sum of their individual transforms. Thus, quite generally, if

$$f(x) = \sum_j \delta(x - x_j)$$

then

$$\mathcal{F}\{f(x)\} = \sum_j e^{ikx_j}$$

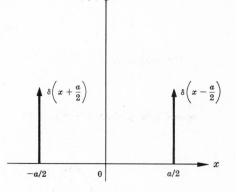

Fig. 8-19

In particular, here

$$f(x) = \delta\left(x - \frac{a}{2}\right) + \delta\left(x + \frac{a}{2}\right)$$

and so

$$\mathcal{F}\{f(x)\} = e^{ika/2} + e^{-ika/2} = 2\cos\frac{ka}{2}$$

Recall Young's experiment. As long as it consisted of two exceedingly narrow slits the idealized interference *field* was cosinusoidal. In other words, if the aperture function corresponds to Fig. 8-19 the irradiance system will be cosine-squared fringes.

8.3 CONVOLUTION

Imagine an object with an irradiance distribution $I_o(y, z)$ followed by an optical system which creates an image $I_i(Y, Z)$. The object information is transformed into the image by a process which can be represented mathematically as

$$I_i(Y, Z) = \iint_{-\infty}^{+\infty} I_o(y, z)\,\mathsf{S}(Y - y, Z - z)\,dy\,dz$$

Here $\mathsf{S}(Y - y, Z - z)$ is called the *point-spread function*. In effect, each source point on the object appears as some sort of blotch of light on the image plane. The exact configuration of the blotch (i.e. the impulse response) is determined by the particular optical system. (In the case of a perfect lens system, for example, S would be an Airy pattern.) The inte-gral merely sums up all of these spots of light thereby yielding the resultant image. Thus, if we put a piece of tape on the front of a camera lens, that would certainly change the image of a point source, i.e. the spread function, and any other photo thereafter would be affected accordingly.

The above expression represents a *convolution integral* in two dimensions, and one speaks of the process as "convolving" the two functions. More succinctly, the integral can be written as

$$I_i(Y, Z) = I_o(y, z) \circledast S(y, z)$$

Geometrically, one can view the convolution as the volume under the product surface $I_o(y, z) S(Y - y, Z - z)$ encompassing the region of overlap of the two functions. And in one dimension the convolution, $f(x) \circledast h(x)$, is the overlap area under the product of the two functions.

SOLVED PROBLEMS

8.15. The *cylinder* or *top hat function* $P(r)$ shown in Fig. 8-20 is a two-dimensional extension of the *unit pulse*. Calculate the self-convolution, $g(R) = P(r) \circledast P(r)$, for $R \geq 0$ and plot it versus the position variable R.

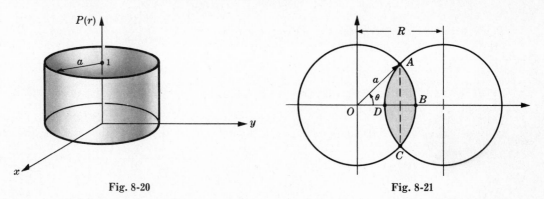

Fig. 8-20   Fig. 8-21

Since $P(r)$ has unit height, the convolution is simply the area of overlap in Fig. 8-21 as a function of R. The area of either segment ($ACBA$ or $ACDA$) is equal to the area of sector $OABCO$ minus the area of triangle AOC, i.e.

$$\frac{1}{2} a^2 (2\theta) - \frac{1}{2} a^2 \sin 2\theta$$

Twice this is the overlap area A:

$$A = a^2(2\theta - \sin 2\theta) = a^2(2\theta - 2 \sin \theta \cos \theta)$$

To express A as a function of R use

$$R = 2a \cos \theta$$

which is valid for $R \leq 2a$. Thus, for $R \leq 2a$,

$$g(R) = a^2 \left[2 \cos^{-1}\left(\frac{R}{2a}\right) - 2\left(1 - \frac{R^2}{4a^2}\right)^{1/2} \frac{R}{2a} \right]$$

When $R = 0$, $g(R) = \pi a^2$, while a value of $R \geq 2a$ yields $g(R) = 0$ (no overlap). The function is graphed in Fig. 8-22.

This is a particularly important calculation since it is closely related to the optical transfer function of an ideal lens.

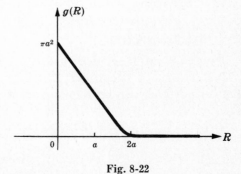

Fig. 8-22

8.16. The *convolution theorem* states that where $g(X) = f(x) \circledast h(x)$,

$$\mathcal{F}\{g\} = \mathcal{F}\{f \circledast h\} = \mathcal{F}\{f\} \cdot \mathcal{F}\{h\}$$

i.e. the transform of the convolution of two functions is the product of the transforms of the individual functions. Prove that this is indeed the case.

The convolution g is a function of X and so its transform is

$$\mathcal{F}\{g\} = \int_{-\infty}^{+\infty} g(X)\,e^{ikX}\,dX$$

$$= \int_{-\infty}^{+\infty} \left[\int_{-\infty}^{+\infty} f(x)\,h(X-x)\,dx\right] e^{ikX}\,dX$$

$$= \int_{-\infty}^{+\infty} \left[\int_{-\infty}^{+\infty} h(X-x)\,e^{ikX}\,dX\right] f(x)\,dx$$

Letting $X - x$ equal w, we have $dw = dX$, $e^{ikX} = e^{ikw}\,e^{ikx}$, and so

$$\mathcal{F}\{g\} = \int_{-\infty}^{+\infty} f(x)\,e^{ikx}\,dx \int_{-\infty}^{+\infty} h(w)\,e^{ikw}\,dw = \mathcal{F}\{f\}\cdot\mathcal{F}\{h\}$$

8.17. Figure 8-23 depicts a cosine function $f(x)$ and its Fourier transform $F(k)$ (see Problem 8.14). Graphically form the self-convolution of $F(k)$.

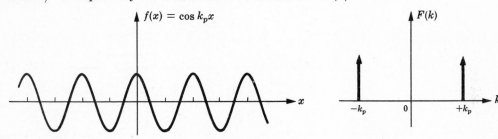

Fig. 8-23

In general, to construct $g = f \circledast h$, we imagine f to be composed of a series of delta functions, each of which is then spread out in the form of h with its *origin* at the position of the delta function. The sum of the contributions for all the delta functions is then g. This is particularly easy to do here, since we are dealing with a function which is in fact composed of delta functions. Hence, we picture $\delta(k + k_p)$ spread out into two delta functions as in Fig. 8-24(b), and similarly $\delta(k - k_p)$ is spread out as in (c). The sum of (b) and (c) is the self-convolution shown in (d).

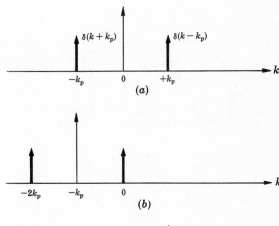

8.18. Use the approach discussed in Problem 8.17 to self-convolve a square pulse of width d.

It follows from the sifting property of the delta function (Problem 8.14) that a function can be expressed as a linear sum of impulses. The left side of Fig. 8-25, page 220, depicts the square pulse as represented by a convenient number of delta functions. Each of these, in turn, serves as the center of a square pulse on the right side of the figure. The sum of all of these is then the convolution. In other words, each delta function constituting f is spread out into a square pulse corresponding to h. Here, of course, f and h are identical.

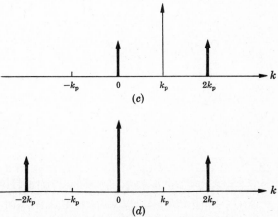

Fig. 8-24

Proceeding to the limit of infinitely many delta functions, we see that the convolution g is triangular with a base of $2d$ and a height d-times the height of the square pulse.

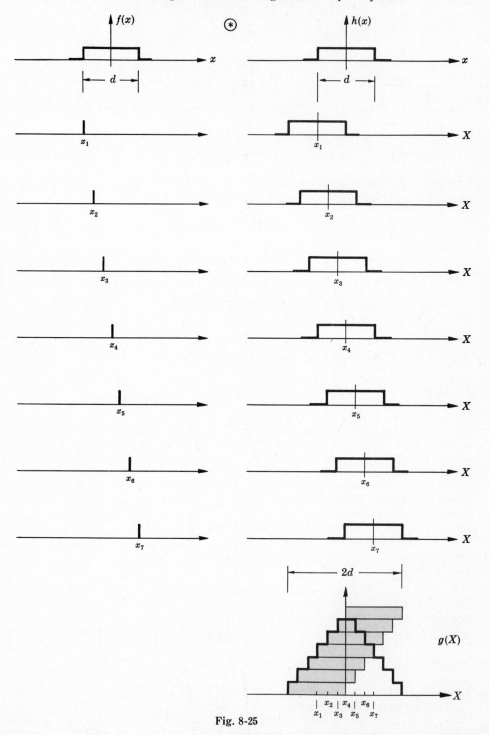

Fig. 8-25

8.19. Use the convolution theorem to verify the results of Problem 8.13, i.e. to show that sinc-squared is the transform of a triangular pulse.

From Problem 8.18 we know that the self-convolution of a square pulse is a triangle. Furthermore, the convolution theorem leads to

$$\mathcal{F}\{f \circledast f\} = [\mathcal{F}\{f\}]^2$$

As we saw in Problem 8.8, the transform of a square pulse is a sinc function. Hence the transform of a triangular pulse is the product of two identical sinc functions, as indicated in Fig. 8-26.

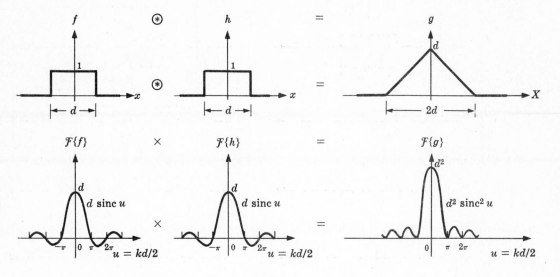

Fig. 8-26

8.20. Suppose we amplitude modulate a sinusoidal function $f(x)$ with a sinusoid $g(x)$ to yield $h(x) = f(x)\,g(x)$, as in Fig. 8-27. Make a sketch of $\mathcal{F}\{h(x)\}$.

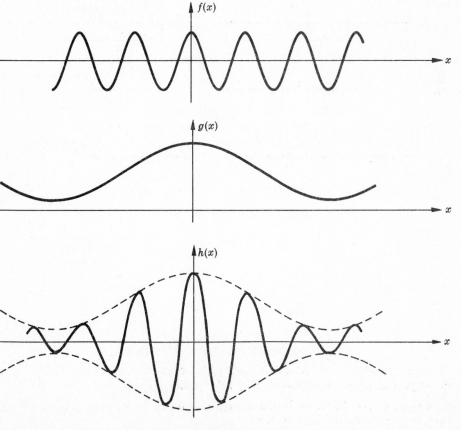

Fig. 8-27

We are looking for $\mathcal{F}\{f(x)\,g(x)\}$, which, from the frequency convolution theorem (Problem 8.37), equals the convolution of the corresponding transforms. The calculation is indicated in Fig. 8-28.

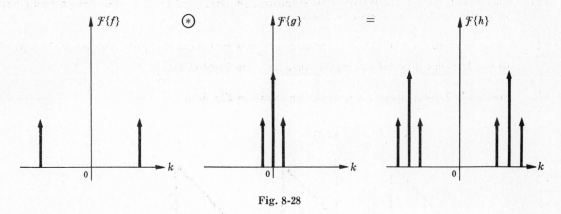

Fig. 8-28

Notice how the dc, or zero-frequency, contribution raises $g(x)$ above the x-axis. No such component is present in the spectrum of either $f(x)$ or $h(x)$.

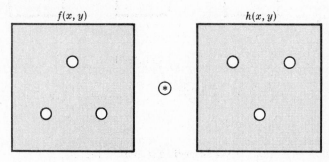

Fig. 8-29

8.21. There are several simple means of optically generating the convolution of two-dimensional functions. Generally one uses opaque screens with appropriate apertures. Determine the convolution of the two functions $f(x, y)$ and $h(x, y)$, each represented by three small holes in a mask, as shown in Fig. 8-29.

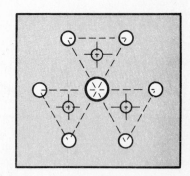

Each aperture constituting $f(x, y)$ serves, in turn, as the impulse which is then spread out into three spots corresponding to $h(x, y)$. Figure 8-30 shows the holes of $f(x, y)$ serving as centers of the spread function. The result is a hexagonal pattern of six equally bright spots surrounding a strong central spot.

Fig. 8-30

Supplementary Problems

PERIODIC WAVES AND FOURIER SERIES

8.22. Show that

$$f(x) \;=\; C_0 \,+\, \sum_{m=1}^{\infty} C_m \cos\!\left(\frac{2\pi}{\lambda/m}x + \varepsilon_m\right)$$

is equivalent to the more usual sine and cosine expression for the Fourier series.

8.23. Show that a function possesing screw symmetry, i.e. $f(x) = -f(x - \lambda/2)$, will have nonzero Fourier series contributions only when m is odd.

8.24. Under what circumstances will a function have only even harmonics in its Fourier series expansion?

Ans. If it has a period of π rather than 2π. See Problem 8.29.

8.25. Derive the Fourier series for the function shown in Fig. 8-31.

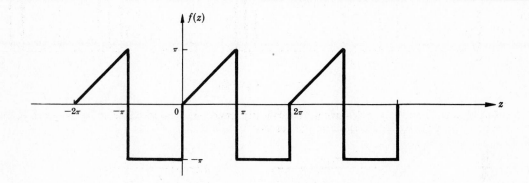

Fig. 8-31

Ans. $f(z) = -\dfrac{\pi}{4} - \dfrac{2}{\pi}\dfrac{\cos z}{1^2} - \dfrac{2}{\pi}\dfrac{\cos 3z}{3^2} - \cdots + 3\sin z - \dfrac{1}{2}(\sin 2z) + \dfrac{1}{3}(3\sin 3z) - \cdots$

where $k = 1$ since $\lambda = 2\pi$.

8.26. Compute the Fourier series for the periodic function of Fig. 8-32.

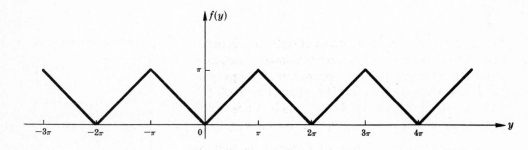

Fig. 8-32

Ans. $f(y) = \dfrac{\pi}{2} - \dfrac{4}{\pi}\left(\dfrac{\cos y}{1^2} + \dfrac{\cos 3y}{3^2} + \dfrac{\cos 5y}{5^2} + \cdots\right)$

where $k = 1$ since $\lambda = 2\pi$.

8.27. Generalize Problem 8.26 by finding the Fourier series equivalent to the function in Fig. 8-33.

Ans. $f(y) = \dfrac{H}{2} - \dfrac{4H}{\pi^2}\left[\dfrac{\cos(2\pi y/L)}{1^2} + \dfrac{\cos(6\pi y/L)}{3^2} + \cdots\right]$

where $k = 2\pi/L$.

8.28. By shifting axes in Problem 8.26, determine the Fourier series for the periodic function depicted in Fig. 8-34.

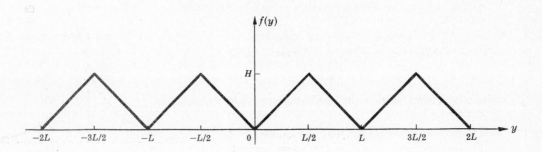

Fig. 8-33

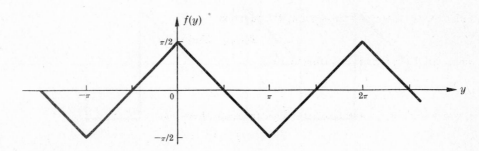

Fig. 8-34

Ans. $\quad f(y) \; = \; \dfrac{4}{\pi}\left(\dfrac{\cos y}{1^2} + \dfrac{\cos 3y}{3^2} + \cdots \right)$

8.29. Derive the Fourier series representation in the time domain of a full-wave rectified sine function of amplitude E_0 having a 1-second period.

Ans. $\quad f(t) \; = \; \dfrac{2E_0}{\pi} - \dfrac{4E_0}{\pi}\left(\dfrac{1}{3}\cos 2\pi t + \dfrac{1}{15}\cos 4\pi t + \cdots \right)$

Recall Problem 8.24.

FOURIER TRANSFORMS

8.30. Recompute the transform of the square pulse of Fig. 8-10, this time using the complex exponential formulation.

8.31. Compute the Fourier transform of the product of the unit square pulse and a sine function of spatial frequency k_p, i.e. $E(x) = P(x)E_0 \sin k_p x$, or,

$$E(x) \; = \; \begin{cases} E_0 \sin k_p x & \text{when } |x| \le L \\ 0 & \text{when } |x| > L \end{cases}$$

Ans. $\quad F(k) \; = \; iE_0L[\text{sinc}\,(k - k_p)L - \text{sinc}\,(k + k_p)L]$

8.32. Determine the Fourier transform of

$$E(x) \; = \; P(x)E_0 \sin^2 k_p x$$

where $P(x)$ is the unit square pulse ranging from $x = -L$ to $x = +L$.

Ans. $\quad \mathcal{F}\{E(x)\} \; = \; E_0L[\text{sinc}\,kL - \tfrac{1}{2}\text{sinc}\,(k + 2k_p)L - \tfrac{1}{2}\text{sinc}\,(k - 2k_p)L]$

8.33. Determine the Fourier transform of the function depicted in Fig. 8-35. Do it first by calculating $\mathcal{F}_C\{f(x)\}$ directly and then by using the even part of the transform of Problem 8.12.

Ans. $\mathcal{F}_C\{f(x)\} = \dfrac{2a}{a^2 + k^2}$

8.34.   Find the Fourier transform of the Gaussian function

$$f(x) = \sqrt{a/\pi}\, e^{-ax^2}$$

Why might this function be used to determine an apodization mask?

Ans. $F(k) = e^{-k^2/4a}$, which is also Gaussian. If a lens were coated so that the aperture function were Gaussian, the Airy rings in the diffraction pattern would vanish.

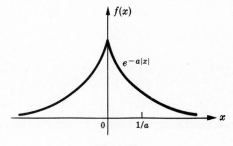

Fig. 8-35

8.35.   Calculate the Fourier transform of the function

$$E(x) = U(x)\, xe^{-ax}$$

where $U(x)$ is the unit step function.

Ans. $\mathcal{F}\{E(x)\} = 1/(a - ik)^2$

8.36.   Using the sifting property of the delta function, show that

$$\mathcal{F}\{\delta(x)\} = 1$$

Now determine $\mathcal{F}\{1\}$.

Ans. $\mathcal{F}\{1\} = 2\pi\, \delta(k)$

CONVOLUTION

8.37.   Prove the *frequency convolution theorem*:

$$\mathcal{F}\{f \cdot h\} = \frac{1}{2\pi}\, \mathcal{F}\{f\} \circledast \mathcal{F}\{h\}$$

8.38.   Use the result of Problem 8.37 to find $\mathcal{F}\{\cos^2 k_0 x\}$.

Ans.   See Fig. 8-24(*d*).

8.39.   Prove that $f(x) \circledast h(x) = h(x) \circledast f(x)$.

8.40.   Construct the self-convolution of the function $f(x)$ depicted in Fig. 8-36.

Ans.   See Fig. 8-37.

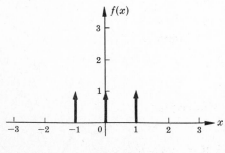

Fig. 8-36

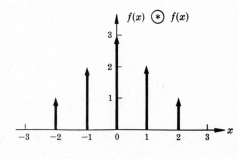

Fig. 8-37

8.41.   With Problem 8.40 in mind, use the fact that

$$\delta(x-a) \;\circledast\; \delta(x-b) \;=\; \delta[x-(a+b)]$$

to compute $f \circledast f$, where

$$f(x) \;=\; \delta(x-1) + \delta(x) + \delta(x+1)$$

Ans. $f(x) \circledast f(x) \;=\; \delta(x-2) + 2\delta(x-1) + 3\delta(x) + 2\delta(x+1) + \delta(x+2)$

8.42.   Construct the self-convolution of the function shown in Fig. 8-38. This is the frequency spectrum of a disturbance consisting of two cosine waves of differing wavelengths.

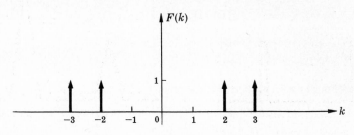

Fig. 8-38

Ans.   See Fig. 8-39.

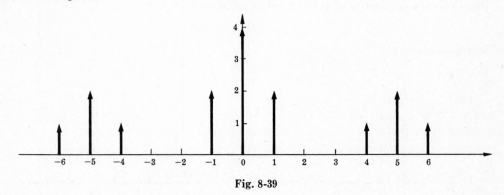

Fig. 8-39

8.43.   Convolve the two functions in Fig. 8-40 to get the function depicted in Fig. 8-11. Then, using the convolution theorem, find its transform (Fig. 8-12).

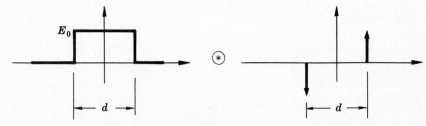

Fig. 8-40

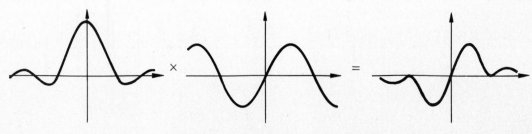

Fig. 8-41

Ans.   See Fig. 8-41.

8.44.   Determine the self-convolution of the function shown in Fig. 8-42.   It corresponds to the aperture function for a double slit.

Ans.   See Fig. 8-43.

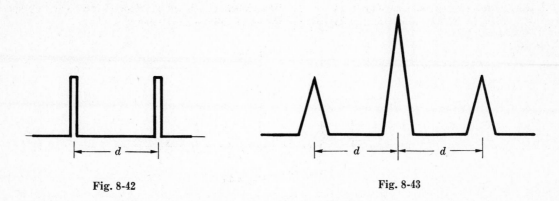

Fig. 8-42   Fig. 8-43

8.45.   Construct the convolution of the two functions shown in Fig. 8-44.

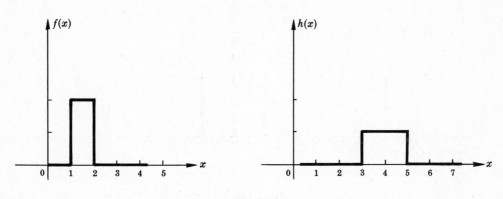

Fig. 8-44

Ans.   Remember to place the *origin* for h at the location of each δ-function making up f.   See Fig. 8-45.

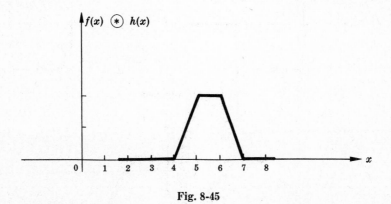

Fig. 8-45

8.46.   Construct the convolution indicated in Fig. 8-46, page 228.

Ans.   See Fig. 8-47.

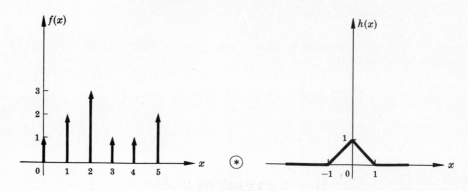

Fig. 8-46

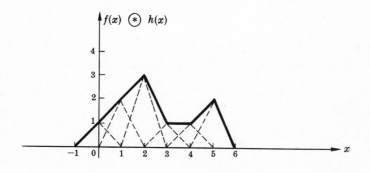

Fig. 8-47

8.47. Figure 8-48 shows a mask with six circular holes representing a two-dimensional function $f(x, y)$. Determine $f(x, y) \circledast f(x, y)$.

Ans. Figure 8-49 depicts the nineteen hexagonally arrayed spots.

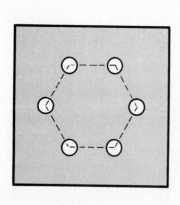

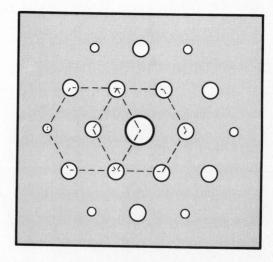

Fig. 8-48

Fig. 8-49

APPENDIX

Appendix

Values of the Function $(\sin u)/u \equiv \operatorname{sinc} u$

u	0.00	0.01	0.02	0.03	0.04	0.05	0.06	0.07	0.08	0.09
0.0	1.000000	0.999983	0.999933	0.999850	0.999733	0.999583	0.999400	0.999184	0.998934	0.998651
0.1	0.998334	0.997985	0.997602	0.997186	0.996737	0.996254	0.995739	0.995190	0.994609	0.993994
0.2	0.993347	0.992666	0.991953	0.991207	0.990428	0.989616	0.988771	0.987894	0.986984	0.986042
0.3	0.985067	0.984060	0.983020	0.981949	0.980844	0.979708	0.978540	0.977339	0.976106	0.974842
0.4	0.973546	0.972218	0.970858	0.969467	0.968044	0.966590	0.965105	0.963588	0.962040	0.960461
0.5	0.958851	0.957210	0.955539	0.953836	0.952104	0.950340	0.948547	0.946723	0.944869	0.942985
0.6	0.941071	0.939127	0.937153	0.935150	0.933118	0.931056	0.928965	0.926845	0.924696	0.922518
0.7	0.920311	0.918076	0.915812	0.913520	0.911200	0.908852	0.906476	0.904072	0.901640	0.899181
0.8	0.896695	0.894182	0.891641	0.889074	0.886480	0.883859	0.881212	0.878539	0.875840	0.873114
0.9	0.870363	0.867587	0.864784	0.861957	0.859104	0.856227	0.853325	0.850398	0.847446	0.844471
1.0	0.841471	0.838447	0.835400	0.832329	0.829235	0.826117	0.822977	0.819814	0.816628	0.813419
1.1	0.810189	0.806936	0.803661	0.800365	0.797047	0.793708	0.790348	0.786966	0.783564	0.780142
1.2	0.776699	0.773236	0.769754	0.766251	0.762729	0.759188	0.755627	0.752048	0.748450	0.744833
1.3	0.741199	0.737546	0.733875	0.730187	0.726481	0.722758	0.719018	0.715261	0.711488	0.707698
1.4	0.703893	0.700071	0.696234	0.692381	0.688513	0.684630	0.680732	0.676819	0.672892	0.668952
1.5	0.664997	0.661028	0.657046	0.653051	0.649043	0.645022	0.640988	0.636942	0.632885	0.628815
1.6	0.624734	0.620641	0.616537	0.612422	0.608297	0.604161	0.600014	0.595858	0.591692	0.587517
1.7	0.583332	0.579138	0.574936	0.570725	0.566505	0.562278	0.558042	0.553799	0.549549	0.545291
1.8	0.541026	0.536755	0.532478	0.528194	0.523904	0.519608	0.515307	0.511001	0.506689	0.502373
1.9	0.498053	0.493728	0.489399	0.485066	0.480729	0.476390	0.472047	0.467701	0.463353	0.459002
2.0	0.454649	0.450294	0.445937	0.441579	0.437220	0.432860	0.428499	0.424137	0.419775	0.415414
2.1	0.411052	0.406691	0.402330	0.397971	0.393612	0.389255	0.384900	0.380546	0.376194	0.371845
2.2	0.367498	0.363154	0.358813	0.354475	0.350141	0.345810	0.341483	0.337161	0.332842	0.328529
2.3	0.324220	0.319916	0.315617	0.311324	0.307036	0.302755	0.298479	0.294210	0.289947	0.285692
2.4	0.281443	0.277202	0.272967	0.268741	0.264523	0.260312	0.256110	0.251916	0.247732	0.243556
2.5	0.239389	0.235231	0.231084	0.226946	0.222817	0.218700	0.214592	0.210495	0.206409	0.202334
2.6	0.198270	0.194217	0.190176	0.186147	0.182130	0.178125	0.174132	0.170152	0.166185	0.162230
2.7	0.158289	0.154361	0.150446	0.146546	0.142659	0.138786	0.134927	0.131083	0.127253	0.123439
2.8	0.119639	0.115854	0.112084	0.108330	0.104592	0.100869	0.097163	0.093473	0.089798	0.086141
2.9	0.082500	0.078876	0.075268	0.071678	0.068105	0.064550	0.061012	0.057492	0.053990	0.050506
3.0	0.047040	0.043592	0.040163	0.036753	0.033361	0.029988	0.026635	0.023300	0.019985	0.016689
3.1	0.013413	0.010157	0.006920	0.003704	0.000507	-0.002669	-0.005825	-0.008960	-0.012075	-0.015169
3.2	-0.018242	-0.021294	-0.024325	-0.027335	-0.030324	-0.033291	-0.036236	-0.039160	-0.042063	-0.044943
3.3	-0.047802	-0.050638	-0.053453	-0.056245	-0.059014	-0.061762	-0.064487	-0.067189	-0.069868	-0.072525
3.4	-0.075159	-0.077770	-0.080358	-0.082923	-0.085465	-0.087983	-0.090478	-0.092950	-0.095398	-0.097823
3.5	-0.100224	-0.102601	-0.104955	-0.107285	-0.109591	-0.111873	-0.114131	-0.116365	-0.118575	-0.120761
3.6	-0.122922	-0.125060	-0.127173	-0.129262	-0.131326	-0.133366	-0.135382	-0.137373	-0.139339	-0.141282
3.7	-0.143199	-0.145092	-0.146960	-0.148803	-0.150622	-0.152416	-0.154186	-0.155930	-0.157650	-0.159345
3.8	-0.161015	-0.162661	-0.164281	-0.165877	-0.167448	-0.168994	-0.170515	-0.172011	-0.173482	-0.174929
3.9	-0.176350	-0.177747	-0.179119	-0.180466	-0.181788	-0.183086	-0.184358	-0.185606	-0.186829	-0.188027
4.0	-0.189201	-0.190349	-0.191473	-0.192573	-0.193647	-0.194698	-0.195723	-0.196724	-0.197700	-0.198652
4.1	-0.199580	-0.200483	-0.201361	-0.202216	-0.203046	-0.203851	-0.204633	-0.205390	-0.206124	-0.206833
4.2	-0.207518	-0.208179	-0.208817	-0.209430	-0.210020	-0.210586	-0.211128	-0.211647	-0.212142	-0.212614
4.3	-0.213062	-0.213487	-0.213888	-0.214267	-0.214622	-0.214955	-0.215264	-0.215550	-0.215814	-0.216055
4.4	-0.216273	-0.216469	-0.216642	-0.216793	-0.216921	-0.217028	-0.217112	-0.217174	-0.217214	-0.217232
4.5	-0.217229	-0.217204	-0.217157	-0.217089	-0.217000	-0.216889	-0.216757	-0.216604	-0.216430	-0.216235
4.6	-0.216020	-0.215784	-0.215527	-0.215250	-0.214953	-0.214635	-0.214298	-0.213940	-0.213563	-0.213166
4.7	-0.212750	-0.212314	-0.211858	-0.211384	-0.210890	-0.210377	-0.209846	-0.209296	-0.208727	-0.208140
4.8	-0.207534	-0.206911	-0.206269	-0.205609	-0.204932	-0.204236	-0.203524	-0.202794	-0.202046	-0.201282
4.9	-0.200501	-0.199702	-0.198887	-0.198056	-0.197208	-0.196344	-0.195464	-0.194568	-0.193656	-0.192728

u	0.00	0.01	0.02	0.03	0.04	0.05	0.06	0.07	0.08	0.09
5.0	−0.191785	−0.190826	−0.189853	−0.188864	−0.187860	−0.186841	−0.185808	−0.184760	−0.183699	−0.182622
5.1	−0.181532	−0.180428	−0.179311	−0.178179	−0.177035	−0.175877	−0.174706	−0.173522	−0.172326	−0.171117
5.2	−0.169895	−0.168661	−0.167415	−0.166158	−0.164888	−0.163607	−0.162314	−0.161010	−0.159695	−0.158369
5.3	−0.157032	−0.155684	−0.154326	−0.152958	−0.151579	−0.150191	−0.148792	−0.147384	−0.145967	−0.144540
5.4	−0.143105	−0.141660	−0.140206	−0.138744	−0.137273	−0.135794	−0.134307	−0.132812	−0.131309	−0.129798
5.5	−0.128280	−0.126755	−0.125222	−0.123683	−0.122137	−0.120584	−0.119024	−0.117459	−0.115887	−0.114310
5.6	−0.112726	−0.111137	−0.109543	−0.107943	−0.106338	−0.104728	−0.103114	−0.101495	−0.099871	−0.098243
5.7	−0.096611	−0.094976	−0.093336	−0.091693	−0.090046	−0.088396	−0.086743	−0.085087	−0.083429	−0.081768
5.8	−0.080104	−0.078438	−0.076770	−0.075100	−0.073428	−0.071755	−0.070080	−0.068404	−0.066726	−0.065048
5.9	−0.063369	−0.061689	−0.060009	−0.058329	−0.056648	−0.054967	−0.053287	−0.051606	−0.049927	−0.048248
6.0	−0.046569	−0.044892	−0.043216	−0.041540	−0.039867	−0.038195	−0.036524	−0.034856	−0.033189	−0.031525
6.1	−0.029863	−0.028203	−0.026546	−0.024892	−0.023240	−0.021592	−0.019947	−0.018305	−0.016667	−0.015032
6.2	−0.013402	−0.011775	−0.010152	−0.008533	−0.006919	−0.005309	−0.003703	−0.002103	−0.000507	0.001083
6.3	0.002669	0.004249	0.005824	0.007393	0.008956	0.010514	0.012066	0.013612	0.015151	0.016684
6.4	0.018211	0.019731	0.021244	0.022751	0.024250	0.025743	0.027228	0.028706	0.030177	0.031640
6.5	0.033095	0.034543	0.035983	0.037414	0.038838	0.040253	0.041661	0.043059	0.044449	0.045831
6.6	0.047203	0.048567	0.049922	0.051268	0.052604	0.053931	0.055249	0.056558	0.057857	0.059146
6.7	0.060425	0.061695	0.062955	0.064204	0.065444	0.066673	0.067892	0.069101	0.070299	0.071487
6.8	0.072664	0.073830	0.074986	0.076130	0.077264	0.078386	0.079498	0.080598	0.081688	0.082765
6.9	0.083832	0.084887	0.085930	0.086962	0.087982	0.088991	0.089987	0.090972	0.091945	0.092906
7.0	0.093855	0.094792	0.095717	0.096629	0.097530	0.098418	0.099293	0.100157	0.101008	0.101846
7.1	0.102672	0.103485	0.104286	0.105074	0.105849	0.106611	0.107361	0.108098	0.108822	0.109533
7.2	0.110232	0.110917	0.111589	0.112249	0.112895	0.113528	0.114149	0.114756	0.115350	0.115931
7.3	0.116498	0.117053	0.117594	0.118122	0.118637	0.119138	0.119627	0.120102	0.120563	0.121012
7.4	0.121447	0.121869	0.122277	0.122673	0.123055	0.123423	0.123779	0.124121	0.124449	0.124765
7.5	0.125067	0.125355	0.125631	0.125893	0.126142	0.126378	0.126600	0.126809	0.127005	0.127188
7.6	0.127358	0.127514	0.127658	0.127788	0.127905	0.128009	0.128100	0.128178	0.128243	0.128295
7.7	0.128334	0.128360	0.128373	0.128373	0.128361	0.128335	0.128297	0.128247	0.128183	0.128107
7.8	0.128018	0.127917	0.127803	0.127677	0.127539	0.127388	0.127224	0.127049	0.126861	0.126661
7.9	0.126448	0.126224	0.125988	0.125739	0.125479	0.125207	0.124923	0.124627	0.124320	0.124000
8.0	0.123670	0.123328	0.122974	0.122609	0.122232	0.121845	0.121446	0.121036	0.120615	0.120183
8.1	0.119739	0.119286	0.118821	0.118345	0.117859	0.117363	0.116855	0.116338	0.115810	0.115272
8.2	0.114723	0.114165	0.113596	0.113018	0.112429	0.111831	0.111223	0.110605	0.109978	0.109341
8.3	0.108695	0.108040	0.107376	0.106702	0.106019	0.105327	0.104627	0.103918	0.103200	0.102473
8.4	0.101738	0.100994	0.100243	0.099483	0.098714	0.097938	0.097154	0.096362	0.095562	0.094755
8.5	0.093940	0.093117	0.092287	0.091450	0.090606	0.089755	0.088896	0.088031	0.087159	0.086280
8.6	0.085395	0.084503	0.083605	0.082701	0.081790	0.080874	0.079951	0.079023	0.078089	0.077149
8.7	0.076203	0.075253	0.074296	0.073335	0.072369	0.071397	0.070421	0.069439	0.068453	0.067463
8.8	0.066468	0.065468	0.064465	0.063457	0.062445	0.061429	0.060410	0.059386	0.058359	0.057328
8.9	0.056294	0.055257	0.054217	0.053173	0.052127	0.051077	0.050025	0.048970	0.047913	0.046853
9.0	0.045791	0.044727	0.043660	0.042592	0.041521	0.040449	0.039375	0.038300	0.037223	0.036145
9.1	0.035066	0.033985	0.032904	0.031821	0.030738	0.029654	0.028569	0.027484	0.026399	0.025313
9.2	0.024227	0.023141	0.022055	0.020970	0.019884	0.018799	0.017714	0.016630	0.015547	0.014464
9.3	0.013382	0.012301	0.011222	0.010143	0.009066	0.007990	0.006916	0.005843	0.004772	0.003703
9.4	0.002636	0.001570	0.000507	−0.000554	−0.001612	−0.002669	−0.003722	−0.004774	−0.005822	−0.006868
9.5	−0.007911	−0.008950	−0.009987	−0.011021	−0.012051	−0.013078	−0.014101	−0.015121	−0.016138	−0.017150
9.6	−0.018159	−0.019164	−0.020165	−0.021161	−0.022154	−0.023142	−0.024126	−0.025106	−0.026081	−0.027051
9.7	−0.028017	−0.028977	−0.029933	−0.030884	−0.031830	−0.032771	−0.033707	−0.034637	−0.035562	−0.036482
9.8	−0.037396	−0.038304	−0.039207	−0.040104	−0.040995	−0.041881	−0.042760	−0.043633	−0.044500	−0.045361
9.9	−0.046216	−0.047064	−0.047906	−0.048741	−0.049570	−0.050392	−0.051208	−0.052017	−0.052819	−0.053614

u	0.00	0.01	0.02	0.03	0.04	0.05	0.06	0.07	0.08	0.09
10.0	−0.054402	−0.055183	−0.055957	−0.056724	−0.057484	−0.058237	−0.058982	−0.059720	−0.060450	−0.061173
10.1	−0.061888	−0.062596	−0.063296	−0.063988	−0.064673	−0.065350	−0.066019	−0.066680	−0.067333	−0.067978
10.2	−0.068615	−0.069244	−0.069865	−0.070477	−0.076143	−0.076663	−0.077174	−0.077677	−0.078170	−0.078655
10.3	−0.074533	−0.075078	−0.075615	−0.076143	−0.076663	−0.077174	−0.077677	−0.078170	−0.078655	−0.079131
10.4	−0.079599	−0.080057	−0.080507	−0.080947	−0.081379	−0.081802	−0.082216	−0.082620	−0.083016	−0.083403
10.5	−0.083781	−0.084149	−0.084509	−0.084859	−0.085200	−0.085532	−0.085855	−0.086169	−0.086473	−0.086768
10.6	−0.087054	−0.087331	−0.087599	−0.087857	−0.088106	−0.088346	−0.088576	−0.088797	−0.089009	−0.089212
10.7	−0.089405	−0.089589	−0.089764	−0.089929	−0.090085	−0.090232	−0.090370	−0.090498	−0.090617	−0.090727
10.8	−0.090827	−0.090919	−0.091001	−0.091073	−0.091137	−0.091191	−0.091236	−0.091272	−0.091299	−0.091316
10.9	−0.091324	−0.091324	−0.091314	−0.091295	−0.091267	−0.091229	−0.091183	−0.091128	−0.091064	−0.090990
11.0	−0.090908	−0.090817	−0.090717	−0.090608	−0.090490	−0.090364	−0.090228	−0.090084	−0.089931	−0.089770
11.1	−0.089599	−0.089420	−0.089233	−0.089037	−0.088832	−0.088619	−0.088397	−0.088167	−0.087929	−0.087682
11.2	−0.087427	−0.087163	−0.086891	−0.086612	−0.086324	−0.086027	−0.085723	−0.085411	−0.085091	−0.084763
11.3	−0.084426	−0.084083	−0.083731	−0.083371	−0.083004	−0.082630	−0.082247	−0.081857	−0.081460	−0.081055
11.4	−0.080643	−0.080223	−0.079796	−0.079362	−0.078921	−0.078473	−0.078017	−0.077555	−0.077086	−0.076609
11.5	−0.076126	−0.075636	−0.075140	−0.074637	−0.074127	−0.073611	−0.073088	−0.072559	−0.072023	−0.071481
11.6	−0.070934	−0.070379	−0.069819	−0.069253	−0.068681	−0.068103	−0.067519	−0.066929	−0.066334	−0.065733
11.7	−0.065127	−0.064515	−0.063898	−0.063275	−0.062647	−0.062014	−0.061376	−0.060733	−0.060084	−0.059431
11.8	−0.058773	−0.058111	−0.057443	−0.056771	−0.056095	−0.055414	−0.054728	−0.054039	−0.053345	−0.052646
11.9	−0.051944	−0.051238	−0.050528	−0.049814	−0.049096	−0.048375	−0.047650	−0.046921	−0.046189	−0.045453
12.0	−0.044714	−0.043972	−0.043227	−0.042479	−0.041727	−0.040973	−0.040216	−0.039456	−0.038694	−0.037929
12.1	−0.037161	−0.036391	−0.035618	−0.034844	−0.034067	−0.033288	−0.032506	−0.031723	−0.030938	−0.030152
12.2	−0.029363	−0.028573	−0.027781	−0.026988	−0.026193	−0.025398	−0.024600	−0.023802	−0.023003	−0.022202
12.3	−0.021401	−0.020599	−0.019796	−0.018992	−0.018188	−0.017384	−0.016578	−0.015773	−0.014967	−0.014161
12.4	−0.013355	−0.012549	−0.011743	−0.010937	−0.010131	−0.009326	−0.008521	−0.007716	−0.006912	−0.006109
12.5	−0.005306	−0.004504	−0.003702	−0.002902	−0.002103	−0.001304	−0.000507	0.000289	0.001083	0.001877
12.6	0.002668	0.003459	0.004248	0.005035	0.005820	0.006603	0.007385	0.008164	0.008942	0.009717
12.7	0.010491	0.011262	0.012030	0.012797	0.013560	0.014312	0.015080	0.015836	0.016589	0.017339
12.8	0.018087	0.018831	0.019572	0.020311	0.021046	0.021778	0.022506	0.023231	0.023953	0.024671
12.9	0.025386	0.026097	0.026804	0.027507	0.028207	0.028903	0.029594	0.030282	0.030966	0.031645
13.0	0.032321	0.032992	0.033658	0.034321	0.034978	0.035632	0.036281	0.036925	0.037564	0.038199
13.1	0.038829	0.039454	0.040075	0.040690	0.041300	0.041905	0.042506	0.043101	0.043690	0.044275
13.2	0.044854	0.045428	0.045996	0.046559	0.047117	0.047669	0.048215	0.048756	0.049291	0.049820
13.3	0.050344	0.050861	0.051373	0.051879	0.052379	0.052873	0.053361	0.053843	0.054319	0.054788
13.4	0.055252	0.055709	0.056160	0.056605	0.057043	0.057476	0.057901	0.058321	0.058733	0.059140
13.5	0.059540	0.059933	0.060320	0.060700	0.061073	0.061440	0.061800	0.062154	0.062500	0.062840
13.6	0.063174	0.063500	0.063820	0.064132	0.064438	0.064737	0.065029	0.065314	0.065593	0.065864
13.7	0.066128	0.066385	0.066636	0.066879	0.067115	0.067344	0.067566	0.067781	0.067989	0.068190
13.8	0.068384	0.068570	0.068750	0.068922	0.069087	0.069245	0.069396	0.069540	0.069677	0.069806
13.9	0.069929	0.070044	0.070152	0.070253	0.070346	0.070433	0.070512	0.070584	0.070649	0.070707
14.0	0.070758	0.070801	0.070838	0.070867	0.070889	0.070904	0.070912	0.070913	0.070907	0.070893
14.1	0.070873	0.070846	0.070811	0.070770	0.070721	0.770666	0.070603	0.070534	0.070457	0.070374
14.2	0.070284	0.070186	0.070082	0.069971	0.069854	0.069729	0.069598	0.069460	0.069315	0.069163
14.3	0.069005	0.068840	0.068668	0.068490	0.068305	0.068114	0.067916	0.067712	0.067501	0.067283
14.4	0.067060	0.066829	0.066593	0.066350	0.066101	0.065845	0.065584	0.065316	0.065042	0.064762
14.5	0.064476	0.064183	0.063885	0.063581	0.063271	0.062954	0.062633	0.062305	0.061971	0.061632
14.6	0.061287	0.060936	0.060580	0.060218	0.059851	0.059478	0.059100	0.058717	0.058328	0.057933
14.7	0.057534	0.057129	0.056719	0.056304	0.055884	0.055459	0.055029	0.054594	0.054154	0.053710
14.8	0.053260	0.052806	0.052347	0.051884	0.051416	0.050944	0.050467	0.049985	0.049500	0.049010
14.9	0.048516	0.048017	0.047515	0.047008	0.046497	0.045983	0.045464	0.044942	0.044416	0.043886

u	0.00	0.01	0.02	0.03	0.04	0.05	0.06	0.07	0.08	0.09
15.0	0.043353	0.042815	0.042275	0.041730	0.041183	0.040632	0.040077	0.039520	0.038959	0.038395
15.1	0.037828	0.037257	0.036684	0.036108	0.035529	0.034948	0.034363	0.033776	0.033187	0.032595
15.2	0.032000	0.031403	0.030803	0.030202	0.029598	0.028992	0.028383	0.027773	0.027161	0.026547
15.3	0.025931	0.025313	0.024693	0.024072	0.023450	0.022825	0.022199	0.021572	0.020944	0.020314
15.4	0.019683	0.019051	0.018418	0.017783	0.017148	0.016512	0.015875	0.015237	0.014599	0.013960
15.5	0.013320	0.012680	0.012040	0.011399	0.010758	0.010116	0.009475	0.008833	0.008191	0.007549
15.6	0.006907	0.006266	0.005624	0.004983	0.004342	0.003702	0.003062	0.002422	0.001783	0.001145
15.7	0.000507	−0.000130	−0.000766	−0.001401	−0.002035	−0.002668	−0.003300	−0.003931	−0.004561	−0.005190
15.8	−0.005817	−0.006443	−0.007067	−0.007690	−0.008311	−0.008931	−0.009549	−0.010166	−0.010780	−0.011393
15.9	−0.012004	−0.012613	−0.013219	−0.013824	−0.014427	−0.015027	−0.015625	−0.016221	−0.016814	−0.017405
16.0	−0.017994	−0.018580	−0.019163	−0.019744	−0.020322	−0.020898	−0.021470	−0.022040	−0.022607	−0.023170
16.1	−0.023731	−0.024289	−0.024843	−0.025395	−0.025943	−0.026488	−0.027030	−0.027568	−0.028103	−0.028634
16.2	−0.029162	−0.029686	−0.030207	−0.030724	−0.031237	−0.031747	−0.032252	−0.032754	−0.033252	−0.033746
16.3	−0.034236	−0.034722	−0.035204	−0.035682	−0.036156	−0.036626	−0.037091	−0.037552	−0.038009	−0.038461
16.4	−0.038909	−0.039352	−0.039792	−0.040226	−0.040656	−0.041081	−0.041502	−0.041918	−0.042330	−0.042737
16.5	−0.043139	−0.043536	−0.043928	−0.044315	−0.044698	−0.045076	−0.045448	−0.045816	−0.046179	−0.046536
16.6	−0.046889	−0.047236	−0.047578	−0.047915	−0.048247	−0.048574	−0.048895	−0.049212	−0.049522	−0.049828
16.7	−0.050128	−0.050423	−0.050713	−0.050997	−0.051275	−0.051548	−0.051816	−0.052078	−0.052335	−0.052586
16.8	−0.052831	−0.053071	−0.053306	−0.053535	−0.053758	−0.053975	−0.054187	−0.054393	−0.054594	0.054789
16.9	−0.054978	−0.055161	−0.055339	−0.055511	−0.055677	−0.055837	−0.055992	−0.056141	−0.056284	−0.056421
17.0	−0.056553	−0.056678	−0.056798	−0.056912	−0.057021	−0.057123	−0.057220	−0.057310	−0.057395	−0.057474
17.1	−0.057548	−0.057615	−0.057677	−0.057732	−0.057782	−0.057826	−0.057865	−0.057897	−0.057924	−0.057944
17.2	−0.057959	−0.057968	−0.057972	−0.057969	−0.057961	−0.057947	−0.057927	−0.057902	−0.057870	−0.057833
17.3	−0.057790	−0.057742	−0.057688	−0.057628	−0.057562	−0.057491	−0.057414	−0.057331	−0.057243	−0.057149
17.4	−0.057049	−0.056944	−0.056834	−0.056717	−0.056596	−0.056468	−0.056336	−0.056197	−0.056054	−0.055905
17.5	−0.055750	−0.055590	−0.055425	−0.055254	−0.055078	−0.054897	−0.054710	−0.054518	−0.054321	−0.054119
17.6	−0.053912	−0.053699	−0.053481	−0.053258	−0.053031	−0.052798	−0.052560	−0.052317	−0.052069	−0.051816
17.7	−0.051558	−0.051296	−0.051028	−0.050756	−0.050479	−0.050198	−0.049911	−0.049620	−0.049324	−0.049024
17.8	−0.048719	−0.048410	−0.048096	−0.047778	−0.047455	−0.047128	−0.046796	−0.046461	−0.046121	−0.045776
17.9	−0.045428	−0.045075	−0.044718	−0.044358	−0.043993	−0.043624	−0.043251	−0.042875	−0.042494	−0.042110
18.0	−0.041722	−0.041330	−0.040934	−0.040535	−0.040132	−0.039726	−0.039316	−0.038902	−0.038485	−0.038065
18.1	−0.037642	−0.037215	−0.036785	−0.036351	−0.035915	−0.035475	−0.035033	−0.034587	−0.034139	−0.033687
18.2	−0.033233	−0.032775	−0.032315	−0.031853	−0.031387	−0.030919	−0.030449	−0.029976	−0.029500	−0.029022
18.3	−0.028541	−0.028059	−0.027574	−0.027086	−0.026597	−0.026105	−0.025612	−0.025116	−0.024619	−0.024119
18.4	−0.023618	−0.023114	−0.022610	−0.022103	−0.021594	−0.021085	−0.020573	−0.020060	−0.019546	−0.019030
18.5	−0.018512	−0.017994	−0.017474	−0.016953	−0.016431	−0.015908	−0.015384	−0.014859	−0.014333	−0.013806
18.6	−0.013278	−0.012750	−0.012220	−0.011691	−0.011160	−0.010629	−0.010098	−0.009566	−0.009033	−0.008501
18.7	−0.007968	−0.007435	−0.006901	−0.006368	−0.005834	−0.005301	−0.004767	−0.004234	−0.003701	−0.003168
18.8	−0.002635	−0.002102	−0.001570	−0.001038	−0.000507	0.000024	0.000554	0.001083	0.001612	0.002140
18.9	0.002668	0.003194	0.003720	0.004245	0.004769	0.005292	0.005813	0.006334	0.006853	0.007371
19.0	0.007888	0.008404	0.008918	0.009431	0.009942	0.010452	0.010960	0.011466	0.011971	0.012474
19.1	0.012976	0.013475	0.013973	0.014468	0.014962	0.015454	0.015944	0.016431	0.016917	0.017400
19.2	0.017881	0.018360	0.018836	0.019310	0.019782	0.020251	0.020717	0.021181	0.021643	0.022102
19.3	0.022558	0.023011	0.023462	0.023910	0.024355	0.024797	0.025236	0.025672	0.026105	0.026535
19.4	0.026962	0.027386	0.027807	0.028224	0.028638	0.029049	0.029457	0.029861	0.030262	0.030659
19.5	0.031053	0.031444	0.031831	0.032214	0.032594	0.032970	0.033342	0.033711	0.034076	0.034437
19.6	0.034794	0.035148	0.035497	0.035843	0.036185	0.036522	0.036856	0.037186	0.037512	0.037833
19.7	0.038151	0.038464	0.038774	0.039079	0.039379	0.039676	0.039968	0.040256	0.040540	0.040820
19.8	0.041095	0.041365	0.041632	0.041893	0.042151	0.042404	0.042652	0.042896	0.043135	0.043370
19.9	0.043600	0.043826	0.044047	0.044263	0.044475	0.044682	0.044885	0.045082	0.045275	0.045464

INDEX

Index

Catalog

If you are interested in a list of SCHAUM'S
OUTLINE SERIES in Science, Mathematics,
Engineering and other subjects, send your name
and address, requesting your free catalog, to:

SCHAUM'S OUTLINE SERIES, Dept. C
McGRAW-HILL BOOK COMPANY
1221 Avenue of Americas
New York, N.Y. 10020